Climate Change Management

Series editor

Walter Leal Filho, Hamburg, Germany

More information about this series at http://www.springer.com/series/8740

Walter Leal Filho · Jesse M. Keenan
Editors

Climate Change Adaptation in North America

Fostering Resilience and the Regional Capacity to Adapt

Springer

Editors
Walter Leal Filho
Faculty of Life Sciences
Hamburg University of Applied Sciences
Hamburg
Germany

Jesse M. Keenan
Graduate School of Design
Harvard University
Cambridge, MA
USA

ISSN 1610-2010 ISSN 1610-2002 (electronic)
Climate Change Management
ISBN 978-3-319-53741-2 ISBN 978-3-319-53742-9 (eBook)
DOI 10.1007/978-3-319-53742-9

Library of Congress Control Number: 2017933436

Printed on acid-free paper

This Springer imprint is published by Springer Nature
The registered company is Springer International Publishing AG
The registered company address is: Gewerbestrasse 11, 6330 Cham, Switzerland

Preface

North America and the Caribbean are affected by extreme weather and climate change at a variety of scales and within the context of an even greater diversity of geographies, ecologies and institutions. Whereas droughts affect the western part of the USA and Canada, the eastern portion of the continent is particularly prone to flooding and sea-level rise. Even in the Caribbean, where tropical cyclones have been the primary risk factor for generations, a persistent drought is leading to severe ecological stresses that are driving unprecedented transformations in economy and society.

According to the Fifth Assessment Report (AR5) produced by the Intergovernmental Panel on Climate Change (IPCC), recent climate variations and individual extreme events demonstrate both impacts of climate-related stresses and the vulnerabilities of exposed systems. Many climate stresses that carry risk—particularly related to severe heat, heavy precipitation and declining snowpack—will increase in frequency and/or severity in North America in the coming decades. AR5 also states that current and future climate-related drivers of risk for small islands during the twenty-first century, such as those in the Caribbean region, will include sea-level rise (SLR), tropical and extratropical cyclones, increasing air and sea surface temperatures, and changing rainfall patterns. In addition, these patterns are likely to persist in some of the most advanced urban environments in the world, including Miami, Washington, D.C., New York and Boston.

Among other things, AR5 states that adaptation to climate change generates greater benefits when delivered in conjunction with other development activities, such as disaster risk reduction and community-based approaches to development. Whether it is a sparsely populated Caribbean island or a major continental urban region, adaptation processes are increasingly be recognized as critical steps where conventional modes of consumption, production and risk mitigation are unsustainable. The above state of affairs illustrates the need for a better understanding of how climate change affects North America and for the identification of processes, methods and tools that may help countries and communities to develop an adaptive capacity. There is also a critical need to showcase successful examples of how to manage the social, economic and political complexities posed by climate change, so that lessons can be learned and best practices may be disseminated.

This book serves the purpose of showcasing experiences from research, field projects and best practice in climate change adaptation in North America that may be useful or implemented in other countries and regions. A further aim of this book is to document and disseminate the wealth of experiences available today. Part I describes experiences on climate adaptation management in rural and urban areas, including elements related to community deliberations and the influences of policy and governance. Part II focuses on climate change and the built environment, also emphasizing aspects of planning. Part III includes a set of papers with an emphasis on adaptation, resilience and multi-hazard mitigation. Part IV puts an emphasis on information, communication, education and training on climate change. Part V entails elements related to climate change, planning and health, as well as two examples from other regions. A final chapter offers a cross-disciplinary perspective on the factors shaping North American adaptation research.

We thank the authors for their willingness to share their knowledge, know-how and experiences, as well as the many peer reviewers, which have helped us to ensure the quality of the manuscripts.

Hamburg, Germany
Cambridge, MA, USA
Spring 2017

Walter Leal Filho
Jesse M. Keenan

Contents

About the Editors

Professor Walter Leal (BSc, PhD, DSc, DPhil, DL, DLitt, DEd) is a Senior Professor and Head of the Research and Transfer Centre "Applications of Life Sciences" at the Hamburg University of Applied Sciences in Germany and at Manchester Metropolitan University. He has in excess of 250 publications to his credit, among which ground-breaking books such as "Handbook of Climate Change Management" and others. He teaches on information, education, communication on climate change at various European universities. He has over 20 years of research experience and has a particular interest on the connections between climate and human behavior.

Professor Jesse M. Keenan (A.B., M.Sc., Ph.D., J.D., LL.M., A.M., ASCE) is a member of the faculty of the Graduate School of Design at Harvard University where he teaches courses and conducts research in resilience and adaptation science. Keenan has served as Vice-Chair of the U.S. Community Resilience Panel for Building and Infrastructure Systems under two White House administrations and as Editor of the Built Environment at climate.gov. Keenan has conducted climate adaptation research with various cities around the globe including Amsterdam, Boston, Miami, New York City (NYC), Rotterdam, Sao Paulo, Rio de Janeiro, Stockholm and Tokyo. Keenan's many publications include, "Blue Dunes: Climate Change By Design."

Chapter 1
Climate Change Adaptation in North America: A Short Review of Priorities

Walter Leal Filho

Abstract This short papers offers an overview of some of the priorities to foster climate change adaptation in North America. It is meant to outline some areas where the impacts of climate change can be better addressed, and in the context of which adaptation strategies may be implemented.

Keywords Climate change · Priorities · North America · Adaptation

Introduction

There are a few regions in the world as prepared to cope with climate change as North America. Apart from Mexico, which is still regarded as a developing country and as such as certain limitations in respect of financial resources and access to technologies, both the United States and Canada are well resourced, and hence better able to adapt, than many other countries round the world. Mearns et al. (2009) produced a regional climate change assessment program for North America.

The latest report produced by the Intergovernmental Panel on Climate Change (IPCC) contains a chapter on North America which outlines the particularities of the region, as well as outlines aspects related to its vulnerability (Romero-Lankao et al. 2014).

Even though the cause of handling the impacts of climate change equally involves action in the mitigation and adaptation fronts-both are equally important-and despite the fact that many economic, social and political aspects are associated with them (Leal Filho 2011), this short overview focuses on adaptation, being consistent with the engagement of the author in international climate change adaptation initiatives. For purposes of this chapter, climate change adaptation is perceived as processes of designing, updating and implementing strategies to take

W. Leal Filho (✉)
Faculty of Life Sciences, Hamburg University of Applied Sciences,
Ulmenliet 20, 21033 Hamburg, Germany
e-mail: walter.leal@haw-hamburg.de

W. Leal Filho and J.M. Keenan (eds.), *Climate Change Adaptation in North America*, Climate Change Management,
DOI 10.1007/978-3-319-53742-9_1

Table 1.1 Preparedness of North American countries to implement climate change adaptation

Item	Availability	Impacts
Technologies	High	Greater availability of tools and methods to adapt
Financing	High	Possibilities to fund large initiatives and projects
Know-how	High	Wide knowledge of techniques and methods in support of adaptive action
Documentation	High	A wide range of data, documents and materials available, which may guide adaptation initiatives
Governance	High	Real commitment from decision-makers on what needs to be done
Public awareness	Medium	A wide-but not full-public awareness on the impacts of climate change

into account and cope with the impacts of climate change, to ensure the best action required to address them is taken.

Table 1.1 outlines the reasons why North America is well suited to engage in climate change adaptation initiatives. Even though its capacity is high in most areas, some deficiencies are seen in respect of public awareness, which shows the need for concerted efforts in this field.

Consistent with this overall high degree of preparedness, both in the United States and in Canada, there is a variety of regional initiatives focusing on climate change adaptation. The U.S. Department of Agriculture via the U.S. Forest Service, for instance, produced a national roadmap for responding to climate change (USDA 2010), which describes a variety of areas where action is needed. In addition, the Wildlife Conservation Society (WSC) has undertaken the Adaptation for Conservation Targets (ACT) framework, which has been designed to "motivate collaborative, scientifically defensible climate change planning for specific landscapes or seascapes by a multidisciplinary group of scientists and practitioners". The framework entails elements of conceptual modeling, scenario-based planning, and adaptive management, with a focus on addressing climate change matters.

In addition, the Conservation Biology Institute has coordinated the "AdaptWest" scheme, a spatial database and synthesis of methods for conservation planning aimed at enhancing resilience and adaptation potential of natural systems under climate change.

A further example of a regional initiative is the Nature Conservacy's **climate adaptation case studies**. This, as the name implies, is aimed at explaining how including future climate change considerations in our project planning strengthens and advances our overall conservation investments.

In Canada, the Government is helping Canadians adapt to the challenges posed by climate change, by emphasizing the need to make adjustments in decisions, activities, and thinking because of observed or expected changes in climate, in order to reduce harm or take advantage of new opportunities (Government of Canada 2016).

Adaptation actions can be in anticipation of, or in response to the impacts of a changing climate. Examples of adaptation measures include the development of more stringent building standards for areas where heavier snowfall is expected, or limiting development in coastal areas where sea level is projected to rise. Black et al. (2010) produced a risk-based guide for local governments in British Columbia, which was well accepted by a number of municipalities in that Province, as well as elsewhere in the country, whereas Richardson (2010) published a guidebook for Canadian municipalities. The usefulness of information and technical materials approaching risks and disasters management is well proven (Leal Filho 2013) and their availability does offer valuable support to local agencies.

Still in Canada, Bizikova et al. (2008) produced a document titled "Canadian communities' guidebook for adaptation to climate change. Including an approach to generate mitigation co-benefits in the context of sustainable development", which outlines a variety of action that can be taken, in order to foster climate change adaptation in the country. Richardson (2010) on the other hand, produced a handbook which shows some useful insights into how Canadian municipalities may adapt to climate change.

A Matter of Prioritising

There is little doubt in relation to the fact that adaptation is vital to attempts made by North American countries to cope with climate change. The noticeable shifts in average conditions (e.g. temperature, precipitations and sea level rise), accompanied by changes in climate variability and the frequency of extreme weather and climate events indicate the pressing need for solid and well defined adaptation efforts. In an important document outlining a framework for responding to climate change (USDA 2008), the USDA and US Department of Forestry describe some key areas where action is necessary.

So, in moving forward and in order to ensure duly emphasis is given to the most essential features of the climate change adaptation process, this paper defends the view that four main priority areas should be considered.

Priority 1—Reducing vulnerability and risks, by investing on changes and/or improvements in infra-structure. This may, for instance, include more sea walls in coastal areas, or better flood control instruments in cities. Such enablers foster risk reduction and help to protect exposed systems.

Priority 2—Increasing resilience by acting in the interface between physically defined hazards and their impacts on specific sectors. For instance, the negative consequences of draughts may be reduced, by modernizing water systems—especially water supply routes-, especially in rural areas.

Priority 3—Raising Awareness and improving information, education and communication on climate change. In order to yield long-term benefits, there is a perceived need to foster a better learning from past events and disasters on the one hand, and to engage the population more actively in coping with future events. Indeed, raising capacity among the population is regarded as vital to raising their capability to handle climate change (UN/ISDR 2004).

Priority 4—Preventing maladaptation by reflecting very carefully on the types and nature of any investments on adaptation. Apart from the fact that adaptation programmes which are not well reflected and considered upon are very costly, they often cause more harm than any good. Projects should be weighted not only in respect of cost-benefits under the specific circumstances at a given time when they are started, but also in terms of their long-term sustainability.

Whereas government will still play a central role in handling the impacts of climate change and in spearheading adaptation, it is vital that the communities are duly engaged.

This list of priorities is not meant to be comprehensive, nor it is meant to be hierarchical, i.e. the listed measures all bear equal relevance. But it does serve the purpose of illustrating some key areas where immediate action is needed.

Climate change is a problem which global in nature, but quite local as far as its impacts are concerned. Therefore, operationally, even though most of the climate change adaptation initiatives are coordinated by the central or regional government level, it is essential that municipalities are involved. This is so because they are uniquely placed to realise the implementation of adaptation plans, especially in respect of land use planning, management of their areas and territories (especially in coastal areas) and in terms of information, awareness raising and communication.

Conclusions

The North American region is well prepared to cope with the many challenges climate change poses to it, even though there is a constant need to monitor expected or possible impacts to human beings, to the physical environment and to private and public property and infra-structure. One lesson which can be useful to other regions in the world, is the need to anticipate the effects of climate change and taking preventive actions, before major impacts occur. In this particular field, the North American region offers many useful lessons and examples, whose replication in other parts of the world could be quite useful. This may entail setting up and implementing effective strategies to monitor, manage and reduce climate risks and increase a given community's overall resilience.

References

Bizikova L, Neale T, Burton I (2008) Canadian communities' guidebook for adaptation to climate change. Including an approach to generate mitigation co-benefits in the context of sustainable development, 1st edn. Environment Canada and University of British Columbia, Vancouver

Black RA, Bruce JP, Egener M (2010) Adapting to climate change: a risk-based guide for local governments in British Columbia. Summit Enterprises International Inc, Ottawa, Ont

Government of Canada (2016) Helping Canadians adapt to climate change. http://www.climatechange.gc.ca/default.asp?lang=En&n=2B2A953E-1

Leal Filho W (ed) (2011) The economic, social and political aspects of climate change. Springer, Berlin
Leal Filho W (ed) (2013) Climate change and disasters management. ISBN Springer, Berlin
Mearns LO, Gutowski W, Jones R, Leung R, McGinnis S, Nunes A, Qian Y (2009) A regional climate change assessment program for North America. Geological and Atmospheric Sciences at Digital Repository, Iowa State University
Richardson GRA (2010) Adapting to climate change: an introduction for Canadian municipalities. Natural Resources Canada, Ottawa, Ont
Romero-Lankao P, Smith JB, Davidson DJ, Diffenbaugh NS, Kinney PL, Kirshen P, Kovacs P, Villers Ruiz L (2014) North America. In: Climate Change 2014: impacts, adaptation, and vulnerability. Part B: regional aspects. contribution of working group II to the fifth assessment report of the intergovernmental panel on climate change [Barros VR, Field CB, Dokken DJ, Mastrandrea MD, Mach KJ, Bilir TE, Chatterjee M, Ebi KL, Estrada YO, Genova RC, Girma B, Kissel ES, Levy AN, MacCracken S, Mastrandrea PR, White LL (eds)]. Cambridge University Press, Cambridge, United Kingdom and New York, NY, USA, pp 1439–1498
United Nations Inter-Agency Secretariat of the International Strategy for Disaster Reduction (UN/ISDR) (2004) Living with risk: a global review of disaster risk reduction initiatives. United Nations Publication Office, Geneva
U.S. Department of Agriculture (USDA), U.S. Forest Service (2008) Strategic framework for responding to climate change. http://www.fs.fed.us/climatechange/message.shtml. Accessed 24 Aug 2016
U.S. Department of Agriculture (USDA), U.S. Forest Service (2010) National roadmap for responding to climate change. http://www.fs.fed.us/climatechange/climate--update.shtml. Accessed 24 Aug 2016

Part I
Climate Adaptation Management in Rural and Urban Areas

Chapter 2
Community Deliberation to Build Local Capacity for Climate Change Adaptation: The Rural Climate Dialogues Program

C. Daniel Myers, Tara Ritter and Andrew Rockway

Abstract Apathy and skepticism about climate change make mobilizing collective action for adaptation difficult in rural areas of the US. This paper evaluates the potential for deliberative public engagement to overcome these obstacles through a case study of the Rural Climate Dialogues (RCD) program. A Rural Climate Dialogue (RCD) convenes a demographically and politically representative group of residents for three days of deliberation about the local impacts of climate change and about how their community can adapt. Following the Citizens Jury model, participants spend three days hearing expert testimony, deliberating together to identify elements of their community that are threatened by climate change, and devising recommendations for individual and community actions that can enhance their community's climate resilience. Drawing on case studies of RCDs in three Minnesota communities, this evaluation finds that participating in an RCD reduces skepticism about climate change and increases beliefs that the local community can and should take action. Further, these dialogues spur collective action by setting clear, public goals and building support for direct involvement from community leaders and public officials. This success suggests that deliberative public engagement can be a useful tool for adaptation planning in rural communities and other areas where apathy and skepticism are significant barriers.

Keywords Climate change adaptation · Deliberation · Public engagement · Citizens Jury · Rural climate adaptation

C.D. Myers (✉)
Department of Political Science, University of Minnesota, 1414 Social Sciences, Minneapolis, MN 55409, USA
e-mail: cdmyers@umn.edu

T. Ritter
Institute for Agriculture and Trade Policy, Minneapolis, MN, USA
e-mail: tritter@iatp.org

A. Rockway
Jefferson Center, St. Paul, MN, USA
e-mail: arockway@jefferson-center.org

W. Leal Filho and J.M. Keenan (eds.), *Climate Change Adaptation in North America*, Climate Change Management,
DOI 10.1007/978-3-319-53742-9_2

Introduction

Rural areas in the United States face unique challenges in adapting to climate change. The economies of rural areas, which "have limited economic diversity and relatively high dependence on climate-sensitive sectors" (IPCC 2014, p. 1471), are particularly threatened by changes in temperature and rainfall patterns (Hales et al. 2014). Rural residents are also more likely to be dependent on carbon-intensive forms of transportation. Despite this vulnerability, a range of factors make adaptation planning difficult in rural areas. Public opinion in rural areas tends to be highly skeptical about climate change (Howe et al. 2015), and rural residents see little place for themselves in media discourses about climate change, which often treat climate change as a global issue whose primary impacts will be on urban and coastal areas (Moser 2014). Additionally, rural areas tend to have lower government capacity (McGuire et al. 1994; Hall 2008) and receive less attention from private philanthropy (Newstead and Wu 2009).

This constellation of factors create a conundrum—rural areas face some of the greatest adaptation challenges, but also the greatest barriers to mobilizing collective action to address these challenges. The Rural Climate Dialogue (RCD) program was developed by the Institute for Agriculture and Trade Policy and the Jefferson Center to provide a way for rural areas to mobilize action on climate change adaptation through a process of deliberative public engagement. In contrast with many forms of public engagement, which seek to educate or inform the public, deliberative public engagement combines education with the active involvement of citizens in decision-making. While deliberative processes include education by technical experts, the centerpiece of this form of engagement is discussion among lay-citizens about the challenges facing their community and ultimately the production of citizen-driven recommendations or findings. The Rural Climate Dialogues process builds on one model of deliberative public engagement, the Citizens Jury (Crosby and Nethercutt 2005), in which a small group of demographically and ideologically representative community members deliberate intensively for an extended period of time to produce recommendations for their community. By directly involving citizens in the adaptation planning process, the RCD program aims to overcome the apathy, skepticism, and lack of capacity that make adaptation planning difficult in rural areas.

This paper describes and evaluates the RCD model using case studies of three RCDs conducted in the rural Minnesota communities of Morris (June 2014), Itasca County (May 2015), and Winona County (March 2016). These communities were selected to reflect the economic and cultural diversity of rural Minnesota. This evaluation finds that the RCD program has contributed to local adaptive capacity by producing a series of recommendations for individual and community action that has served as a focal point for local adaptation planning, by changing attitudes about climate change and the need for action on climate change adaptation among participants, and by helping to build networks among local groups for future action. However, the nature and degree of these contributions depends on contextual factors about the communities.

The Importance of Rural Engagement

While public engagement is important to all climate adaptation efforts, it plays a particularly important role in rural areas. Rural communities are particularly susceptible to climate change impacts on many levels. Rural communities are more likely than urban or suburban communities to have natural resource-based economies. These industries, including agriculture, forestry, and fishing, will become less predictable in the face of more frequent extreme weather events, temperature changes, droughts, floods, wildfires, and increases in weeds, diseases and other pests. As a result, rural economies based on these industries will become less stable as climate change intensifies (Hales et al. 2014).

This increased instability occurs amidst existing economic insecurity. In 2014, the rural poverty rate was just over 18%, compared to the national average of 15% (USDA Economic Research Service 2016). Rural households have lower incomes and older housing stock on average as compared to urban households (Cutter et al. 2003). This means that most rural residents spend a larger percentage of their income on energy costs and often use more energy to heat and cool energy-inefficient spaces. Rural residents will be disproportionately impacted by energy costs as heating and cooling needs change in the face of more extreme temperatures.

Though the stakes are high in rural America, support for climate action among rural residents is lower than in the general population. Environmental concern in general, and climate change concern specifically, has been found to be higher among urban than rural residents (Safford et al. 2012; Howe et al. 2015). The polarization of climate change attitudes along cultural lines can make discussing climate change difficult in these environments, as the science of climate change is overwhelmed by questions of identity and group membership (Kahan 2012, 2015). Rural engagement on climate change must confront this skepticism by providing a space where citizens can discuss the local impacts of climate change without triggering identity threat. The perspectives that arise from these conversations can form the basis for local adaptation planning and help guide state and national policy to ensure that rural voices are included in policy solutions.

This is particularly important because many of the interventions needed to address climate change will come from rural communities. According to the 2010 Census, rural America encompasses nearly 75% of the land area and 19% of the population in the United States. The rural landscape—forests, farms, and rangelands—has exceptional potential to capture carbon and generate wind, solar and other renewable energy, with the people and ingenuity to oversee the transition to a low carbon economy. Although rural America will be disproportionately impacted by climate change, it has much to gain by undertaking climate change adaptation and mitigation efforts.

Deliberative Public Engagement

One tool to address this engagement challenge is deliberative public engagement. Advocates of deliberative public engagement argue that it is a way to both improve policy and democratize the policy-making process.[1] In contrast to forms of public engagement that focus on a one-way process of educating the public, deliberative public engagement aims to engage members of the public in a two-way conversation about important public issues. Citizens are not seen just as an audience to educate, but instead as experts in their own right whose values and situated knowledge are important inputs into the policy process. Deliberative public engagement aims to create situations where citizens can learn from experts, from each other, and come to collective decisions through (generally face-to-face) discussion.

Often, deliberative engagement takes the form of mini-publics, in which "citizens representing different viewpoints are gathered together to deliberate on a particular issue in small groups" (Grönland et al. 2014). While mini-publics take a variety of forms, all contain some form of the following three elements: *education* about the issue under discussion, *deliberation* in which citizens discuss the issue in a structured fashion, and *recommendations* agreed to by the forum's participants, which are sometimes actual policy decisions but more frequently a report or series of findings that are treated as inputs into a broader policy process. Mini-publics usually involve face-to-face interaction within an intensive, but time-bound period (e.g. a few hours or days). Proponents of mini-publics argue that education and deliberation with a diverse group of fellow citizens produces "refined" public opinion, and thus adds democratic legitimacy to policy-making processes (Fishkin 2009). This is particularly true on highly polarized issues where citizens are unlikely to talk with others who hold different views in the normal course of political life, or on technically complex issues where even highly informed citizens cannot be expected to hold well-reasoned opinions (Warren and Gastil 2015).

Climate change adaptation—an issue that is complex, poorly understood in the mass public, and politically divisive (Moser 2014)—would thus seem to be an ideal candidate for deliberative public engagement.[2] Indeed, a number of existing studies report efforts to include deliberative public engagement as part of adaptation

---

[1]See, for example, Fishkin (2009) and Myers and Mendelberg (2013) in political science, Gastil and Black (2007) in communications, Forester (1999) in planning, and De Vries et al. (2011) in health policy.

[2]For a related argument see Brulle (2010).

planning processes (e.g. Few et al. 2007; Milligan et al. 2009; Heberle et al. 2014; Phadke et al. 2015). Notably, these examples all deal with adaptation in urban or coastal areas. Nevertheless, they contain important lessons for deliberative engagement about climate adaptation in general. Since climate change is an issue where a small number of citizens are highly engaged, recruitment for deliberative fora must be conducted carefully in order to include a range of participants, not merely the usual activists (Few et al. 2007). The complexity of climate change adaptation means that fora must be carefully designed to allow for meaningful discussion of technical issues among non-experts (Milligan et al. 2009; Sheppard et al. 2011). Engagement neither begins nor ends with the deliberative forum itself; instead the process of organizing and conducting the forum should be seen as an opportunity to build trust and social capital around the issue to drive future action (Heberle et al. 2014; Phadke et al. 2015). Perhaps most importantly, deliberative engagement must put real power over the forum's outcomes in the hands of citizen deliberators (Few et al. 2007). Processes that are structured to produce a predetermined conclusion produce backlash against adaptation planning efforts; organizers need to support citizens' recommendations even if these decisions "get it wrong" from the perspective of technical or government elites.

To address these challenges, the Rural Climate Dialogues were designed based on the Citizens Jury model of deliberation (Crosby and Nethercutt 2005). A Citizens Jury provides citizens the opportunity to study an issue intensively over a number of days, deliberate together with a diverse group of their peers, and develop solutions to challenging public issues. The recommendations of a Citizens Jury provide insight for policymakers and the broader public into the informed opinions and priorities of a community. Citizens Juries have been used for a range of purposes including evaluating political candidates, proposing reforms for state electoral processes, assessing health care reform proposals, and evaluating ballot initiatives (Crosby and Nethercutt 2005; Knobloch et al. 2013; Munno and Nabatchi 2014).

A Citizens Jury consists of a randomly selected and stratified group of participants that, as nearly as possible, resembles the demographic and attitudinal makeup of their community. This includes political identification and attitudes towards the issue under discussion; for deliberation about climate change this means recruiting a group that includes Democrats, Republicans, and Independents as well as those who believe in anthropogenic climate change and those who deny or are skeptics of it. The ideal jury is large enough to reflect the demographic and cognitive diversity of the community, but not so large jurors are unable to engage in productive deliberation with one another; most are in the range of 15–24 jurors. To limit barriers to participation and ensure that jurors reflect a community's varying levels of engagement on the issue, jurors are paid a stipend and receive reimbursement for travel and childcare costs. This recruitment strategy ensures the Citizens Jury serves as a microcosm of community perspectives, a key element in the forum's claim to democratic legitimacy.

Citizens Juries usually deliberate in a concentrated, intensive fashion, generally meeting eight hours a day for several days. A jury begins with introductory exercises intended to help jurors get comfortable with each other and with the process of deliberation. During the first several days jurors alternate hearing testimony from a diverse set of experts on the issue with deliberating in small groups about the testimony. Experts are typically instructed to provide only background information on an issue or topic to inform the jury's deliberation without unduly biasing that deliberation. After an expert testifies, jurors deliberate among themselves with the goal of identifying those elements of the information provided that are most relevant to the question facing their community, as well any lingering questions or doubts about the information provided. Once expert testimony concludes, participants draw on the information presented as well as their collective knowledge to develop recommendations that address the issue.

The Citizens Jury process of public deliberation allows non-expert community members to influence policy and community action. The structure of expert testimony and the educative nature of the deliberation allow jurors to make informed recommendations on behalf of their community. The diversity of participants also lends broader legitimacy to the recommendations. Since many rural municipalities lack the technical and financial resources to adequately explore the local impacts of climate change and develop an adaptation plan in response, the work of empowered citizens can extend local governmental capacity and help diffuse tension associated with top-down approaches to addressing climate change.

The Rural Climate Dialogues Process

The Rural Climate Dialogues aim to galvanize leadership in rural communities by connecting diverse citizens and community groups and create a space for rural citizens to directly influence climate policy at the state and national levels by identifying key challenges facing their communities. To achieve these aims, staff pursue three distinct phases in the Rural Climate Dialogues process: networking and relationship building to form community coalitions, Citizens Jury-style public deliberation to produce climate resilience recommendations, and sustained community organizing to support implementation of actions and projects identified through deliberation.

The first phase of the RCD process aims to build community support for action on climate change and extreme weather. Staff meet with a diverse cross-section of local leaders from government, education (K-12 and higher education), business, and community organizations to discuss the most pressing issues in the community and to identify connections between their work and the challenges prompted by climate change. Staff form an ad hoc advisory committee of interested leaders who help select the issues most important for the Citizens Jury in their community to consider, identify speakers to address those issues, and begin forming an ongoing coalition of leaders and organizations committed to advancing the work of the

Citizens Jury. This phase of the process can vary in length, but usually takes three to eight months.

The next phase, engaging the public in deliberative dialogue, serves as the creative focal point of the Rural Climate Dialogues process. The Citizens Jury model, described above, engages eighteen people from the community to study, discuss, and outline courses of action to address the local impacts of extreme weather and climate change.

To recruit a diverse group of participants, five thousand randomly selected individuals from the community are sent invitations to participate in the dialogue. Interested individuals apply by answering a questionnaire, also available online and over the phone, to assess demographic and attitudinal characteristics. Direct mail recruitment is supplemented with online advertising through Craigslist and Facebook, media releases, and word of mouth. Applicant data is anonymized and aggregated in a potential pool of participants. Eighteen jurors and three alternates are selected from this pool to reflect the demographics of the community/county, including political affiliation and attitude toward climate change. Individuals unable to participate are replaced with an applicant closely matching their demographic and attitudinal profile.

The three-day Citizens Jury convenes for eight hours per day over a Thursday, Friday, and Saturday. Over the course of three days, participants are asked to set priorities in three categories: challenges pertinent to climate change that pose a threat to the long-term well-being of the community, opportunities to strengthen the community in the face of climate change, and action steps to address challenges and realize opportunities. This framework helps participants and the community set clear, actionable priorities without being overwhelmed by the scale of the climate change problem. Figure 2.1 shows the schedule for the Winona RCD; while the list of expert speakers was slightly different for each jury, the overall schedule was similar.

At the start of the jury, participants familiarize themselves with discussion guidelines and the dynamic of group deliberation by engaging in a simulation exercise focused on a public challenge a fictional community faces. The challenge encourages participants to think through risk mitigation in the face of uncertainty while practicing discussion skills. In small groups, participants assess information and develop a course of action for the fictional community. Each small group shares their course of action and describes their process for arriving at the recommendation. Participants are also asked to share their feelings about the process of deliberation. Importantly, the public challenge used in this exercise is not related to climate change. This gives jurors a chance to learn how to work as a group in a low-stakes environment where the political divisions that might become salient in a discussion about climate change are not relevant.

During the remainder of the first day and most of the second day the event alternates between expert presentations related to local climate change and extreme weather and small group deliberation about the information presented by these experts. To frame the overall discussion, the first presentation focuses on local weather and climate trends, describing the magnitude and effects of change in the

WINONA COUNTY CLIMATE DIALOGUE

Thursday, March 3rd – Introduction, Issue Expert Testimony, Deliberation

8:00am-12:00pm: Introductions, practicing the deliberative dialogue process

12:00pm-12:50pm: Lunch for participants

1:00pm-2:50pm: WEATHER TRENDS (*Mark Seeley, University of Minnesota climatologist and meteorologist*), idea and question generation, Q&A, debrief

2:50pm-3:35pm: ENERGY (*Lynn Hinkle, MN Solar Energy Industry Association & Chris Meyer, SE Clean Energy Resource Team Coordinator*), idea and question generation, Q&A

3:50pm-5:00pm: Speaker debrief & wrap up work

Friday, March 4th – Issue Expert Testimony, Deliberation

8:30am-8:50am: Morning introductions

8:50am-11am: WATER (*Josh Eash, Regional Hydrologist with Upper Mississippi River Fish and Wildlife Refuge & Jennifer Biederman, Professor of Biology at Winona State University*), idea and question generation, Q&A, debrief

11:00am-12:15pm: INSURANCE (*Mark Kulda, Insurance Federation of Minnesota*), idea and question generation, Q&A, debrief

12:15pm-1:15pm: Lunch for participants

1:15pm-2:30pm: PUBLIC HEALTH (*Bruce Snyder, University of Minnesota*), idea and question generation, Q&A, debrief

2:30pm-3:45pm: AGRICULTURE *Jake Overgaard, University of Minnesota Extension - Winona County Agriculture Production Systems Specialist*), idea and question generation, Q&A, debrief

3:45pm-4:40pm: Making connections between presentations

4:40pm-5:00pm: Wrap up work

Saturday, March 5th – Final Recommendations for Neighbors

8:30am-9:00am: Morning introductions

9:00am-9:50am: Drafting

10:05am-11:00am: Choosing top challenges

11:00am-12:00pm: Choosing top opportunities

12:00pm-12:45pm: Lunch for participants

12:45pm-3:10pm: Identifying: What should our neighbors know?

3:10pm-5:00pm: Identifying: What should be done?

5:00pm: Adjourn

Fig. 2.1 Winona County climate dialogue schedule

historical record. Participants deliberate about which information from the presentation is most important to share with neighbors in understanding climate change, prioritizing five to ten key "facts" to include in their final report. The next presentations, five in total, focus on specific topics relevant to the community and the impacts of climate change and extreme weather on each. Topics vary by community, but can include agriculture, public health, local infrastructure, water resources, energy systems, insurance, wildlife and habitat, tourism, and recreation. As shown in Fig. 2.1, local experts, such as professors at local universities or agents of the local agricultural extension office, are used whenever possible. Each topic presentation is followed by small group deliberation to discuss challenges, opportunities, and action steps and evaluate the trade-offs in pursuing one course of action over another.

The third and final day involves extensive deliberation in small and large groups to produce a final report for their community. The day begins with an assessment of the top challenges and opportunities posed by climate change, and continues in the afternoon with discussion of the actions that are most critical to address these as well as the information that is most important to transmit to other residents of the community. From the list created over the first two days, participants consider trade-offs involved in each challenge, opportunity, and action before voting to identify priorities for the community. These priority lists form the bulk of the information participants share with their neighbors. Participants are also asked to assess whether actions are best taken by individuals or by the community collectively. Finally, participants draft a brief statement for their neighbors outlining their experience in the dialogue process, the reasons for selecting certain challenges/opportunities/actions over others, and the importance of acting to address extreme weather and climate change.[3] Importantly, the Rural Climate Dialogues process does not necessarily aim to build a unified consensus around how the community should adapt to climate change; instead, the final recommendations reflect a range of options for actions that reflect the community's diversity. The difference in values and opinions add strength to the overall recommendations, providing varied foci for diverse actors and "stakeholders" to coalesce behind when thinking about and working toward local adaptation efforts.

While drafting this statement marks the end of the Citizens Jury, it serves as the beginning of the next stage of engagement. Following public deliberation, participants and community leaders identified through pre-Citizens Jury relationship building are empowered to work on the action steps together. With assistance from project staff, community leaders, jury participants, and other community members seek and share resources to implement community action recommendations. High School students develop community-based service learning opportunities to act on the priorities they identified. Through peer-to-peer networking, community members share climate change information with their neighbors and friends, using the Citizens Jury report as a starting point for deeper conversation and movement

[3]Though the community jury is the main public deliberative activity of the Rural Climate Dialogues process, high school students are also engaged in deliberation to advance the perspectives of young people in the community. In an abbreviated deliberative process over the course of many class periods, students hear from experts and develop their own priorities for addressing climate change. Depending on the timing of student deliberation, their priorities are either presented to the community jury or incorporated into the community report.

towards community action. Post-jury organizing work in each of the three Minnesota communities are discussed in detail in the following sections.

Results of the Process

To evaluate the success of the RCD process this section looks at three different kinds of outcomes. The first is the impact of the deliberative process on individual participants, in terms of changes in individual attitudes as well as participants' satisfaction with the deliberative process. This is examined by testing for change in participants' responses to pre- and post-jury surveys. The second outcome is the substantive content of the jurors' conversations and recommendations, evaluated through a thematic analysis of the final recommendations produced by the juries that draws together themes common to the three juries while also noting differences across them. The final outcome is post-jury organizing, evaluated based on the extent to which deliberative engagement served to spur future community action, and the factors that might contribute or detract from deliberative success.

Quality of Deliberation and Impact of Process on Participants

To evaluate jurors' perceptions of the quality of the process, as well as the effect of the jury on participants' attitudes, all jurors completed a pre-deliberation survey at the start of the first day as well as a post-deliberation survey at the end of the final day. This section reports mean responses to questions about deliberative quality on the post-deliberation surveys, and test of attitude change using paired one-sided *t*-tests comparing pre-deliberation responses to post-deliberation responses.

As is commonly found in studies of deliberative public engagement (see Myers and Mendelberg 2013, p. 709), most deliberators reported being highly satisfied with the citizen jury process. The post-deliberation survey measured perceived deliberative quality using a five-item index drawn from Esterling et al. (2015), where each item asked participants to agree or disagree with a statement about the event, where agreement indicates a positive evaluation of the event. On a 5-point scale where 5 indicates strong agreement the average score on the index was 4.52, showing that deliberators were highly satisfied with the quality of discussion at the event. Focusing on specific aspects of the process, jurors were highly satisfied with the information presented, with a mean response of 4.3 on a 5-point scale from "very unsatisfied" to "very satisfied," and with the work of the discussion moderators, with a mean response of 4.3 on a 5-point scale from "very ineffective" to "very effective" on a three-item scale. Jurors reported high levels of agreement with their groups' recommendations (mean response of 4.2 on a 5-point strongly agree-strongly disagree scale), and also agreement with the statement "I can live with the recommendations produced at this meeting, including any that I disagree

Table 2.1 Effect of Citizens Jury on support for action

Who?	Can take action		Should take action	
	Pre-deliberation	Post-deliberation	Pre-deliberation	Post-deliberation
Myself	**3.3**	**4.2**	**3.6**	**4.3**
Community	**3.9**	**4.4**	**3.9**	**4.4**
Local government	**3.8**	**4.3**	4	4.2
State government	**3.9**	**4.4**	**4.0**	**4.4**

Bolded numbers indicate a statistically significant difference between pre- and post-deliberation measures (1-sided t-test, $p = 0.05$)

with" (mean response of 4.4). Asked how much they agreed or disagreed with the statement "I would participate in an event like this again" all but one participant agreed or strongly agreed, with 44% agreeing and 53% strongly agreeing.

Participating in the event significantly increased participants' expectations that climate change would have an impact on their communities. Both pre- and post-deliberation surveys asked how likely it was that their community would see an increase in the number of extreme weather events and major shifts in climate patterns in the coming years a 5-point scale from "very likely" to "very unlikely." The mean response to the question about extreme weather increased from 3.9 to 4.4, while the mean response to the question about climate patterns increased from 3.8 to 4.3; both changes are statistically significant ($p < 0.001$).

Stronger beliefs that climate change would have an effect on their communities was accompanied by increased support for action at the individual, community, and state level. Participants were asked whether they agreed or disagreed with the statements about whether four different entities can take action to address changes in climate as well as whether these entities should take action. Responses were on a five-point scale from "strongly disagree" to "strongly agree." Table 2.1 shows the result. Belief that action is possible as well as support for action increased for all four entities, with the strongest effect at the individual and community level.

Themes and Recommendations from Jury Deliberation

To summarize the substance of jurors' recommendations for adaptation in their communities team members conducted a thematic analysis of the final recommendation documents produced by the three juries. Two team members independently reviewed the final recommendations produced by all three juries, noting commonalities as well as differences across these documents. They then collaborated to produce a final list of themes that both observed in all three communities, and that might thus shed light on the adaptation priorities in rural communities more generally.

Jurors in all three initial Climate Dialogues shared common assessments of the major challenges and opportunities presented by climate change.[4] In each community, participants highlighted the critical importance of their local and regional natural resources base as drivers of economic activity and local quality of life. Recommendations focused on managing land-intensive activities (like agriculture and forestry) by introducing diversity into those systems to both add resilience in the event of extreme weather "shocks" and to provide data to evaluate the success of these changes as the overall climate continues to change. In Morris, jurors were concerned with the susceptibility of monoculture farming to climate change and highlighted the opportunity to "sustain and strengthen [our] agricultural economy" by introducing "diversity in farming" that develops new businesses and supports the interests of younger generations of farmers. In Itasca County, jurors recommended the community "manage forests so that they're more adaptable in the face of changing conditions" by evaluating native species and non-native species in areas with climate conditions similar to those projected in Itasca, by thinning dense pine forests, and by replacing ash trees susceptible to pests. Winona County jurors proposed "adopting agricultural best management practices," like "planting perennials and forages," introducing buffer strips, and planting "pollinator habitat, native plants, and prairie grasses" to improve water quality and reduce soil and nutrient loss while "maintaining production and profitability" for farmers.

Jurors also emphasized local water resources as a "canary in the coal mine" of unsustainable practices that threaten drinking water supplies, industrial and agricultural water use, habitat degradation, and more. For each community, water quality and quantity serve as highly visible markers of progress toward or regression away from sustainable activity. Winona County jurors noted "high intensity precipitation events may lead to short-term increases in water temperature, higher magnitude flooding, erosion, runoff of sediments and pollution, and degraded stream habitat for coldwater fish and other aquatic invertebrates." Jurors in Itasca County prioritized the impact of extreme events on "the life of capital assets" and "operational disruptions for public infrastructure." All three communities recommended actions related to green stormwater infrastructure, including "ecosystem restoration" (Winona County); "reduc[ing] imperviousness and allow[ing] water to infiltrate into the ground, … adapt[ing] stormwater infrastructure to hold higher volumes, and … maintain[ing] riparian buffers and forest cover, using natural features that slow or retain water" (Itasca County); and "us[ing] water channeling and drainage … and, where possible, captur[ing] water for other uses" (Morris) that closely reflected local ecological and economic features.

Finally, jurors shared the assessment that others in the community, including elected officials and other policymakers, lack adequate understanding of the threat presented by climate change. Their recommendations focused on the need for community education efforts to expand climate change awareness. They also

[4]For the full list of findings and recommendations from each jury, see http://www.ruralclimatenetwork.org/content/rural-climate-dialogues.

promoted the need for broader decision making authority, including through better public participation processes, to allow for more voices and perspectives to shape policy and community action. Morris jurors cited "the lack of education on these issues overall, particularly among public officials who are responsible for advocating change to the general public, undermines the ability to make changes." They recommended "building and reinforcing community relationships through discussion," "involving social and local media and others in promoting sustainable and energy-efficient practices and habits", "implement[ing] discussion of climate change into K-12 education," and "hold[ing] town meetings where government officials, agricultural producers, utility providers, human services, and the public can generate and discuss new ideas so voters and consumers can make more informed decisions." Jurors in Itasca County noted that "information is power," "information [needs to be] accessible," and "decision-makers at all levels—including individuals, government, and businesses—need to be informed and engaged concerning how changes in climate affect our natural resources and economy." Jurors in Winona County recommended supporting "local organizations [that] can provide community members with resources to help successfully implement [action] ideas," as well as research and outreach efforts specific to each priority topic area.

Post-jury Organizing

Despite much common discussion around issues of land use, water resources, and public education, the organizing efforts that developed from the jury recommendations led in unique directions for each community. Organizing is an ongoing process and the long-term effects of this organizing should continue for several years to come. This section reports the initial outcomes of post-jury organizing in Morris and Itasca County which were held roughly two years and one year, respectively, prior to this writing. Organizing after the Winona jury, held just a month prior to this writing, shows promise but is at a very early stage.

In Morris, jury recommendations helped shift the thinking of a skeptical City Manager, who took to heart the jury's encouragement to strengthen community resilience around climate threats. In particular, the City of Morris is exploring new methods of managing stormwater and generating local renewable energy as an economic driver. Toward the latter point, the City Council signed a climate protection technical assistance agreement with the City of Saerbeck, Germany to outline opportunities for Morris to develop and sell renewable energy generated locally. The agreement was hailed as "unprecedented" and "unlikely to happen without the Dialogue" by community members and other public officials. The City of Morris is also installing energy efficiency improvements within public buildings and on public streets. Other community members are helping to expand awareness of climate change in the community by pursuing meetings with neighbors, hosting movie nights that focus on the broader issue of global climate change, hosting discussions in their churches, and convening additional deliberative events.

In Itasca County, the site of the second dialogue, public action has come less quickly. City officials are exploring additional green stormwater infrastructure to help manage runoff. High school teachers and students are working to secure funding to install rain gardens near a local lake to reduce runoff. However, many of the recommendations from the Itasca Climate Dialogue focused on individual-level action, such as planting native grasses and pollinator habitat or implementing best practices to mitigate stormwater runoff, actions about which it is difficult to collect data.

Conclusion

Rural areas in the United States face a double-bind: unique adaptation challenges on the one hand, and a skeptical public and limited resources on the other. Deliberative public engagement presents a way to address these challenges by placing open, informed conversation among average citizens at the center of adaptation planning and building on that conversation to coordinate action. Engaging citizens of all backgrounds in a two-way conversation gives legitimacy to adaptation planning it might otherwise lack, while the highly local focus contrasts with the popular discourse of climate change as dire changes that are happening "someplace else." Meanwhile, the recommendations of deliberative processes can provide a focal point around which informal networks of actors can coordinate. The Citizens Jury is not the only method of deliberative public engagement. However, the successes of the Rural Climate Dialogues described here suggest that the Citizens Jury's combination of a small, diverse group of deliberators, an intensive educative component, and a focus on developing a set of recommendations that all participants can live with makes it particularly useful for engagement on this complex, potentially polarizing issue.

The recommendations of the three Minnesota Rural Climate Dialogues may provide insight into rural climate adaptation more generally. Residents of rural areas expressed anxiety about the impact of climate change on natural resource-based industries, but beyond this economic focus worry about the impact on quality-of-life factors ranging from the aesthetic (pristine landscapes) to instrumental (recreation). These concerns were often tied together by discussions about water—threats to water quality, dangers of drought as well as floods and other extreme precipitation events. However, residents also saw opportunities for their communities, such as clean energy jobs, local and regional food initiatives, and improved management of runoff and stormwater. Across all three juries, residents saw these opportunities as ways to prepare for climate change while improving the local economy and reaffirming their community's basic values.

While all three of the cases examined here succeeded at producing a useful set of recommendations for their communities, implementation efforts in the first case,

Morris, have been more successful than in the second case, Itasca County. While it is difficult to generalize from two cases, the analysis presented here suggests some elements that might influence community action. Morris, unlike Itasca County, is home to a public university that has served as an anchor and coordinating body for community adaptation efforts. The University of Minnesota-Morris has devoted staff and student resources to writing grants and conducting community education events. The City of Morris, through the City Manager, was also initially invested in the Citizens Jury process, presenting about city infrastructure despite being personally skeptical about climate change. That commitment has translated to ongoing support for the jury recommendations. Public officials in Itasca County, while initially interested in the Rural Climate Dialogues process, were less involved in the design and delivery of the deliberation. And, pertinent to the lack of philanthropic and other resources in rural communities, Morris has received grant funding to implement community organizing and education efforts, while Itasca County, despite numerous applications, has not. In sum, deliberative engagement can play an important role in addressing the social and informational aspects of building adaptive capacity, but is not a panacea—public institutions in rural areas play an important role in shaping and supporting community action, particularly when resources that can enable quick wins to spur ongoing action are scarce.

In addition to these general limits on what can be accomplished by deliberative public engagement, some limitations to the generalizability of the current research should be kept in mind. While the similarities across the RCDs provide some confidence that the positive results will generalize, this analysis is still limited to only three cases. Further, all of these cases take place in Minnesota, a state whose political culture may make it particularly well suited to deliberation. Finally, the Citizens Juries at the center of the RCD process are a relatively intensive form of deliberation, requiring a significant commitment of time and resources by organizers and participants alike. It is unclear whether less intensive forms of deliberation, lasting perhaps hours instead of days, would be as effective at producing productive conversation across lines of political and demographic difference.

Acknowledgements The authors acknowledge the funding of the McKnight Foundation and the Carolyn Foundation in support of this project.

Appendix 1: Survey Items

Items measuring deliberative quality (all responses on 1–5 scale "Strongly Disagree"—"Disagree"—"Neither Agree Nor Disagree"—"Agree"—"Strongly Agree")

To what extend do you agree or disagree with each statement below:

1. I am more informed about the challenges and options for addressing this issue.
2. People at this meeting listened to one another respectfully and courteously.

3. Other participants seemed to hear and understand my views.
4. The meeting today was fair and unbiased. No particular view was favored.
5. Even when I disagreed with them, most people made reasonable points and tried to make serious arguments.

Items measuring agreement with recommendations (all responses on 1–5 scale "Strongly Disagree"—"Disagree"—"Neither Agree Nor Disagree"—"Agree"—"Strongly Agree")

1. I personally agree with the recommendations produced at this meeting.
2. I can live with the recommendations produced at this meeting, including any that I disagree with.

Item measuring satisfaction with information (1–5 scale "Very Dissatisfied"—"Dissatisfied"—"Neutral"—"Satisfied"—"Very Satisfied"

To what extent are you satisfied with the information presented overall?

Items measuring effectiveness of facilitation (all responses on 1–5 scale "Very ineffective"—"Somewhat ineffective"—"Neither"—"Somewhat effective"—"Very effective")

How effective were the facilitators for the Climate Dialogue at:

1. Keeping the group on task.
2. Making sure that everyone was heard.
3. Remaining neutral (not expressing their opinions).

Items measuring views on likelihood of climate change (all responses on 1–5 scale "Extremely Unlikely"—"Unlikely"—"Neither Likely nor Unlikely"—"Likely"—"Extremely Likely")

How likely do you think it is that the following things will happen?

4. The number of extreme weather events in <location> will increase in the coming years.
5. Climate patterns in <location> will experience major shifts in the coming years.

Items measuring whether entities can and should take action (all responses on 1–5 scale "Strongly Disagree"—"Disagree"—"Neither Agree Nor Disagree"—"Agree"—"Strongly Agree")

Please indicate the extent to which you agree with each statement.

1. There are actions that <entity> can take that will address these risks.
2. <Entity> should take action to address these risks.

Repeated for "I," "the <location> community," "local government in <location>," and "Minnesota state government."

References

Brulle RJ (2010) From environmental campaigns to advancing the public dialog: environmental communication for civic engagement. Environ Commun 4(1):82–98
Cutter S, Boruff B, Shirley L (2003) Social vulnerability to environmental hazards. Soc Sci Q 84:242–261
Crosby N, Nethercutt D (2005) Citizens juries: creating a trustworthy voice of the people. In: Gastil J, Levine P (eds) The deliberative democracy handbook: strategies for effective civic engagement in the twenty-first century. Jossey-Bass, San Francisco
Vries De, Stanczyk AE, Ryan KA, Kim SYH (2011) A framework for assessing the quality of democratic deliberation: enhancing deliberation as a tool for bioethics. J Empirical Res Hum Res Ethics 6(3):3–17
Esterling KM, Fung A, Lee T (2015) How much disagreement is good for democratic deliberation? Polit Commun 32(4):529–551
Fishkin J (2009) When the people speak: deliberative democracy and public consultation. Oxford University Press, New York, NY
Few R, Brown K, Tompkins EL (2007) Public participation and climate change adaptation: avoiding the illusion of inclusion. Clim Policy 7(1):46–59
Forester J (1999) The deliberative practitioner: encouraging participatory planning processes. MIT Press, Cambridge, Mass
Gastil J, Black L (2007) Public deliberation as the organizing principle of political communication research. J Publ Deliberation 4(1). http://www.publicdeliberation.net/jpd/vol4/iss1/art3
Grönlund K, Bächtiger A, Setälä M (2014) Introduction. Deliberative mini-publics: involving citizens in the democratic process, studies in European political science. ECPR Press, Colchester, pp 1–7
Hales D, Hohenstein W, Bidwell MD, Landry C, McGranahan D, Molnar J, Morton LW, Vasquez M, Jadin J (2014) Rural communities. In: Melillo JM, Richmond TT, Yohe GW (eds) Climate change impacts in the United States: the third national climate assessment. U.S. Global Change Research Program, Washington, DC, USA, pp 333–349
Hall JL (2008) The forgotten regional organizations: creating capacity for economic development. Publ Adm Rev 68(1):110–125
Heberle LC, Merrill S, Keeley CH, Lloyd S (2014) Local knowledge and participatory climate change planning in the Northeastern U.S. In: Leal Filho W, Alves F, Caeiro S, Azeiteiro UM (eds) International perspectives on climate change. Springer International Publishing, Cham, CH, pp 239–52
Howe PD, Mildenberger M, Marlon JR, Leiserowitz A (2015) Geographic variation in opinions on climate change at state and local scales in the USA. Nat Clim Change 5(6):596–603
IPCC (2014) Climate change 2014: impacts, adaptation and vulnerability. Cambridge University Press, Cambridge, United Kingdom and New York, NY, USA
Kahan D (2012) Why we are poles apart on climate change. Nature 488(7411):255–255
Kahan DM (2015) Climate-science communication and the measurement problem. Adv Polit Psychol 1(1):1–43
Knobloch KR, Gastil J, Reedy J, Walsh KC (2013) Did they deliberate? Applying an evaluative model of democratic deliberation to the oregon citizens initiative review. J Appl Commun Res 41(2):105–125
McGuire M, Rubin B, Agranoff R, Richards C (1994) Building development capacity in nonmetropolitan communities. Publ Adm Rev 54(5):426–433
Milligan J, O'Riordan T, Nicholson-Cole SA, Watkinson AR (2009) Nature conservation for future sustainable shorelines: lessons from Seeking to involve the public. Land Use Policy 26 (2):203–213
Moser SC (2014) Communicating adaptation to climate change: the art and science of public engagement when climate change comes home. Wiley Interdisc Rev: Clim Change 5(3): 337–358

Munno G, Nabatchi T (2014) Public deliberation and co-production in the political and electoral arena: a citizens' Jury Approach. J Publ Deliberation 10(2)
Myers CD, Mendelberg T (2013) Public deliberation. In: Huddy L, Sears DO, Levy JS (eds) Oxford handbook of political psychology, 2nd edn. Oxford University Press, New York
Newstead B, Wu P (2009) Nonprofits in rural America: overcoming the resource gap. Bridgespan Group, Boston, MA
Phadke R, Manning C, Burlager S (2015) Making it personal: diversity and deliberation in climate adaptation planning. Clim Risk Manage 9:62–76
Safford TG, Cutler M, Henly M, Norman K, Levin P (2012) Urban-rural differences in concern about the environment and jobs in the Puget Sound region. The Carsey School of public policy at the scholars' repository. Paper 185. Carsey Institute, University of New Hampshire, Durham, N.H
Sheppard SRJ et al (2011) Future visioning of local climate change: a framework for community engagement and planning with scenarios and visualisation. Futures 43(4):400–412
USDA Economic Research Service (2016) Rural at a glance, 2015th edn. United States Department of Agriculture, Washington, DC, USA
Warren ME, Gastil J (2015) Can deliberative minipublics address the cognitive challenges of democratic citizenship? J Polit 77(2):562–574

Chapter 3
A Participatory Process to Design Climate Change Adaptation Measures for the Carmen-Pajonal-Machona Lagoon System in Mexico

Valentina Giannini, Emiliano Ramieri, Margarita Caso Chávez, Karina Santos del Prado Gasca and Karla Priego Martínez

Abstract The participatory process described is part of a project for the design of climate change adaptation measures to be implemented in the Carmen-Pajonal-Machona coastal lagoon socio-ecosystem, situated in the Mexican state of Tabasco. Since these measures should be implemented and managed by the local communities impacted by climate change, they were designed based on the active engagement of the local communities, international, national and local institutions, academics, i.e. on knowledge provided by local stakeholders. The project was organized around a set of different activities, a multidisciplinary effort drafted to create synergies among researchers of the Consortium. Two workshops were designed and organized as part of this. The first workshop's (March 2015) aim was to elicit local knowledge on impacts and vulnerabilities, and ideas on what measures are needed to address them. Impacts and vulnerabilities identified in the first workshop and in the diagnostic were used to identify measures adequate to address climate and anthropogenic changes in the lagoon socio-ecosystem. The preliminary list of measures identified was refined through the second workshop (June 2015), in which participants provided qualitative information needed to identify the five most promising measures to address impacts using a multi-criteria decision support system: Mulino Decision Support System (mDSS).

Keywords Climate change adaptation · Community based adaptation · Participatory process · Tabasco · Carmen-Pajonal-Machona Lagoon system

---

V. Giannini (✉)
Centro Euro-Mediterraneo sui Cambiamenti Climatici (CMCC),
Isola di San Giorgio Maggiore 8, 30124 Venice, Italy
e-mail: valentina.giannini@cmcc.it

E. Ramieri
Thetis SpA, Castello 2737/F, 30122 Venice, Italy

M. Caso Chávez · K. Santos del Prado Gasca · K. Priego Martínez
Instituto Nacional de Ecología y Cambio Climático (INECC), Periférico 5000, Col. Insurgentes Cuicuilco, Delegación Coyoacán 04530, D.F., Mexico

W. Leal Filho and J.M. Keenan (eds.), *Climate Change Adaptation in North America*, Climate Change Management,
DOI 10.1007/978-3-319-53742-9_3

Introduction

The Carmen-Pajonal-Machona (CPM) lagoon system is located along the Gulf of Mexico in the State of Tabasco, about 90 km from Villahermosa, Tabasco's capital. It is a low-depth (0.9 m on average) coastal wetland of about 190 km^2, which is separated from the ocean by a fragile and highly vulnerable sand bar (Fig. 3.1). It communicates with the sea through two inlets: Boca Santana of natural origin and currently affected by erosion, and Boca Panteones, artificially created in 1975 and currently under sedimentation. The CPM lagoon system consists of two lagoons (Carmen on the west and Machona on the east) which are connected by the Pajonal narrower water body. Two main rivers flow into the CPM lagoon system: Rio San Felipe in the Carmen lagoon and Rio Santa Ana in the Machona lagoon. The whole CPM lagoon system has a high biological and ecological relevance. The area around the lagoons hosts important mangrove habitats, that have been highly impacted in the last decades, due to land-use changes (mainly into agriculture land and grassland for animal breeding), illegal cut for wood consumption, and erosion of the mangrove banks. Three mangrove species are present in the area: *Avicennia germinans*, *Rhizophora mangle*, and *Laguncularia racemosa*. The area around the lagoon system belongs to the municipalities of Cárdenas, Comalcalco and Paraíso and hosts medium-small human communities (the largest being the Sanchez-Magallanes settlement on the western littoral), strongly relying on the CPM lagoon system services for their subsistence. The main productive activities are agriculture, livestock and, in areas closer to the lagoons, fisheries and aquaculture. The whole area is also characterised by the strong presence of oil extraction.

Between December 2014 and February 2016, the Consortium formed by Thetis, Centro Euro-Mediterraneo sui Cambiamenti Climatici (CMCC) and Coastal Environments (hereafter consortium) run a project "Design of adaptation measures to reduce vulnerability of the pilot site Carmen-Pajonal-Machona Lagoon System, Tabasco, to the impacts from climate change and anthropogenic activities" (hereafter CPM project). The project aimed at assessing current and future vulnerability to human activities and climate change of the CPM lagoon system and, consequently, identifying adaptation measures, which could be implemented and managed by the local communities living in the area around the lagoon. The project is part of the wider macro-project "Adaptation of coastal wetlands of Gulf of Mexico to climate change impacts" (hereafter macro-project) funded by the World Bank and implemented by the Instituto Nacional de Ecología y Cambio Climático (INECC) and the Instituto Mexicano de Tecnología del Agua (IMTA). In addition to the CPM lagoon system, this macro-project focuses on other two pilot sites: (i) Río Papaloapan—Laguna de Alvarado in the Veracruz State, and (ii) Humedal Punta Allen: Reserva de la Biosfera de Sian Ka'an in the Quintana Roo State.

The whole macro-project, including the specific CPM project, applies a holistic approach to the analysis of current and future problems, and to the identification of adaptation responses. The CPM lagoon system is seen as a unique and interlinked

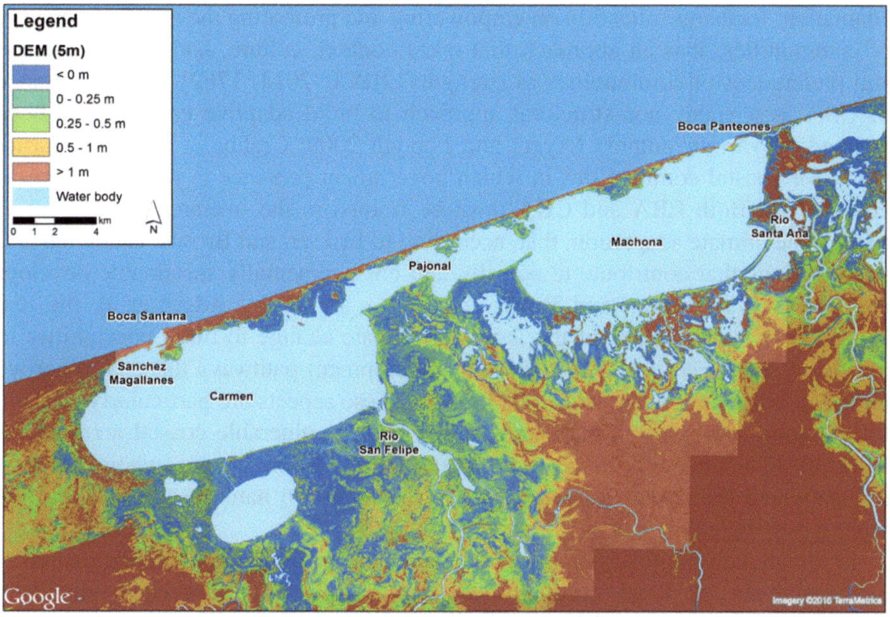

Fig. 3.1 The Carmen-Pajonal-Machona Lagoon system. The map shows the land elevation of the area around CPM lagoon system (source of DEM data: LIDAR 5 m INEGI 2012)

socio-ecosystem, recognising that complex interactions among social, economic and environmental factors operating on different spatial and temporal scales are essential in determining local vulnerability, and in shaping adaptation capacity to current climate variability and long-term climate changes (Thomalla et al. 2006). De facto, social and ecological systems act as strongly coupled, complex and evolving integrated systems; treating them independently in the approach towards improved resilience and adaptation capacity can be a fundamental error (Folke et al. 2002). As other natural ecosystems (Carpenter et al. 2009), CPM provides crucial services (in particular provisioning and regulating ones) for the basic existence of local communities in the area; the integrity of this ecosystem and the services it provides are affected by both current human pressures and future climate change impacts.

Core element of the adopted approach is the ecosystem-based adaptation (EBA) concept, that can be defined as the protection and improvement of biodiversity and ecosystem services (i.e. of the lagoon itself, the littoral bar, the mangrove habitats, etc.) as part of a wider adaptation strategy, for which the role of ecosystem services is highly relevant, aiming to help local people adapt to adverse effects of climate change and climate variability (CBD 2009; Hale et al. 2009; Lhumeau and Cordero 2012). In this perspective, local communities are seen as the main beneficiaries of adaptation strategies.

Within the CPM project the application of EBA is therefore strictly connected to the concept of community-based adaptation (CBA), i.e. local, community-driven adaptation, focusing "attention on empowering and promoting the adaptive capacity of communities. It is an approach that takes context, culture, knowledge, agency, and preferences of communities as strengths" (IPCC 2014: 1762). The importance of CBA as a viable non-structural approach to build adaptive capacity has been stressed by various authors (Ayers and Forsyth 2009; Crabbe et al. 2009) particularly for coastal communities in which government presence is weak (Villamizar et al. 2016). Both EBA and CBA must be framed in the overarching concept of sustainable climate adaptation, that, according to Eriksen and Brown, is defined as a set of actions that contribute to socially and environmentally sustainable development pathways (Eriksen and Brown 2011). As argued by Adger et al. the real challenge is "to make use of the issues of climate change to find opportunities to transform social-ecological systems into development pathways that may improve human conditions" (Adger et al. 2011: 765). These aspects are particularly relevant for communities living in marginalised and highly vulnerable coastal areas of the world (Villamizar et al. 2016), such as the CPM lagoon system, where people's existence and their adaptation capacity strongly rely on natural resources and the integrity of the environment.

In the above described perspective engagement of local stakeholders, including representatives of local communities, was therefore considered strategically important for the successful identification of adaptation measures to be implemented at the community level, with the involvement of local people in their future possible design and management. CBA is, in fact, becoming a prevalent practice where participation by stakeholders informs adaptation measures design and implementation (Mimura et al. 2014). Local communities, including vulnerable groups, should therefore be involved in the design and implementation of adaptation measures (Wong et al. 2014; Noble et al. 2014). Moreover, involving communities in risk and capacity assessment can foster cooperation between a community and an outside institution (van Aalst et al. 2008), contributing local knowledge on impacts of climate (Zubair 2004; Adger et al. 2005; Fabricius et al. 2007). Ultimately the engagement of local communities in management can support adaptation by building adaptive capacity, by introducing new institutional arrangements (Plummer 2013), and by increasing resilience (Fabricius et al. 2007).

The diagnostic studies developed in the CPM project highlighted that the main current problems affecting the lagoon system are: erosion and high vulnerability to extreme sea storms of the littoral bar, high social and physical vulnerability of the areas around the lagoon to flooding, lack of wastewater treatment and bacteria contamination of the lagoon, loss of mangrove habitats, saltwater intrusion in freshwater systems, salinization of soils and consequent impacts on agriculture, lack of facilities for proper collection and management of waste, and high vulnerability of local communities in general and to climate variability in particular. Climate change will exacerbate some of these impacts (e.g. increased erosion of the littoral bar, increased frequency and intensity of flooding events, expansion of areas affected by soil salinization, etc.) and induce other effects (e.g. risk of rupture of the littoral bar, mangrove inland migration, potential diffusion of vector-borne diseases, introduction of new alien species, etc.).

The present article focuses on the role that the participatory process had in identifying community-based adaptation measures for the CPM lagoon system. It has a twofold goal:

- From a research perspective, it illustrates the methodology that was ad hoc developed and tested in the first workshop (mainly focusing on vulnerability evaluation and assessment of climate change impacts), and discusses the application of a more consolidated multi-criteria approach based on the Mulino Decision Support System software tool (Giupponi 2014), which was applied in the second workshop (focusing on adaptation).
- From a practitioner's perspective, it presents the key outcome of the two workshops, that had important implications for other activities of the CPM project. Results of the first workshop complemented the scientifically-based analysis of impacts due to current human activities and future climate change run in the CPM project, that made use of previously existing data, new data acquired in the field, climate change scenarios and projections, mathematical and conceptual models, GIS, statistical analysis, etc. Results of the second workshop crucially contributed to the final selection of the community-based adaptation measures that were designed in the last activity of the project. The second workshop was also extremely helpful in providing local knowledge about highly vulnerable and/or valuable sites where the adaptation measures should be implemented.

The article is organized as follows. The methodology and the results are presented in the same section, in chronological sequence, to enable better understanding of the process. Then the results are discussed and some conclusions are drafted with respect to their potential interpretation for use.

Methodology and Results

Within the broader scope of the CPM project a participatory process consisting of two workshops (marked in dark yellow in the flow chart—Fig. 3.2) was designed to enable involvement of local stakeholders to the design of climate change adaptation measures (marked in orange in the flow chart).

The CPM project consisted of three main activities, carried out independently but in synergy with the two workshops (marked in pale yellow in the flow chart): (1) field surveys and field visits, (2) a desk study on current and future vulnerability of the CPM lagoon system to climate change and human activities, considering a wide range of sectors, i.e. littoral dynamic, hydrology, environmental contamination, soil degradation, biodiversity and habitat protection and restoration, infrastructures, social aspects, local economy (including in particular agriculture and fishery), (3) analysis of adaptation options, which could be implemented to decrease impacts from climate change and anthropogenic pressures and improve resilience. Three outputs (marked in green in the flow chart) were the result of these three

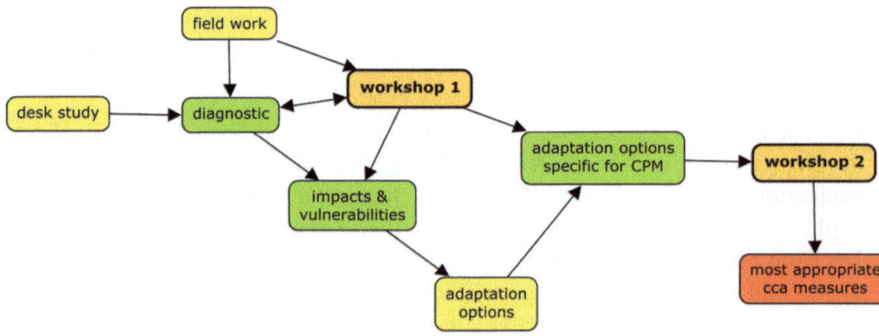

Fig. 3.2 Flow chart of project

activities: (1) a diagnostic of the present state of the lagoon socio-ecosystem, (2) a detailed analysis of impacts and vulnerabilities due to future climate change (based on climate change projections specifically generated for this project) and current anthropic activities, and (3) the identification of good adaptation practices to be potentially applied to local specificities of the CPM lagoon system.

More specifically, the **goal of the first workshop** was to investigate problems associated with impacts from climate and from anthropogenic activities in the CPM lagoon system, describing the socio-ecosystem and its vulnerability to impacts from climate change. Information on issues related to the environmental, social and economic spheres, as well as to cultural and ecological systems was collected and integrated in a diagnostic report. This information was then used in the last part of the workshops to identify possible adaptation measures, which should reduce vulnerability.

The **goal of the second workshop** was to evaluate relative effectiveness of proposed adaptation measures. These have been identified based on the impacts and vulnerabilities described for the lagoons, both through the diagnostic analysis and in the first workshop. Also, resources for their implementation were identified. Moreover, during the second workshop a list of key sites, hot spots to be preserved, where the adaptation measures should be implemented, was also created.

Then the **selection of workshop participants** was carried out. Having defined the goals for the two workshops, we decided that the best option would be for the same people to participate in both, so we could strengthen the process by building on previously shared knowledge. We identified participants to be invited following recommendations from INECC and with reference to the project goals. Therefore, we thought it would be appropriate to involve local stakeholders, who would share with us their knowledge on the CPM lagoon system. Stakeholders identified belonged to different sectors: academia, local, national and international institutions, non-governmental organization, and local communities. Due to the heterogeneity of the backgrounds, and to the diverse capabilities of these people we decided to divide each workshop in two days: **day one** for academia and institutions, and **day two** for local communities and civil society organizations.

The most relevant criteria to identify participants from academia and institutions were to select people knowledgeable on climate change impacts, and, in particular, on how climate change affects the CPM lagoon system. Professors and researcher from the local Universities as well as civil servants from the three levels of government (Federal, State, and local) were selected. First the institutions with mandates related to the goal of the workshops were listed, then one or more representative from each institution was invited, specifically requesting people who could contribute to the workshop. The selected people were formally invited by INECC.

The most relevant criteria to select local communities and civil society organizations were:

1. communities from the municipalities of Cárdenas, Comalcalco, and Paraíso surrounding the CPM lagoon system,
2. communities that are vulnerable to climate change impacts and anthropogenic pressures,
3. communities with civil society organizations, who represent women, fishers, and farmers.

After the communities and civil society organizations were selected the representatives of these were invited to participate, making sure these were people who were willing to share their knowledge and experience on climate change vulnerabilities and anthropogenic pressures on the lagoons. The selected people were formally invited by INECC, and this invitation was followed-up during face-to-face interviews carried out as part of the activities of the socio-economic diagnostic analysis, making sure invitees would understand what kind of opportunity participating in this workshop would bring for their communities.

The methodologies, and their implementation, were designed taking into consideration the goals of the workshops, and the differences among the groups of participants (including illiteracy of the local communities). Day one of both workshops was dedicated to academic and institutional stakeholders, day two to community representatives and civil society organizations.

The **first workshop**'s (March 2015) specific aim was to elicit local knowledge on impacts and vulnerabilities of the lagoon socio-ecosystem to both current human activities and climate change, as well as to discuss initial ideas on what measures are needed to address them (Ziccardi Contigiani 2004).

During **day one** (27 March 2015) 37 out of the 72 people that were invited participated. Academic and institutional stakeholders were first given presentations to introduce: the macro-project "Adaptation of coastal wetlands of Gulf of Mexico to climate change impacts", the concepts of climate change vulnerability and adaptation in general, the expected impacts on the CPM lagoon system, possible activities and strategies to contrast climate impacts, and the specific objectives and activities of the project on the CPM lagoon system. Then participants were asked to fill matrices individually to identify anthropogenic and climatic pressures on the lagoon socio-ecosystem, and to list measures to address them. In particular, participant worked on the following matrixes:

- Matrices 1A and 1B: assessing contribution of human activities to the impacts of climate change (Matrix 1A—Table 3.1), and to anthropogenic impacts (Matrix 1B—Table 3.2)
- Matrix 2: identifying climate change impacts on human activities (Table 3.3)
- Matrix 3: identifying possible adaptation measures in the CPM lagoon system.

The agenda allowed for individual time and discussion time, to start synthetizing concepts and ideas divided in four groups, and ultimately to present the preliminary results in a plenary session.

After the end of the workshop all the matrices were analyzed and processed using a spreadsheet, summing values given individually. The number of matrices filled was as follows:

- Matrix 1A: 13
- Matrix 1B: 12
- Matrix 2: 24
- Matrix 3: 22

Looking at graphs and charts relative to Matrices 1A and 1B, the human activities that are perceived as those causing the greatest impacts are: agriculture, infrastructure, industry and human settlements. Moreover, the human activities with the lowest impacts are: aquaculture, fisheries and tourism. Even if fisheries are one of the activities with the lowest impacts, it contributes the most in changes of fish species distribution (Fig. 3.3), and in overexploitation of resources associated (Fig. 3.4). The other sector with the least impact is tourism, which is close to being currently absent in the area, but it must be noted that this sector could cause high impact on one of the most problematic issue: water resources availability (Fig. 3.4).

Looking at the graphs and tables that synthetize Matrix 2 (impacts of climate change on human activities), the increase in hydro-meteorological events is the highest perceived impact. Other relevant impacts are: droughts, precipitation increase and temperature increase. The least perceived relevant impacts are: ocean acidification, and sea temperature increase. Moreover, the most impacted sectors are: rainfed agriculture, water availability for human consumption, human settlements, health, and livestock. Each sector was then analysed individually, the most relevant impacts by sectors are as follows:

- rainfed agriculture: drought
- water availability for human consumption: drought
- human settlements: sea level increase, precipitation increase, hydro-meteorological events increase
- health: drought
- livestock: drought
- fisheries: hydro-meteorological events increase

Table 3.1 Matrix 1A: how human activities (rows) contribute to impacts from climate change (columns)

	Sea level increase	Air temperature increase	Change in precipitation patterns	Extreme hydro-meteorological events	Water availability	Change in species distribution
Livestock						
Fishery						
Agriculture						
Tourism						
Aquaculture						
Infrastructure						
Forestry						
Industry						
Human settlements						

Table 3.2 Matrix 1B: how human activities (rows) contribute to anthropogenic impacts (columns)

	Fresh water contamination	Salt water contamination	Upper watershed deforestation	Local deforestation	Invasive species	Resources overexploitation	Costal infrastructure	Upper watershed infrastructure	Road and communication infrastructure	Hydrological network modification
Livestock										
Fishery										
Agriculture										
Tourism										
Aquaculture										
Infrastructure										
Forestry										
Industry										
Human settlements										

Table 3.3 Matrix 2: identification of climate change impacts (rows) on human activities (columns)

	Livestock	Fishery	Aquaculture	Agriculture		Tourism	Subsistence	Infrastructure	Forestry	Industry
				Rainfed	Irrigated					
Seal level rise										
Ocean acidification										
Sea temperature increase										
Air temperature increase										
Precipitation increase										
Increase in hydro-meteorological events										
Drought										
Forest fires										

	Biodiversity			Water bodies		Coastal morphology	Water available for human consumption	Human settlements	Health	Employment	Education	Organization
	land	Marine	Coastal	Quality	Quantity							
Seal level rise												
Ocean acidification												
Sea temperature increase												
Air temperature increase												
Precipitation increase												
Increase in hydro-meteorological events												
Drought												
Forest fires												

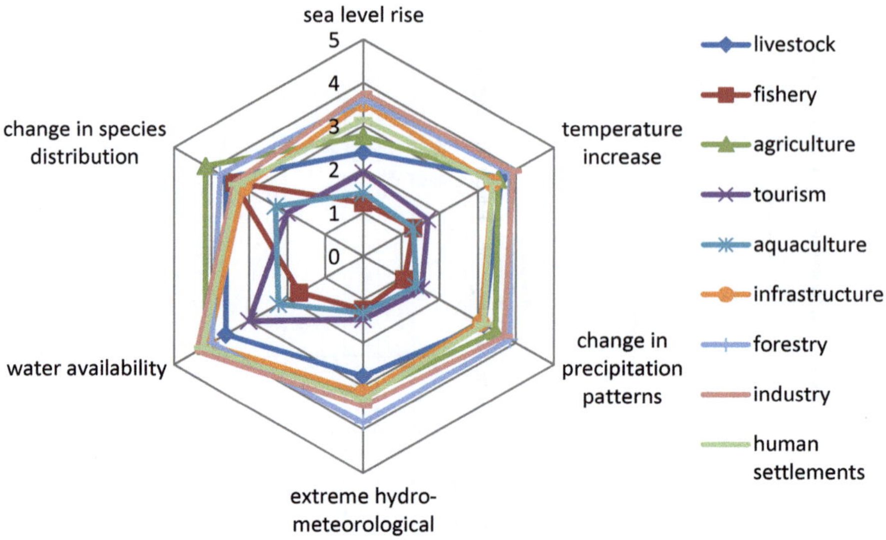

Fig. 3.3 Matrix 1A: how human activities contribute to the impacts from climate change

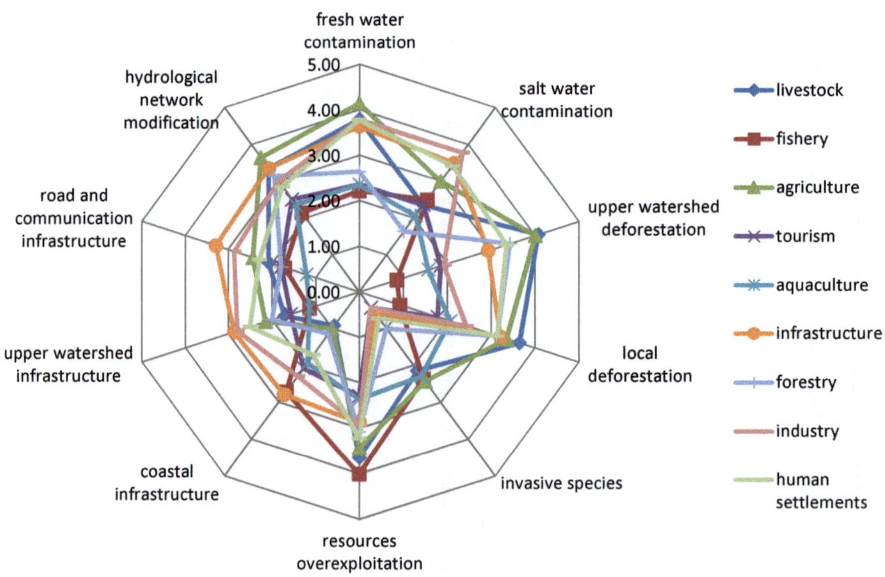

Fig. 3.4 Matrix 1B: how human activities contribute to the anthropogenic impacts

The last activity for the day was filling Matrix 3 to identify initial proposals of adaptation measures for the CPM lagoon system. This was achieved based on the activities and discussions carried out earlier during the day. To close the workshop each of the four groups developed measures, which were then presented in a plenary session.

On **day two** (28 March 2015) 28 out of the 65 people who were invited participated. Representatives of local communities and of civil society organizations first listened to presentations, which introduced: the macro-project "Adaptation of coastal wetlands of Gulf of Mexico to climate change impacts", the concepts related to climate change, and the specific objectives and activities of the project on the CPM lagoon system. Following this an interactive activity to learn and discuss about key concepts was carried out. Participants were divided into three random groups and each group was asked to define each of the concepts: climate change, adaptation, and vulnerability. The discussion on the three concepts made us aware of the fact that participants have some sense and idea of what we would be talking about during the workshop.

Then, for the last part of activities, participants were divided into three homogeneous groups: women, fishers, and farmers. Each participant, supported by facilitators, filled a set of coloured cards individually to identify impacts of climate change on the different economic sectors or community levels (Table 3.4). These cards were gradually added to a board on the wall, so that workshop facilitators could start synthetizing the information capturing the most relevant impacts and their consequences (Table 3.4). To end this activity each group discussed about impacts and identified possible measures, which should be implemented to address these impacts. The workshop finished with representatives from each group presenting impacts and measures in a plenary session.

After the workshop the cards were systematized using the following criteria:

- deletion of opinions which are not related to the described impact, e.g. the impact "sea water temperature increase" and "fertile soil loss" do not appear to have any relation,
- selection of opinions specifically related to the described impact,
- grouping of similar opinions,
- highlighting of the most recurring opinions.

All opinions were recorded, even when they seem to contradict each other, thinking they might be all true depending on the community and place it refers to. For example, one participant said that during drought fish landings increase, another they decrease: this might be related to fished species and specific fishing place. Generally speaking, however, opinions are repeated, giving strength to their truthfulness.

The women's group highlighted scarcity and/or low quality in most resources needed: fisheries, soil, water, yield, energy, wood, livestock. This has consequences on the social-ecological system: migration, unemployment and job scarcity, as well as health issues, which have also been identified.

The fishers' group identified mostly problems related to fisheries: species and fisheries decline, and temperature changes for the water of the lagoons. However, they also identified other general problems: decrease in water quality, soil

Table 3.4 Matrix 1: number of cards filled in each group

	Community			Agriculture sector			Livestock sector			Fisheries sector			Family			
Impacts	P	A	M	P	A	M	P	A	M	P	A	M	P	A	M	Total
Sea level rise	6	8	6	4	8	7	0	4	6	4	0	0	6	8	6	73
Sea water temperature increase	3	7	7	0	0	2	0	0	2	4	3	7	0	0	3	38
Temperature increase	4	6	7	2	6	7	2	7	8	5	1	7	6	7	7	82
Precipitation increase	0	1	7	2	7	7	4	6	6	5	0	7	6	6	7	71
Extreme events increase (hurricanes, Nortes)	4	4	7	4	8	7	1	7	8	4	0	7	6	10	7	84
Droughts	5	7	7	1	6	6	0	5	1	4	1	8	6	6	7	70
Total	22	33	41	13	35	36	7	29	31	26	5	36	30	37	37	**418**

P fishers; *A* farmers; *M* women

salinization, health issues, livestock loss, job scarcity, yield loss. They also lamented lack of support from authorities, e.g. fisheries closures do not take into account local biological cycles.

The farmers' group reaffirmed what the other two groups have also identified: soil and well salinization, water scarcity, yield decrease and loss, livestock suffering, mangroves loss, human health, unemployment.

Based on these opinions adaptation measures to decrease vulnerabilities from climate change while benefiting the socio-ecosystem, were identified and discussed.

To conclude the description of the first workshop, we present below a summary of the most relevant information, which formed one of the bases upon which the consortium identified possible adaptation measures.

- Human activities affected by highest impact: agriculture, infrastructure, industry, and human settlements.
- Human activities affected by lowest impact: aquaculture, fisheries, and tourism.
- Highest climate change direct impacts: increase of extreme events, followed by droughts, changes in precipitation patterns and temperature increase.
- Lowest climate change direct impacts: ocean acidification, and sea water temperature increase.
- Sectors and activities perceived as more affected by climate change impacts: rainfed agriculture, availability of water for human consumption and livestock.
- Other relevant local problems: soil and water contamination due to oil extraction and to agrochemicals, garbage disposal, deforestation, upstream pollution, emigration, water and soil salinization, increase in morbidity, fisheries decline, decrease in agriculture production, estate and assets loss.

Moreover, the measures identified are related to the following themes: capacity building and organization in the communities, green infrastructure, and new technologies for agriculture and livestock.

At this point the consortium elaborated and integrated information coming from the workshop with information coming from the other research efforts of the project, in particular taking into account local vulnerabilities of the CPM lagoon system, and the concrete feasibility of possible adaptation measures. This process led to identify a list of 32 possible **adaptation measures**, considered to be the most feasible to apply in the area (Fig. 3.2). For the specific purpose of the second workshop, this list was further refined using a set of criteria (about 20) defined by the consortium.

The nine selected measures to evaluate in the second workshop were:

- **M1** Micro-credit system to strengthen local micro-industry, and sustainable socio-economic activities, which could convert problems caused by climate change into opportunities
- **M2** Drainage system to control water levels during precipitation events
- **M3** Renewable energy: wind mills and solar panels
- **M4** Garbage disposal plan based on community involvement
- **M5** Management plan for mangroves with innovative strategies: reforestation and banks stabilization, biofences
- **M6** Waste water treatment with micro-scale phytodepuration
- **M7** Water abstraction from deep wells
- **M8** Management plan for oyster harvesting and processing to improve fishery resource quality
- **M9** Adaptation of agricultural practices to improve resilience to climate change (e.g. climate resilient farming, agro-forestry, use of climate resistant crops, etc.)

The **second workshop**'s goal was to present and discuss the results of the first workshop as well as of the analysis on current and future impacts of human activities and climate change; to present, discuss and evaluate the refined list of nine measures; to identify resources for their implementation; and to identify key sites for the implementation of the most promising measures.

In the two days of the second workshop (June 2015) all participants first discussed the benefit, and possible weaknesses, of the nine measures, then provided qualitative evaluations to identify the five most promising measures to address impacts using a multi-criteria decision support system software: Mulino Decision Support System[1] (mDSS) (Feás Vázquez 2008; Giupponi 2014). Only during day one an activity was carried out to identify possible financial resources to support the realization of those measures, and key sites, where the measures could be implemented.[2]

---

[1]http://www.netsymod.eu/mdss/ accessed June 2016.

[2]This activity will not be reported in this article.

For the multi-criteria analysis the first exercise was to weight the relative importance of the criteria. The Simple Additive Weighting (SAW) rule was used, therefore each participant had a fixed number of points to distribute (100 the first day, 20 the second). A list of ten criteria was developed by the consortium with reference to present and future environmental, social and economic impacts, feasibility of implementation, and resources needed. The criteria used in both days were similar, but allowed to account for the specificity of the participants (also in terms of criteria descriptions). The ten criteria were then used to evaluate the nine measures. Taking into account the different background of the participants, in day one (academic and institutional stakeholders) matrices were used to carry out the exercise, while in day two (local communities representatives and civil society organizations) a questionnaire was provided. The evaluation was done using a qualitative scale from one (worse performance) to five (best performance).

The core of the workshop was the evaluation of the measure with respect to their effectiveness in addressing impacts from climate and anthropogenic change. For this each participant filled in individually a table to weight criteria. Then on day one each participant filled in a matrix to evaluate effectiveness of measures, where the ten criteria are presented in rows, and the nine measures are presented in columns. While on day two each participant selected the five most promising measures among the nine presented, and then answered the questionnaire to evaluate relative effectiveness of only these five measures.

On **day one** (19 June 2015) 40 participants came to the workshop, while the invitations were sent to the same 72 people of the first workshop. The activities were divided into three sections. The first was dedicated to presentations on: (1) update on the macro-project "Adaptation of coastal wetlands of Gulf of Mexico to climate change impacts"; (2) results of the analysis of the current state of the CPM lagoon system and assessment of expected climate change impacts performed by the consortium experts; (3) results of the first workshop. The second section focused on: (a) the two exercises related to the multi-criteria analysis, (b) the identification of resources for measures' implementation, and (c) the identification of key sites of the CPM lagoon system.[3] This section included the explanation of multi-criteria analysis, the presentation and discussion of the criteria and, most importantly, the illustration of the nine pre-identified adaptation measures. In the last part of the workshop some preliminary results of the multi-criteria analysis were presented in the plenary session.

On **day two** (20 June 2015) 48 participants came to the workshop, while the invitations were sent to the same 65 of the first workshop. Similarly to day one, the activities were divided into three sections. The first section was dedicated to presentations on: (1) update about the macro-project "Adaptation of coastal wetlands of Gulf of Mexico to climate change impacts", (2) disaster risk management in communities, (3) results of the analysis of the current state of the CPM lagoon system and assessment of expected climate change impacts performed by the

[3](b) and (c) will not be reported in this article.

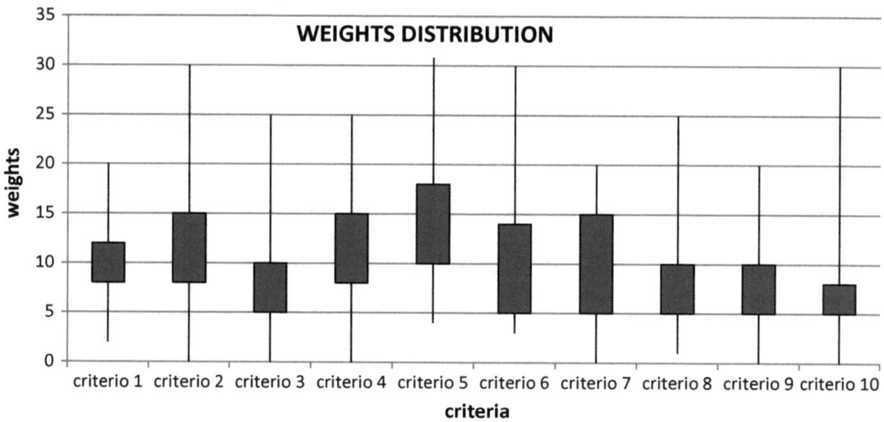

Fig. 3.5 Weights distribution from day one (central box is between 25th and 75th percentile)

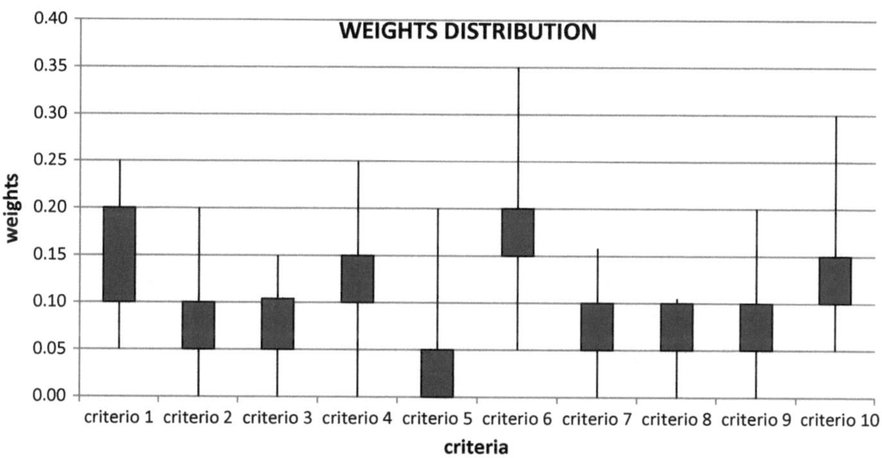

Fig. 3.6 Weights distribution from day two (central box is between 25th and 75th percentile)

consortium experts, (4) results of the first workshop. The second section focused on the two exercises related to the multi-criteria analysis. This section included the explanation of multi-criteria analysis, the presentation and discussion of the criteria and, most importantly, the illustration of the nine pre-identified adaptation measures. To end the workshop a discussion on the measures, their effectiveness, and positive and negative impacts took place: participants were divided into the same three groups as the first workshop (women, fishers, and farmers).

Comparing **criteria weights** from day one to those of day two we can say that there was a higher degree of consensus on day one, and that on day one generally criteria related to feasibility were given less importance (Figs. 3.5 and 3.6). Moreover, criteria weights can give us some insights on which problems the adaptation measures should address (Tables 3.5 and 3.6). In both days, in fact, these were the issues expressed by the most important criteria: socio-economic needs of communities (criterion no. 1), decrease impacts from climate change (criterion no. 4), measures should benefit the most disadvantaged groups, e.g. women, children, elderly (criterion no. 6). Last but not least, the criterion on maladaptation (criterion no. 5) was the most important in day one, and the least important in day two. Maladaptation, which is highly relevant for academic and institutional participants, appears not to be so important for local community representatives.

To **evaluate the measures** all criteria weights and matrices (questionnaires from day two were transformed into matrices) were input in mDSS participant by

Table 3.5 Statistics of criteria weights for day one (total points 100)

	Average	Median	Mode
1. The measure addressed socio-economic needs	10.11	10	10
2. The measure decreases vulnerability	11.81	10	10
3. The measure addresses present issues related to climate	9.60	10	10
4. The measure will address future climate change impacts	10.86	10	10
5. The measure will bring medium term concrete benefits (environmental, social, and economic) and will not bring new socio-environmental vulnerabilities (maladaptation actions)	13.60	11	10
6. The measure will benefit the most vulnerable people	10.69	10	5
7. The measure is feasible with respect to public policies and laws	9.78	10	10
8. The measure can be adapted to new socio-environmental and climatic conditions	9.42	9	10
9. The measure is coherent with state and municipal policies	7.50	5	5
10. Your institution will be able to support the measure	6.64	6	5

Table 3.6 Statistics of criteria weights for day two (total rescaled to 100)

	Average	Median	Mode
1. The measure addresses socio-economic problems	14	15	15
2. The measure addresses environmental problems	9	10	10
3. The measure reduces current climate change impacts	9	10	10
4. The measure will address future climate change impacts	12	10	10
5. The measure will cause future damage	5	5	5
6. The measure will help the most vulnerable people	17	15	15
7. Considering your capacities, will it be easy to develop the measure?	9	10	10
8. The measure can be adapted to changing social, economic and environmental conditions	7	5	10
9. The measure takes into account local traditions	7	5	5
10. The measure will be accepted by your community	13	10	10

Table 3.7 Effectiveness ranking of measures

	1	2	3	4	5	6	7	8	9
Day 1	**M5**	**M4**	M9	M6	**M8**	*M2*	M7	*M3*	M1
Day 2	M1	**M5**	**M4**	**M8**	M7	M9	*M3*	M6	*M2*

participant, generating option files, each with an individual relative ranking of the measures, not all participants completed the exercises: 36 out of 40 for day one, and 26 out of 48 for day two. Then all the option files were imported in mDSS and combined with an algorithm corresponding to a decision rule, which aggregates all individual preferences in a consensus ranking (Table 3.7).

Since one of the final goals of the project is to identify five adaptation measures for the CPM lagoon system, the rankings from both days were compared looking at the first five measures in each days' ranking. Three measures, namely **M4**, **M5**, and **M8** (**bold** in Table 3.7) are positioned in the first five ranks in both days, so there is a consensus coming from the workshop to consider those as worth being developed. Moreover, there is a consensus on two measures which should not be taken into consideration, according to local knowledge, namely *M2* and *M3* (*italic* in Table 3.7), as these were not in the first five ranks in neither day. Considering that the communities will be implementing and managing the measures, M1 could be developed (underlined in Table 3.7), even if it ranked last in the first day, because it ranked first in the second day. The fifth measure to be developed should be chosen between M6, M7, and M9, using information coming from other sections of this project, e.g. feasibility, socio-economic impact, resilience, and synergies.

Discussion and Conclusions

The high vulnerability of the CPM lagoon system to climate change, climate variability, including extreme events in particular, and human impacts calls for the design of tailored adaptation measures. Climate change, in fact, will exacerbate some of the already existing problems (e.g. flooding, littoral erosion, soil salinization) and cause new ones (e.g. great variation of lagoon salinity, inland migration of mangroves, spreading of species transmitting diseases to humans as dengue). Moreover, since present and future problems affect the whole socio-ecosystem, the measures should be ecosystem based. Also, to grant success in implementing the measures, they should be community-based, thus involving local people in their realisation and maintenance. To implement these principles a participatory process has been designed within the project for the design of the adaptation measures.

The activities developed and carried out for the **first workshop** (diagnostic workshop) were adequate with respect to the goal and, thus, gave expected results: knowledge on (current and expected) impacts and vulnerabilities in the CPM lagoon system. During day one (involving academic and institutional participants) there was general consensus, as testified by the spread of the opinions as expressed through qualitative values. Also, discussions were useful to gain specific insights. In day two (involving local communities and civil society organizations) all participants shared their knowledge and opinions, identifying most pressing needs of the communities. Moreover, both days were an opportunity for people to meet and possibly have future opportunities to engage and collaborate. Last but not least, the activity to start identifying possible adaptation measures was successful: initial ideas were identified individually in day one, while in day two measures were listed in a group setting. All this information, together with the information coming from the field visits and the wide range of desk studies, was used by the consortium to identify best practices to increase the resilience of the socio-ecological system (Fig. 3.2). Having information coming from different studies and obtained using different methods gave robustness to the findings: one conclusion can be drawn on the necessity to cross-check information using a similar integrated approach as the one used in the CPM project.

The multi-criteria analysis used in the **second workshop** to evaluate the relative effectiveness of the nine proposed measures not only gave promising results, but also was useful to enable discussion on climate change impacts, vulnerabilities and how to increase local resilience. It is important to say that this is coherent with the methodology itself, that, as any DSS tool, it should not be used to achieve "the absolute truth", but rather to enable discussion through comparison. Moreover, the information on performance of measures with respect to needs and expectations, which are expressed by criteria, results in the possibility of finding consensus, rather than conflict, by identifying pressing issues and the relatively best measure to address them. These findings are coherent with those of other similar processes where multi-criteria analysis was used in a participatory setting to identify and develop adaptation measures (Giannini et al. 2011; Giupponi 2014).

The final ranking (Table 3.7) was further analysed going back to the single option files created by the software mDSS. Several possible groupings were hypothesised, but none was in the end judged methodologically correct. The first aggregation was created considering all option files together, but this introduces a bias because we have 36 option files from day one and 26 from day two. The second aggregation was created based on the fact that on day one some institutions were represented by more than one person, so we tried averaging the information from participants belonging to the same institution. The results were again judged not methodologically correct, because of the diversity of opinions expressed by each participant. So we can conclude that the comparison of the two rankings (day one vs. day two) is the best option, therefore M4, M5 and M8 should be selected and developed further, while M2 and M3 should be discarded. The other two measure to be developed should then be chosen among M1, M6, M7 and M9. Considering that the measure will be implemented and managed by the communities it is advisable to choose M1, the preferred measure in day two, even if it was judged the worse in day one. Then, perhaps, the other measure could be decided between M7 and M9, because M6 was the second to last in day two, that were local communities participated. Both M7 and M9 have the possibility of addressing some of the pressing needs and issues, e.g. water availability (M7), and soil salinization and job availability (M9). However, before making a final decision more specific information would need to be gathered, e.g. on feasibility or on economic possibilities.

Referring to some of the comments made during both workshops, the following points and specificities should be taken into consideration while designing the measures:

- measure 1: improve capacity of institutions to cooperate, empower local micro-industry and promote local socio-economic sustainable activities;
- measure 4: address local specificities, e.g. transform solid residues in economic benefit for communities, design a deposit to sell waste;
- measure 8: consider algal blooms, temperature and salinity increase of water;
- measure 9: address soil erosion, organize farmers, identify most resistant plants.

Moreover, it must be noted that the selected measures could be developed trying to enable synergies among them or with other activities, which are useful for the sustainability of the measures themselves. Moreover, the development of measures should also have as a goal that of re-creating social bonding, as it appears that communities are quite disaggregated. All measures, to fulfil needs voiced, could also integrate capacity building and empowerment, community participation and organization, and, last but not least gender issues.

Some issues and difficulties arose and were addressed during the organization of the two workshops. For day one all institutions that were contacted were able to identify and send experts to participate in both workshop, also ensuring some continuity by sending almost the same people to both workshops. For day two, the biggest challenge was due to the difficulty in accessing the location of the workshop. There was discussion on whether it would be more appropriate to have small focus groups, one for each community, but the timing and resources of the CPM project did not allow for this, and, also, grouping all communities in one location had the added value of possibly establishing ties and synergies among local communities, who share similar vulnerabilities and needs, and have similar or complementary capacities, which could foster peer-to-peer learning.

Further research could try to investigate and evaluate the degree of acceptance by local communities of the proposed measures, perhaps focusing on understanding which communities could be mostly benefited by each of the measures. This could give insights, which could be useful for the implementation and management of the measures.

Last but not least, with reference to a generic project cycle, having completed the preliminary phases of problem framing and project design, now the focus should shift towards completing the CPM project by progressing towards the implementation, management, monitoring and assessment, and review phases. The latter could be a useful test to understand more about the effectiveness of participation as a tool to guarantee success of adaptation measures.

Acknowledgements This research is part of the project "Diseño de Medidas de Adaptación para Reducir la Vulnerabilidad del sitio piloto del Sistema lagunar Carmen-Pajonal-Machona, Tabasco, a los Impactos Generados por el Cambio Climático y por Actividades Humanas", funded by the World Bank and implemented by the Instituto Nacional de Ecología y Cambio Climático (INECC) and the Instituto Mexicano de Tecnología del Agua (IMTA).

Valentina Giannini coordinated all work related to the two workshops, developed the methodology for the second workshop, facilitated both workshops, and analysed their results. Emiliano Ramieri coordinated the general CPM project and supervised the methodology for the two workshops. Margarita Caso Chávez, Karina Santos del Prado Gasca, and Karla Priego Martínez managed the project, developed the methodology for the first workshop and supervised the development of the methodology for the second workshop.

A word of thanks goes to the stakeholders who shared their opinions, to the people who gave presentations during both workshops, and to the people from the consortium and from INECC who helped carry out the activities during both workshops.

References

Adger WN, Hughes TP, Folke C et al (2005) Social-ecological resilience to coastal disasters. Science 309:1036–1039. doi:10.1126/science.1112122

Adger WN, Brown K, Nelson DR et al (2011) Resilience implications of policy responses to climate change. WIREs Clim Change 2:757–766. doi:10.1002/wcc.133

Ayers J, Forsyth T (2009) Community-based adaptation to climate change. Environ Sci Policy Sustain Dev 51:22–31. doi:10.3200/ENV.51.4.22-31

Carpenter SR, Mooney HA, Agard J et al (2009) Science for managing ecosystem services: beyond the millennium ecosystem assessment. PNAS 106:1305–1312. doi:10.1073/pnas.0808772106

CBD (2009) Connecting biodiversity and climate change mitigation and adaptation: report of the second ad hoc technical expert group on biodiversity and climate change. [online] https://www.cbd.int/doc/publications/cbd-ts-41-en.pdf. Accessed June 2016

Crabbe MJC, Martinez E, Garcia C et al (2009) Is capacity building important in policy development for sustainability? A case study using action plans for sustainable marine protected areas in Belize. Soc Nat Resour 23:181–190. doi:10.1080/08941920802409593

Eriksen S, Brown K (2011) Sustainable adaptation to climate change. Clim Dev 3:3–6. doi:10.3763/cdev.2010.0064

Fabricius C, Folke C, Cundill G, Schultz L (2007) Powerless spectators, coping actors, and adaptive co-managers: a synthesis of the role of communities in ecosystem management. Ecol Soc 12:29

Feás Vázquez J (2008) Una metodología multicriterio para la gestión sostenible de recursos hídricos. [online] http://dspace.usc.es/handle/10347/2504. Accessed June 2016

Folke C, Carpenter S, Elmqvist T et al (2002) Resilience and sustainable development: building adaptive capacity in a world of transformations. AMBIO: A J Hum Environ 31:437–440. doi:10.1579/0044-7447-31.5.437

Giannini V, Ceccato L, Hutton C et al (2011) Development of responses based on IPCC and"what-if?" IWRM scenarios. Adv Sci Res 7:71–81

Giupponi C (2014) Decision support for mainstreaming climate change adaptation in. Water Resour Manage 28:4795–4808. doi:10.1007/s11269-014-0776-y

Hale LZ, Meliane I, Davidson S et al (2009) Ecosystem-based adaptation in marine and coastal ecosystems. Renew Resour J 25:21–28

IPCC (2014) Glossary, Annex II. Contribution of working group II to the fifth assessment report of the intergovernmental panel on climate change [online] http://ipcc.ch/pdf/assessment-report/ar5/wg2/WGIIAR5-AnnexII_FINAL.pdf. Accessed June 2016

Lhumeau A, Cordero D (2012) Adaptación basada en Ecosistemas: una respuesta al cambio climático. IUCN, Quito, Ecuador 17 pp

Mimura N, Pulwarty RS, Duc DM, Elshinnawy I, Redsteer MH, Huang HQ, Nkem JN, Sanchez Rodriguez RA (2014) Adaptation planning and implementation. In: Field CB, Barros VR, Dokken DJ, Mach KJ, Mastrandrea MD, Bilir TE, Chatterjee M, Ebi KL, Estrada YO, Genova RC, Girma B, Kissel ES, Levy AN, MacCracken S, Mastrandrea PR, White LL (eds) Climate change 2014: impacts, adaptation, and vulnerability. Part a: global and sectoral aspects. Contribution of working group II to the fifth assessment report of the intergovernmental panel on climate change. Cambridge University Press, Cambridge, United Kingdom and New York, USA, pp 869–898

Noble IR, Huq S, Anokhin YA, Carmin J, Goudou D, Lansigan FP, Osman-Elasha B, Villamizar A (2014) Adaptation needs and options. In: Field CB, Barros VR, Dokken DJ, Mach KJ, Mastrandrea MD, Bilir TE, Chatterjee M, Ebi KL, Estrada YO, Genova RC, Girma B, Kissel ES, Levy AN, MacCracken S, Mastrandrea PR, White LL (eds) Climate change 2014: impacts, adaptation, and vulnerability. Part a: global and sectoral aspects. contribution of working group II to the fifth assessment report of the intergovernmental panel on climate change. Cambridge University Press, Cambridge, United Kingdom and New York, USA, pp 833–868

Plummer R (2013) Can adaptive comanagement help to address the challenges of climate change adaptation? Ecol Soc 18(4):2. [online] http://dx.doi.org/10.5751/ES-05699-180402. Accessed June 2016

Thomalla F, Downing T, Spanger-Siegfried E et al (2006) Reducing hazard vulnerability: towards a common approach between disaster risk reduction and climate adaptation. Disasters 30:39–48. doi:10.1111/j.1467-9523.2006.00305.x

van Aalst MK, Cannon T, Burton I (2008) Community level adaptation to climate change: the potential role of participatory community risk assessment. Glob Environ Change 18:165–179

Villamizar A, Gutiérrez ME, Nagy GJ et al (2016) Climate adaptation in South America with emphasis in coastal areas: the state-of-the-art and case studies from Venezuela and Uruguay. Clim Dev 0:1–19. doi:10.1080/17565529.2016.1146120

Wong PP, Losada IJ, Gattuso J-P, Hinkel J, Khattabi A, McInnes KL, Saito Y, Sallenger A (2014) Coastal systems and low-lying areas. In: Field CB, Barros VR, Dokken DJ, Mach KJ, Mastrandrea MD, Bilir TE, Chatterjee M, Ebi KL, Estrada YO, Genova RC, Girma B, Kissel ES, Levy AN, MacCracken S, Mastrandrea PR, White LL (eds) Climate change. Impacts, adaptation, and vulnerability. Part a: global and sectoral aspects. contribution of working group II to the fifth assessment report of the intergovernmental panel on climate change. Cambridge University Press, Cambridge, United Kingdom and New York, USA, pp 361–409

Ziccardi Contigiani AE (2004) Participación ciudadana y políticas sociales en el ámbito local, UNAM, Instituto de Investigaciones Sociales; Instituto Nacional de Desarrollo Social (INDESOL). México, Consejo Mexicano de Ciencias Sociales (COMECSO)

Zubair L (2004) Tiempo climate cyberlibrary, empowering the vulnerable. [online] http://tiempo.sei-international.org/newswatch/xp_feature040625.htm. Accessed June 2016

Chapter 4
Experimentalist Regional Governance for Climate Change Adaptation: A Canadian Case Study

Nicole Klenk, Dragos Flueraru and James Ian MacLellan

Abstract Climate change is affecting the life, livelihoods and survival of individuals and communities in many parts of the world. Moreover, the uncertainties associated with climate change impacts present an unprecedented challenge for adaptation planning. While climate change projections can be developed on the basis of global information and models, adaptation has to be informed by the knowledge of the communities where the consequences are felt. For the maritime province of New Brunswick there is growing necessity for policy-makers to incorporate adaptation considerations into daily decision-making and policies. The Regional Service Commissions (RSCs) are a new governance arrangement put in place to deliver municipal services, to facilitate regional planning and act as decision-making body on cross-boundary issues. As such, this institution may be a driving force for regional adaptation planning and implementation. This chapter aims to answer the following research questions: How can regional planning facilitate cooperation among municipalities with shared water and infrastructure governance issues? How are regional planners integrating and mobilizing local knowledge into regional adaptation planning? What models of environmental governance could inform the further development of RSCs in the context of climate change adaptation planning? In-depth interviews with RSC directors and planners provided the data, which was analyzed with a grounded theory approach. It was found that RSCs play a leadership role for adaptation planning and some have policies for infrastructural adaptation already in place. Institutional barriers to adaptation such as outdated legislation, centralized provincial power, and lack of a clear mandate were found to be common themes among RSCs. We discuss regional planning in light of experimentalist and co-productive models of environmental governance to address these barriers. While we focus on a case study of adaptation planning in New Brunswick, Canada, the insights derived from this case study and its implications for adaptation governance

N. Klenk (✉) · D. Flueraru · J.I. MacLellan
Department of Physical and Environmental Sciences,
University of Toronto, Toronto, ON, Canada
e-mail: Nicole.klenk@utoronto.ca

N. Klenk
1265 Military Trail, Toronto, ON M1C 1A4, Canada

W. Leal Filho and J.M. Keenan (eds.), *Climate Change Adaptation in North America*, Climate Change Management,
DOI 10.1007/978-3-319-53742-9_4

are not limited to this location, but speak to common adaptation planning challenges. In addition to presenting an illustrative case, this article also makes a theoretical contribution to the role of regional organizations in climate change adaptation governance, and understudied focus of climate change adaptation governance (Antonson et al. in Land Use Policy 52:297–305, 2016).

Keywords Regional governance · Experimentalism · Knowledge co-production · Coastal communities

Introduction

In recent decades efforts to mitigate carbon dioxide emissions have largely failed, and adapting to climate change impacts has become a pressing issue for policy and decision-makers across the globe (Adger 2003). Adaptation is broadly defined as the 'adjustment in natural or human systems in response to actual or expected climatic changes or their effects, which moderates harm or exploits beneficial opportunities' (IPCC 2014). Although adapting to changing climate conditions has been a rule rather than an exception for human societies and how they derive ecosystems services from their environment, the rate and magnitude of anthropogenic climate change threatens to push vulnerable ecosystems towards undesirable future states (Folke et al. 2005). Adaptation therefore is more than a technical issue, rather it should be framed as a dynamic social and institutional problem—a governance challenge—that requires a reconsideration of how we arrange and govern social-ecological systems (Hinkel et al. 2013; Van Nieuwaal et al. 2009). In this sense, adaptation represents a shift in the way societies operate, and so institutional change is necessary for policy processes that enable decisions to be made differently and be informed by newly available knowledge (Dovers and Hezri 2010).

By virtue of being along the Atlantic coast the province of New Brunswick has to tackle adaptation to climate change impacts such as: flooding of low-lying areas, coastal erosion, rising sea levels, saltwater leakage, groundwater contamination, higher prevalence of wildfires, and the appearance of invasive species and new pests (Province of New Brunswick 2014a). To address these issues the province has issued two policy documents: the New Brunswick Climate Action Plan and the New Brunswick Flood Reduction strategy. The express goal of these policies is to include considerations of adaptation into decision-making, integrate adaptation into relevant sectors, and to adapt communities and (current and future) infrastructure to climate change (Province of New Brunswick 2014a). Prior to 2008, the local governance system in New Brunswick had failed to deliver services to smaller communities, which had no methods of facilitating regional cooperation on planning or cross-boundary issues and constituted a significant rural population that had no elected representation (35% of the population) (Finn 2008). In response to this problem, the Regional Service Commissions (RSC) were instituted (in 2008) with the mandate to provide municipal services and facilitate regional collaboration and planning. Though the mandate does not include adaptation planning, a number of RSCs have taken initiative to integrate adaptation planning into their policies and practices (Province of New Brunswick 2015). This raises a number of questions as to how RSCs are facilitating regional collaboration and whether they are considering climate change in regional planning. Climate change is a global phenomenon manifesting itself at the local and regional level and as such transcends sectoral lines and impacts many aspects of planning and development (Sanchez-Rodriguez 2009). As urban development and land-use planning are generally facilitated at the regional level, it is of interest whether regional planners from RSCs can play a role in adaptation planning. Furthermore, RSCs may be a candidate to facilitate a novel governance approach: experimentalist governance. Experimentalist governance is a multi-scale governance arrangement that draws in multiple stakeholders in a deliberate polycentric decision-making process (Sabel and Zeitlin 2011).

The objective of this chapter is to better understand how RSCs enable regional collaboration and land-use planning and the extent to which climate change adaptation is included in planning and activities. The barriers and enablers RSCs face will be explored within the context of challenges to local level adaptation governance identified within the literature such as (1) conflicting timescales, (2) substantive, strategic and institutional uncertainty, (3) institutional crowdedness and institutional void, (4) institutional fragmentation, (5) lack of awareness and communication, (6) motives and willingness to act, and (7) lack of resources (Biesbroek et al. 2011). In this sense, close analytical attention will be paid to the relationship between knowledge, policy and practice (Dannevig and Aall 2015). Furthermore, the potential for the RSCs to facilitate regional adaptation governance using the theoretical model of experimentalist governance as a potentially useful scaffold is explored. This article builds upon prior research on multi-level adaptation governance (Betsill and Bulkeley 2007; Cashmore and Wejs 2014; Urwin and Jordan 2008; Knieling and Leal Filho 2013) and contributes an experimentalist institutional design perspective to recent studies on regional adaptation governance studies in Australia, Norway, Sweden, Holland and Canada (Jacobs et al. 2016; Dannevig and Aall 2015; Termeer et al. 2011; Cloutier et al. 2015; Antonson et al. 2016; Hanssen et al. 2013).

Research Questions and Case Study Methods

This study aimed to answer the following research questions: How are RSCs planning for and implementing regional adaptation measures? What are barriers to regional adaptation planning? How are RSCs facilitating regional cooperation? How are RSCs integrating and mobilizing knowledge into adaptation planning? Do RSCs have the potential to support an experimentalist governance approach to adaptation?

A qualitative data method was employed to answer these questions and in-depth interviews with RSC Directors comprise the data. Our case study is set within an inductive and interpretive research tradition in social science and humanities

(Gomm et al. 2000) that seeks to generate knowledge from immersion within a particular context. Emphasis is placed on "general fitness" and "transferability" of knowledge gained rather than the ability to generalize through statistical inference (Lincoln and Guba 2000). Efforts were made to draw out themes from interview transcripts, rather than to impose a theoretical framework for analysis.

Local-Level Adaptation Planning

Adaptation strategies are difficult to pinpoint for three overarching reasons: (i) how much adaptation is necessary given uncertainties implicit in changing climate conditions, (ii) how adaptation strategies may be tailored to the particular local context, and (iii) which adaptation actions are best suited within the contextual institutional capacity (Füssel 2007; Dovers and Hezri 2010). The concept of uncertainty pervades discussions of adaptation. This is due to several sources of inherent uncertainty: uncertainty in climate variables and climate impacts (often discussed in probabilistic terms) and uncertainty of context due to variation in local conditions: climate, economic, and social (Adger et al. 2005). While adaptation planning and implementation has largely had a focus at the national scale, the past decade has seen emphasis on locally based adaptation initiatives (IPCC 2014). In addition to the realization that adaptation at the national scale has largely failed to produce tangible adaptation action, findings suggest that the nature of adaptation is inherently "context-specific" (Measham et al. 2011). Local and regional governance systems are therefore the most on-the-ground and appropriate institutions to address these impacts (Sanchez-Rodriguez 2009). Local institutions play critical roles in climate adaptation: mobilizing responses to locally felt impacts, connecting individual and collective responses, and managing the delivery of resources that enable adaptation (Agrawal 2008). Municipal governments are required to provide a number of services and climate change issues are perceived as added burdens on a crowded agenda and limited resources (Crabbé and Robin 2006). Instead "community based environmental planning" (local, bottom-up approach) has been championed as an effective way to deal with environmental problems (Measham et al. 2011). A key dimension of CBEP is integrating different sources of knowledge: local, experiential, traditional and scientific; however, actors with competing interests have made this integration difficult (Measham et al. 2011).

Previous studies have noted the following constraints to local and regional climate adaptation governance, including competing and contradictory agendas at the local level (Storbjörk 2007; Betsill and Bulkeley 2007; Mukheiber et al. 2013); barriers to local knowledge integration in decision-making (Storbjörk 2007; Measham et al. 2011; Lieske et al. 2015); lack of resources and capacities (Sander-Regier et al. 2009; Craft et al. 2013); event-driven and short term policy horizons (Storbjörk 2007; Naess et al. 2005; Measham et al. 2011; Amundsen et al. 2010; Adger et al. 2009, Dannevig et al. 2013) and institutional constraints involving inadequate structures, processes, and distribution of responsibility across decision-making levels, and a lack in local capacity (Storbjörk 2007; Naess et al. 2005; Measham et al. 2011; Matthews 2013; Amundsen et al. 2010; Adger et al. 2009). From the governance challenges mentioned above, we can identify some of the desired characteristics of a regional climate adaptation governance model: it should reflect the multi-level, polycentric nature of the adaptation problem, and encourage knowledge generation, sharing, and learning from a variety of sources (Emerson and Gerlak 2014; Fossum 2012; Monkelbaan 2015; Knieling and Leal Filho 2013; Cloutier et al. 2015; Hanssen et al. 2013; Dannevig and Aall 2015). Changes to climate occur over timescales of decades (in other words a long-term change) which is difficult to reconcile with the short-term nature of decision-making, policies and political cycles (Adger et al. 2009). As previously mentioned though climate change may be a pressing issue that needs to be addressed, other problems such as economic growth and natural resource development may be perceived to have higher priority on the agenda. As such integrating long-term adaptation planning into new policies may be met with resistance from politicians, policy-makers and citizens (Edvardsson-Bjornberg and Hansson 2011; Hanssen et al. 2013; Antonson et al. 2016; Storbjörk and Hedrén 2011; Nilsson et al. 2012).

Experimentalist Governance

Experimentalist governance is a novel governance model that takes a different approach to the integration of knowledge and decision-making. It has been defined as "a recursive process of provisional goal-setting, and revision based on learning from the comparison of alternative approaches to advancing them in different contexts" (Sabel and Zeitlin 2011). Conceptually it encompasses a multi-scale architecture that brings together multiple stakeholders (public, private) in a polycentric decision-making process. It diverges from the conventional hierarchical model (top-down) but also from the bottom-up approach championed by community-based planning, and advocates for a fluid partnership type governance (or network governance) (Monkelbaan 2015; Sabel and Zeitlin 2011). It is a form of pragmatic and experimenting governance in that assumptions, knowledge and practice are treated with skepticism and doubt such that solutions are viewed as incomplete, and constantly subject to revision (Sabel and Zeitlin 2011). In other words, it advocates the reformulation and readjustment of ends and means through learning from comparison of local efforts to advance a framework of broad goals. In this sense it holds much in common with adaptive governance, as it advances principles of pragmatism and incrementalism (Monkelbaan 2015). It diverges from adaptive governance as it is concerned mainly by producing knowledge through practice and trial-and-error, stemming from decision-makers, rather than tackling the issue of knowledge integration central to adaptive governance regimes (Monkelbaan 2015).

The experimentalist governance architecture sets broad framework goals such as "sustainable agriculture" or "climate adaptation" and incorporates metrics (agreed-upon by multiple stakeholders) to measure success. Lower-level institutions such as civil organizations, regulatory authorities, ministries or municipalities are given wide discretion to achieve agreed-upon goals. These lower-level units must report their performance and participate in a peer-review process in which results from different units are compared. If local units fail to produce results, they are required to make necessary adjustments in consultation with other units. Alternatively, it may involve the redefinition of original goals (Sabel and Zeitlin 2011). In this way goals are revised on the basis of new practices and emerging knowledge. The principles underlying this model are functional rather than structural and as such the realization of this model can take a variety of institutional forms (Overdevest and Zeitlin 2012). One of the virtues of experimentalist governance is that it works as a decentralized structure that accommodates differences in resources and expertise (Fossum 2012). The principles underlying this model are functional rather than structural and as such the realization of this model can take a variety of institutional forms (Overdevest and Zeitlin 2012).

Some of the benefits of experimentalist governance may lend themselves well to the adaptation problem. For one, framework goals are broad such that local units can tailor actions to their particular context. Furthermore, stakeholders and decision-makers take an active role in producing practical and policy-relevant knowledge and therefore facilitate social learning. The polycentric nature of decision-making facilitates a governance arrangement that is transparent, accountable and inclusive, preventing powerful interests from hijacking the process (Monkelbaan 2015). Experimentalist governance may be a useful scaffold to address climate change policy as it depends on strategic uncertainty: the only way to find a solution to a problem is along the process of solving the problem, while the multi-polar nature of the framework prevents any one actor from imposing his particular solution (Overdevest and Zeitlin 2012).

Governance Arrangements in New Brunswick

Canada's system of government is comprised of three hierarchical levels: federal, provincial, and municipal. The policy framework in which local government operates is largely imposed by higher levels of governance, such as provincial, state and national policies. Municipal authorities have no constitutional standing of their own, they are the delegated agents of the province. The main role of municipalities is to be the provider of services for their constituents. Legislation requires that municipalities provide services such as: police protection, development planning and emergency measures. However in practice municipalities also provide solid waste management, transportation planning, land-use planning, and water management among other services dependent on the resource capacity of the municipality.

In New Brunswick, municipalities are subject to the Department of the Environment and Local Government in New Brunswick. The department oversees and addresses issues within land-use planning and management, zoning development and waste management and ensures that municipalities comply with environmental legislation. New Brunswick has already felt some of the effects of climate change. Warmer temperature trends have resulted in the appearance of invasive species and new pests. Snow and river ice is melting earlier in spring, increasing the risk of ice jams and flooding. Precipitation is projected to increase in frequency and intensity resulting in the flooding of low-lying areas, coastal erosion, saltwater leakage and groundwater contamination (Province of New Brunswick 2014b). The Climate Change Secretariat functions within the Department of the Environmental and Local Government and has a mandate to develop, implement and oversee climate change mitigation and adaptation strategies and policies. The secretariat is required to engage multiple stakeholders and work across provincial departments to mainstream adaptation and to provide provincial leadership on adaptation action.

The province of New Brunswick is also facing a number of challenges to local governance (Conteh 2013a, b). The population is shifting towards urban centers, there is limited population growth (0.1% growth according to 2006 census data), and the population profile shows an aging population trend (Finn 2008). The economic profile of the province is also changing. There is a shift away from a rural natural resource economy to a service-based economy concentrated in urban centers (Finn 2008). These two trends, one of a shifting population and economic profile place strain on the way municipalities deliver services. Some municipalities are facing a declining tax base and must deliver services to a plummeting constituency while urban municipalities are straining to provide services under the influx of rural migrants. In addition, the province is faced with governance challenges: the large number of municipalities and Local Service Districts (LSDs) relative to tax base and population, the mismatch of administrative boundaries with community interest and the lack of elected municipal officials for 35% of the province's population. Created in 1967, LSDs consist of an advisory committee brought together informally (for instance at town hall meetings), designed to act as administrative units serviced directly by the provincial government. Therefore LSDs (comprising a third of the provincial population) do not have an elected council and as such were not intended to act as a legitimate local governance structure. Furthermore, the lack of any governance arrangement to facilitate regional cooperation between communities (such as cost-sharing and regional planning) suggested that changes were needed.

The Regional Service Commissions were created in 2008 to try to address these governance problems. The mandate of the RSC is to:

- Facilitate regional collaboration between municipalities in terms of regional planning and delivery of services such as: land-use planning, solid waste management, policing, and emergency measures
- Develop a strategic regional plan

- Facilitate the planning of environmental protection
- Serve as a forum to deal with issues of regional significance
- Advancing the interests of the region.

The regional service commissions provide a gambit of services to their constituent communities such as land-use planning, solid waste management, regional policing services, emergency management, and stormwater management. These services are prime targets for integrating adaptation or "mainstreaming" adaptation at the local level. The adaptation literature champions local-level government as the best opportunity for adaptation action as they are the scale at which planning and development is most likely to occur and be effective, and also because adaptation action can be tailored to the local context (Agrawal 2008). According to Rauken et al. (2013), the concept of mainstreaming means addressing adaptation in all relevant sectors and coordinating the policies of different sectors across scales of government in a synergistic way. Mainstreaming should therefore incorporate both vertical and horizontal integration (Nunan et al. 2012). There is a need for communication and coordination between local and provincial governments (vertical) and between different sectors and departments (horizontal). In New Brunswick however, for each municipalities to have its own adaptation component to planning is largely not an option. Planning is a major challenge for small communities in rural areas, of which there are many: some RSCs have over 20 LSDs in their area. These smaller communities lack the expertise, funding and information required for adaptation. Although not part of their official mandate, all interviewed RSC have adopted adaptation planning to their agenda, and all RSCs have long-term goals that include adaptation programs.

However, the extent of adaptation varies across the RSCs. For example the Southwest RSC has decided as recently as the summer of 2015 that adaptation planning will be added to their agenda, and that they have interest in beginning to plan (though they are not sure what form it may take). RSC 10 has adaptation strategies lead by the municipalities themselves (with no involvement from the RSC) such as Fredericton's comprehensive Climate Change Adaptation Strategy, which addresses adaptation in transportation, infrastructure and planning sectors. By contrast, there are two RSCs that have been planning for adaptation for some time; they have adaptation measures integrated into land-use planning. These adaptation measures are integrated in land-use planning and so consist of the writing of zoning bylaws, setting new benchmarks for construction, building permits, and providing development recommendations, so climate vulnerable areas are avoided.

Though not explicitly mandated by the Minister to include adaptation planning within this plan, all RSC have said that they believe their regional plan will have an adaptation component. Apart from one RSC, this regional plan has yet to be realized. In many ways, challenges to regional development planning are synonymous with adaptation planning. Only one out of twelve RSCs has a completed regional development plan in place, which also happens to be the RSC which has a coastal area with infrastructure fully adapted to flooding and storm surges, with ambitious plans to expand the adaptation implementation in-land and also plan for other facets of adaptation, such as social and economic adjustments for the future. However, all RSCs have expressed profound challenges and difficulties in facilitating the regional development plan. Similarly to regional planning bodies in Sweden and Norway (Antonson et al. 2016; Dannevig and Aall 2015), RSCs can conceptually act as a collaborative entity working with provincial departments (public safety, natural resource management) that proactively mainstream adaptation, and alleviate some of the abject political dimensions of adaptation planning.

What Are Barriers to Regional Adaptation Planning?

For the RSCs challenges to regional adaptation planning comes from a number of sources. A commonly discussed barrier is outdated planning legislation. Planners largely feel that they do not have the tools in place to implement planning and development strategies in a fast-paced modern world. By this planners refer to inadequate legislative powers and authority. Although the goal of the RSC governance arrangement is to initiate a type of institutional change that will allow better planning for the region, old institutions still hold too much sway: adaptation planning in this context suffers from institutional crowdedness. The RSCs are subject to institutions that influence the decision-making process such that adaptation planning is hindered and the process is muddied by: (i) outdated *Community Planning Act* and *Municipalities Act*, (ii) an undemocratic system of representation (LSDs) and (iii) centralized power from Fredericton. According to Measham et al. (2011), a lack of legislative municipal authority is the cause of confusion about roles and responsibilities, uncertain or conflicting goals, and divergent ideas about how the adaptation problem should be solved. Without an infusion of (legislative) power into the RSCs, difficult planning decisions on tradeoffs and adaptation costs cannot be made, and the process will suffer from stagnation.

Another common barrier reported in the literature is policy fragmentation resulting from a lack of collaboration between institutions and policies at different governance levels and geographic and temporal scales (Biesbroek et al. 2011) Such institutional fragmentation is present in New Brunswick– each department (Transportation, Environment and Local Government, Health) pursues its own activities without consultation or integration with one another—even when synergies can lead to a more effective policy outcome (Conteh 2013a, b). One RSC in particular has criticized the provincial government for diluting what the RSC were originally intended to do—fix an outdated and broken local governance structure—to an entity that lacks a clear mandate, does not have the appropriate tools to achieve their mandate and has received no provincial leadership or guidance.

RSCs have also faced a number of hurdles from the municipalities. Essentially they are struggling with (i) the unwillingness of members to consider tradeoffs between climate objectives and development goals, (ii) the politics of regionalization (iii) unqualified Board members and (iv) a lack of public awareness and policy communication. Mayors are not receptive to regulating development in coastal areas vulnerable to climate impacts when there are short-term economic or political gains available while planners try to produce long-term plans. Although some impacts have manifested themselves already, climate change is largely a long-term issue. It is of no surprise that there is a conflict of timescales between adaptation actions that thoroughly address climate change and conventional policy-making, decision-making, planning, and political cycles (Biesbroek et al. 2011). Some RSCs have voiced frustration over working with Board members that do not have a basic understanding of the planning process. Indeed, political awareness and active community participation is often discussed as a prerequisite for local adaptation initiatives (Foster et al. 2011; Cloutier et al. 2015).

Another major issue is the economic situation of the province. Indeed, according to the Conference Board of Canada New Brunswick ranked last among the provinces on overall economic performance and is only one of two provinces with a per capita income of less than 30,000 US dollars (The Conference Board of Canada 2014). Resource limitation is a barrier to adaptation for New Brunswick: the province is faced with an economic crisis, and municipalities are equally strained in terms of where to allocate funds for an already crowded agenda. However, these findings suggest that the biggest obstacle facing the successful adaptation planning is not a lack of a capacity: resources or expertise, rather it is the translation of existing capacity into adaptation action. Outdated institutions such as the *Municipalities Act*, *Community Planning Act* and the lack of clarity as to the mandates of RSCs (from provincial officials) contribute to an institutional path dependency that prevents effective adaptation. As such, facilitating adaptation requires that these path-dependent institutional arrangements be reformed so that existing latent capacity can translate into tangible adaptation activities.

How Are RSCs Facilitating Regional Cooperation?

Part of the mandate of the RSCs is to facilitate regional cooperation between municipalities. The process consists of representatives from all communities (this includes cities, LSDs and unincorporated areas) in the region coming together and discussing regional plans; decisions are made on the basis of votes cast by Board members. The RSC has effectively brought together multiple stakeholders (municipal representatives and regional planners) and engaged experts (consultants) to bring forward a regional plan. According to Amaru and Chhetri (2013), rural adaptation can be strengthened by participation of relevant actors, considering local contexts, considering sustainability and reducing dependency on external intervention. In a sense, regional cooperation as described by RSCs are demonstrating these characteristics by engaging rural communities in their local contexts and doing so in a self-sufficient manner. However this is not a consistent pattern; there is division as to whether RSCs have been effective in facilitating regional cooperation.

Without cooperative representatives or ministry approval, plans cannot take root, stultifying regional initiatives that do not share broad consensus. Essentially, for plans to take shape, consensus is needed across three levels of governance. Some Board members see the process itself as a threat to municipal autonomy; if the RSCs were to be successful, it may lead to the formation of a regional government to which they would have to answer to. Therefore they are deliberately uncooperative. In contrast, as one planner puts it, some municipalities "*see the need for regionalization and the benefit of well-trained and well-experienced staff*". In this sense, regional planners employ some of these tools for collaboration yet most have not realized many of the collaborative tools potentially available to them for adaptation planning (e.g. regional status report, organizing information seminars, scenario-based planning) (Wachsmuth 2015).

Do RSCs Have the Potential to Support an Experimentalist Governance Approach to Adaptation?

As of yet, experimentalist governance has not been applied to climate change adaptation planning, mainly it is a transnational governance model employed in the European Union for sectors such as energy, transportation and finance (Sabel 2012). The multi-scalar architecture of this model and the polycentric decision-making process can conceptually lend itself well to addressing a number of the specific barriers to adaptation planning and implementation in New Brunswick. Essentially this model would bring stakeholders together to set broad framework goals and compare their own contextualized plans (and outcomes) to achieve those goals. The stakeholders would then adjust their plans based on what has worked for their peers, or collectively readjust framework goals. Conceptually, this "fluid network" would entail RSCs and municipalities working as peers within this process, collaborating across jurisdictional divides. If the objective was simply set as "adaptation planning" for example, each municipalities or RSC can be free to pursue their own approach to adaptation, while having the benefit of consultation with peers. Furthermore, actors from across scales can be brought into this iterative collaborative process: municipalities, RSCs, and provincial departments. The poly-centric deliberative decision-making process is a feature of experimentalism that may help to address the problem of power relations in the province: between municipalities and RSCs, and the province. As an institutional architecture, our proposal shares some of the design elements of the New South Wales government's approach to enabling regional adaptation, namely regional capacity building, enhancement of social capital, knowledge dissemination, and research partnerships (Jacobs et al. 2016). However, experimentalist governance arrangements focus less on steering than on supporting local knowledge co-production within an experimental framework.

The experimentalist procedure (applied specifically for adaptation planning) holds no threat to municipal autonomy in the way unilateral regional planning does —a critical factor that has compromised the ability of regional adaptation governance in other countries such as Sweden (Antonson et al. 2016). Theoretically this governance regime can engage the key players in the province together: the climate secretariat, RSCs, municipalities, NGOs and universities. The architecture of this model can be built in multiple settings at different governance scales, or nested within one another across scales (Overdevest and Zeitlin 2012). Therefore, the way this architecture may be conceptually realized in New Brunswick is for provincial departments to collaborate to create one network and for RSCs to create another, with interpenetration between the two. More specifically, the experimentalism for adaptation in New Brunswick may be realized similar to the experimentalist regime created by the Food Safety Modernization Act of 2011 (FSMA). The FSMA requires factories to identify, monitor and manage all areas within the food production process vulnerable to contamination, with the Food Drug Administration periodically inspecting licensed factories (Sabel 2012). If the performance of factories lag behind that of their peers, the frequency of inspections increases, and the facility must reanalyze its risk management plan (Sabel 2012). This method can theoretically translate to adaptation planning within the current New Brunswick governance system.

In this model RSCs can act as autonomous actors perusing the goal of adaptation planning, periodically measuring performance in relation to their peers. The Climate Secretariat could act as the regulatory body, inspecting progress and requesting revision and reformation of adaptation planning that is not meeting peer-established "performance standards". Furthermore, NGOs as well as relevant provincial government departments can be included within this framework as the experimentalist framework does not differentiate between public and private stakeholders (Fossum 2012). The benefit of taking a nested approach, in which provincial departments and RSCs collaborate to make policies together, is that there is the latent potential for mainstreaming adaptation efficiently. Indeed, Sabel and Zeitlin (2011) have defined experimentalism as a "machine for learning from diversity". The decentralized structure of experimental governance allows for tensions between actors to diffuse, as there is no overarching authority. If the power struggle between municipalities and RSCs (which clogs the process of decision-making) is relieved, it may smooth the progress towards a regional development plan. In addition, another characteristics of municipalities in New Brunswick is the large difference in resources and expertise (adaptive capacity). The experimental architecture is intended to produce results in circumstances where capacity is low. The RSCs themselves can participate in their own cycle of deliberation, setting framework goals such as "regional adaptation planning", and comparing and readjusting results.

Nonetheless, there is the question of feasibility: how would these actors be brought together? Given the collaborative nature of experimentalist governance, what would provide the impetus for actors to converge and sustain cooperation without sanctions or mandates remains ambiguous (Fossum 2012). Furthermore, as a province facing financial difficulties and a shifting economy, the political climate may not be favorable to a governance structure dependent on experimenting with policies. Another limitation of this model is its reliance on shared constitutional principles (Fossum 2012). There remains the inherent problem of actors that do not acknowledge adaptation planning as a shared goal preferring instead to pursue vested interests (Antonson et al. 2016). Cultural elements may be a root cause of divergent principles and views from stakeholders within the adaptation process. Indeed, cultural factors—for example differences between Anglophone or predominantly Francophone communities—may be responsible for the disparity in adaptation planning across the RSCs.

Conclusion

RSCs are collaborative decision-making bodies with the potential to facilitate adaptation planning in the region as they have: (i) had some success with regional development planning (one RSC has a regional plan), (ii) integrated climate knowledge into policy (for two RSCs) and (iii) implemented adaptation across municipal boundaries (two RSCs). However, RSCs have qualified their successes as there are many problems to be solved and many challenges ahead. Outdated planning legislation, vague leadership from the province as to the mandate and responsibilities of RSCs, centralized provincial power, and an undemocratic election process (lack of democratic representation for LSDs) are institutional barriers to adaptation. In addition the decision-making process is stultified by conflict between municipalities and RSCs. From the perspective of the RSCs, municipalities often look out for their own economic interests (ergo refusing to make tradeoffs to address climate vulnerabilities) and favor short-term (economic) gain over long-term benefits to resilience. These results offer additional insights on the conflict between local and regional levels of decision-making in adaptation planning (Storbjörk and Hedrén 2011; Nilsson et al. 2012; Antonson et al. 2016).

Experimentalist governance as a polycentric deliberate decision-making process that operates across scales may rectify some issues, and have merits for adaptation planning. The decentralized structure may alleviate problems and impasses (to decision-making) caused by power struggles between municipalities and RSCs, as well as the RSCs and the province. This approach has some affinities with the 'boundary work' that enables a 'hybrid management space' advocated for regional adaptation governance in Norway (Dannevig and Aall 2015), as it can enable the functions of knowledge translation, mediation and communication by facilitating the tailoring of responses to climate change from municipalities within their respective adaptive capacities. Yet, it is recommended that strong provincial leadership may play a key role in adaptation policy and in the future of RSCs. Many RSCs feel that the province has been ambiguous in terms of what the mandate of the RSCs should extend to and whether regionalization should be discouraged or embraced. The provincial Departments of Health, Safety and Transportation should collaborate closely with the RSCs or delegate respective powers for regional planning: the RSC need to be empowered to take a multi-scaled approach to adaptation planning. The regional development plan may represent a vital policy instrument to initiate the kind of change required for effective adaptation planning (Antonson et al. 2016). As RSCs begin to formulate regional development plans future research should explore the way adaptation planning is integrated, and the factors that hinder or enable the process.

References

Adger WN (2003) Social capital, collective action and adaptation to climate change. Econ Geogr 79(4):387–404

Adger WN, Arnell NW, Tompkins EL (2005) Successful adaptation to climate change across scales. Glob Environ Change 15(2):77–86

Adger WN, Dessai S, Goulden M, Hulme M, Lorenzoni I, Nelson DR, Naess LO, Wolf J, Wreford A (2009) Are there social limits to adaptation to climate change? Clim Change 93:335–354

Agrawal, A (2008) The role of local institutions in adaptation to climate change. Paper presented at the social dimensions of climate change, Social Development Department, The World Bank, Washington DC, 5–6 March 2008

Amaru S, Chhetri NB (2013) Climate adaptation: institutional response to environmental constraints, and the need for increased flexibility, participation and integration of approaches. Appl Geogr 39:128–139

Amundsen H, Berglund F, Westskog H (2010) Overcoming barriers to climate change adaptation: a question of multilevel governance? Environ Plann C Gov Policy 2010(28):276–289

Antonson H, Isaksson K, Storbjörk S, Hjerpe M (2016) Negotiating climate change responses: regional and local perspectives on transport and coastal zone planning in South Sweden. Land Use Policy 52:297–305

Betsill M, Bulkeley H (2007) Looking back and thinking ahead: a decade of cities and climate change research. Local Environ 12:447–456

Biesbroek R, Klostermann J, Termeer C, Kabat P (2011) Barriers to climate change adaptation in the Netherlands. Clim Law 2:181–199

Cashmore M, Wejs A (2014) Constructing legitimacy for climate change planning: a study of local government in Denmark. Glob Environ Change 24:203–212

Cloutier G, Joerin F, Bubois C, Labarth M, Legay C, Viens D (2015) Planning adaptation based on local actors' knowledge and participation: a climate governance experiment. Clim Policy 15:458–474

Conteh C (2013a) Transitions in regional development policy implementation in Canada: the cases of New-Brunswick and Manitoba. Brit J Can Stud 26:105–128

Conteh C (2013b) Navigating Canada's federal maze: regional development policy governance in New-Brunswick. Reg Fed Stud 23:129–149

Crabbé P, Robin M (2006) Institutional adaptation of water resource infrastructures to climate change in eastern Ontario. Clim Change 78:103–133

Craft J, Howlett M, Crawford M, McNutt K (2013) Assessing policy capacity for climate change adaptation: governance arrangements, resource deployments, and analytical skills in Canadian Infrastructure Policy making. Rev Policy Res 30:42–65

Dannevig H, Aall C (2015) The regional level as boundary organization? An analysis of climate change adaptation governance in Norway. Environ Sci Policy 54:168–175

Dannevig H, Hovelsrud GK, Husabø IA (2013) Driving the agenda for climate change adaptation in Norwegian Municipalities. Environ Plan C Gov Policy 31:490–505

Dovers S, Hezri R (2010) Institutions and policy processes: the means to the ends of adaptation. Wiley Interdisc Rev Clim Change 1(2):212–231

Edvardsson-Bjornberg K, Hansson SO (2011) Five areas of value judgement in local adaptation to climate change. Local Gov Stud 37(6):671–687

Emerson K, Gerlak AK (2014) Adaptation in collaborative governance regimes. Environ Manage 54:768–781

Finn J-G (2008) Building stronger local governments and regions an action plan for the future of local governance in New Brunswick. Report of the Commissioner on the Future of Local Governance

Folke C, Hahn T, Olsson P, Norberg J (2005) Adaptive governance as social-ecological systems. Ann Rev Environ Resour 30:441–473

Foster J, Winkelman S, Lowe A (2011) Lessons learned on local climate adaptation from the urban leaders adaptation initiative, center for clean air policy (CCAP). DC, USA, Washington 23 pp

Füssel H-M (2007) Adaptation planning for climate change: concepts, assessment approaches, and key lessons. Sustain Sci 2(2):265–275

Fossum JE (2012) Reflections on experimentalist governance. Regul Gov 6:394–400

Gomm R, Hammersley M, Foster P (eds) (2000) Case study method. Sage Publications, London

Hanssen GS, Mydske PK, Dahle E (2013) Multi-level coordination of climate change adaptation: by national hierarchical steering or by regional network governance? Local Environ 18: 869–887

Hinkel J, van Vuuren DP, Nicholls RJ, Klein RJT (2013) The effects of adaptation and mitigation on coastal flood impacts during the 21st century. An application of the DIVA and IMAGE models. Clim Change 117:783–794

IPCC (2014). Summary for policymakers. In: Field CB, Barros VR, Dokken DJ, Mach KJ, Mastrandrea MD, Bilir TE, Chatterjee M, Ebi KL, Estrada YO, Genova RC, Girma B, Kissel ES, Levy AN, MacCracken S, Mastrandrea PR, White LL (eds) Climate change 2014: Impacts, adaptation, and vulnerability. Part A: global and sectoral aspects. Contribution of Working Group II to the fifth assessment report of the intergovernmental panel on climate change. Cambridge University Press, Cambridge, pp 1–32

Jacobs B, Lee C, Watson S, Dunford S, Coutts-Smith A (2016) Adaptation planning process and government adaptation architecture support regional action on climate change in New South Wales, Australia. In: Leal Filho W (ed) Innovation in climate change adaptation. Series in climate change management. Springer, Berlin

Knieling J, Leal Filho W (eds) (2013) Climate change adaptation. Series in climate change management. Springer, Berlin

Lieske DJ, Roness LA, Phillips EA, Fox MJ (2015) Climate change adaptation challenges facing New Brunswick coastal communities: a review of the problems and a synthesis of solutions suggested by regional adaptation research. Revue D'Études Sur Le Nouveau Brunswick 6:32–54

Lincoln YS, Guba EG (2000) The only generalization is: there is no generalization. In: Gomm R, Hammersley M, Foster P (eds) Case study method. Sage Publications, London

Matthews T (2013) Institutional perspectives on operationalizing climate adaptation through planning. Plan Theor Pract 14:198–210

Measham TG, Preston BL, Smith TF, Brooke C, Gorddard R, Withycombe G, Morrison C (2011) Adapting to climate change through local municipal planning: barriers and challenges. Mitig Adapt Strat Glob Change 16(8):889–909

Monkelbaan J (2015) Experimentalist sustainability governance: jazzing up environmental blues. Center for International Sustainable Development Law, Public Participation and Climate Governance Working Paper Series

Mukheiber P, Kuruppu N, Gero A (2013) Overcoming cross-scale challenges to climate change adaptation for local government: a focus on Australia. National Climate Change Adaptation, Research Facility, Gold Coast, 95 pp

Naess L, Bang G, Eriksen S, Vevatne J (2005) Institutional adaptation to climate change: flood response at the municipal level in Norway. Glob Environ Change 15:125–138

Nilsson AE, Gerger Swartling Å, Eckerberg K (2012) Knowledge for local climate change adaptation in Sweden: challenges of multilevel governance. Local Environ 17(6–7):751–767

Nunan F, Campbell A, Foster E (2012) Environmental mainstreaming: the organizational challenges of policy integration. Public Adm Dev 32(3):262–277

Overdevest C, Zeitlin J (2012) Assembling an experimentalist regime: transnational governance interactions in the forest sector. Comparative Research in Law & Political Economy, Research Paper No. 16/2012. Retrieved from http://digitalcommons.osgoode.yorku.ca/clpe/18

Province of New Brunswick (2014a) New Brunswick climate action plan 2014–2020. Retrieved from the department of environment and local government http://www2.gnb.ca/content/dam/gnb/Departments/env/pdf/Climate-Climatiques/ClimateChangeActionPlan2014-2020.pdf

Province of New Brunswick (2014b) New Brunswick's flood risk reduction strategy. Retrieved from the department of environment and local government http://www2.gnb.ca/content/dam/gnb/Departments/env/pdf/Publications/NBFloodRiskReductionStrategy.pdf

Province of New Brunswick (2015) Structure of the new regional service commissions. Retrieved from the department of environment and local government http://www2.gnb.ca/content/gnb/en/departments/elg/local_government/conte

Rauken T, Mydske PK, Winsvold M (2013) Mainstreaming climate change adaptation at the local level. Local Environ 20(4):408–423

Sabel C (2012) Dewey, democracy, and democratic experimentalism. Contemp Pragmatism 9(2):35–55

Sabel C, Zeitlin J (2011) Experimentalist governance. In: Levi-Faur D (ed) The oxford handbook of governance. Oxford University Press, Oxford

Sanchez-Rodriguez R (2009) Learning to adapt to climate change in urban areas. A review of recent contributions. Curr Opin Environ Sustain 1:201–206

Sander-Regier R, McLeman R, Brklacich M, Woodrow M (2009) Planning for climate change in Canadian rural and resource-based communities. Environ J 37:35–57

Storbjörk S (2007) Governing climate adaptation in the local arena: challenges of risk management and planning in Sweden. Local Environ 12:457–469

Storbjörk S, Hedrén J (2011) Institutional capacity-building for targeting sea level rise in the climate adaptation of Swedish coastal zone management. Lessons from Coastby. Ocean Coast Manag 54(3):265–273

The Conference Board of Canada (2014) How Canada performs: provinces and territorial ranking. Retrieved from http://www.conferenceboard.ca/hcp/provincial/economy.aspx

Termeer C, Dewulf A, van Rijswick H, van buuren A, Huitema D, Meijerink S, Rayner T, Wiering M (2011) The regional governance of climate adaptation: a framework for developing legitimate, effective, and resilient governance arrangements. Clim Law 2:159–179

Urwin K, Jordan A (2008) Does public policy support or undermine climate change adaptation? Exploring policy interplay across different scales of governance. Glob Environ Change 18:180–191

Van Nieuwaal K, Driessen P, Spit T, Termeer C (2009) A state of the art governance literature on adaptation to climate change: towards a research agenda. Report N. kfc 003/2009, Dutch National Research Programme Knowledge for Climate, Utrecht University, Wageningen University, The Netherlands. Retrieved from http://dx.doi.org/10.1088/1755-1307/6/36/362019

Wachsmuth J (2015) Cross-sectoral integration in regional adaptation to climate change via participatory scenario development. Clim Change 132:387–400

Chapter 5
The Participative Action Research Approach to Climate Change Adaptation in Atlantic Canadian Coastal Communities

Omer Chouinard, Steve Plante, Sebastian Weissenberger, Mélinda Noblet and Julie Guillemot

Abstract Canada's Atlantic coast has been increasingly affected by inundations and coastal erosion over time. Results presented here stem from participative research action (PRA) projects undertaken in several coastal communities in New-Brunswick and Quebec, using methods such as semi-directed interviews, focus groups, participant observation, public meetings, kitchen assemblies, collaborative mapping, Method of Evaluation by Group Facilitation (MEGF) or participative ranking. The PRA projects were found to deliver tangible short-term results as well as reinforcing the communities' governance and adaptation capacity and resilience over the long term. The engagement of stakeholders, and the

O. Chouinard (✉) · S. Weissenberger
Département de sociologie, Programme de Maîtrise en études de l'environnement, Université de Moncton, Campus de Moncton, Pavillon Léopold-Taillon, 18, avenue Antonine-Maillet, Moncton, NB E1A 3E9, Canada
e-mail: omer.chouinard@umoncton.ca

S. Weissenberger
e-mail: weissenberger.sebastian@uqam.ca

S. Plante
Département sociétés, territoires et développement, Université dU Québec à Rimouski, 300, allée des Ursulines, C.P. 3300, succ. A, Rimouski, QC G5L 3A1, Canada
e-mail: steve_plante@uqar.ca

S. Weissenberger
Institut des sciences de l'environnement, Université du Québec à Montréal, CP 8888, succ. Centre-Ville, Montreal, QC H3C 3P8, Canada

M. Noblet
Faculté de Droit et de Science Politique, Centre Universitaire de Recherches sur l'Action publique et le Politique, Epistémologie et Sciences Sociales (CURAPP-ESS), Université Picardie Jules Verne, Pôle Universitaire Cathédrale, 10, Placette Lafleur, BP 2716, 80027 Amiens Cedex 1, France
e-mail: melinda.noblet@gmail.com

J. Guillemot
Université de Moncton Campus de Shippagan, 218 boul. JD-Gauthier, Shippagan, NB E8S 1P6, Canada
e-mail: julie.guillemot@umoncton.ca

W. Leal Filho and J.M. Keenan (eds.), *Climate Change Adaptation in North America*, Climate Change Management,
DOI 10.1007/978-3-319-53742-9_5

exchange of information between scientists and local actors, led to a better evaluation of vulnerabilities and adaptation options and permit in some cases the co-construction of new knowledge and the coproduction of priorities to build adaptation plans and tools with and for the communities. Thus, reflexive options such as sea walls were sometimes substituted by less costly and more targeted adaptation options, that are better suited to local circumstances and to the values and aspirations of the community. These solutions are more easily accepted within the community as well as by government authorities. However, not all projects led to immediate decision-making and especially the option of coastal retreat remains highly contentious and emotionally charged.

Keywords Climate change · Atlantic canada · Participatory action research · Community based adaptation · Coastal zone

Introduction

Participatory approaches and in particular participatory action research (PAR) has been advocated as a useful approach in tackling climate change adaptation in Africa, Central America, Asia, and in some cases in developed nations in Europe, Australia and the Canadian Arctic by researchers and international agencies such as Care, CGIAR, IDRC, IIED, or IISD (Bizikova et al. 2014; Brown and Harvey 2009; German et al. 2012; Jost et al. 2014; McClymont Peace and Myers 2012; Reid et al. 2012; van Buuren et al. 2015). Most often, PAR is envisaged for developing nations of First Nations, and more rarely for industrialized nations, which might have historic reasons due to the emergence of the PAR approach in South Asia and Latin America. By presenting case studies from Canadian coastal communities, we aim at demonstrating the usefulness (but also the limitations) of this approach in the context of adapting to ongoing climate change in the coastal zone.

In the PAR approach, researchers are seen as participants in the project, but not as project leaders and decision-makers, nor as simple consultants (e.g. Leonard et al. 2004; Mackinson et al. 2015). PAR is a rather heterogeneous ensemble of practices, but guided by some common principles. Projects are oriented towards action and bringing on change rather than just gathering knowledge (Park 2006). The involvement of people is central (Park 2006). Thus, PAR aims at co-constructing a research (intervention) project rather than imposing a predetermined research agenda on a community, as McIntyre (2008) states, "codeveloping research programs *with* rather than *for* people". Researchers play the role of facilitators (Park 2006). It is recognized that local actors are responsible for elaborating their adaptation plan and will have to deal with their consequences, whereas the researchers will only be involved for the duration of the project. Thus, the terms of the project, including goals, expectations and limits are negotiated with the communities prior to engaging in the research process (Plante et al. 2011, 2016). By favouring the implication of local stakeholders, PAR aims at promoting the transfer of information and tools, the sharing of knowledge and ideas, and generally speaking reinforce the communities' adaptation capacity and local governance (Plante et al. 2011).

In this article, we present case studies of climate change adaptation projects undertaken in Canada, in the provinces of Quebec and New Brunswick, using PAR approaches. The coast of New-Brunswick measures about 5500 km and that of the Gaspé peninsula, Quebec, about 800 km. Most of that coast is vulnerable to the impacts of climate change and sea level rise. An increase in sea level (Forbes et al. 2006) as well as in the amplitude of storm surges and extreme precipitations have been observed (Parkes et al. 2006). Several storms in recent years (2000, 2005, 2010) would have previously been qualified as exceptional and caused extensive damage (Vasseur and Catto 2008; Friesinger and Bernatchez 2010). Coastal erosion is present on the majority of the coast, reinforced by a decrease in winter sea ice and more frequent freeze-thaw cycles. Houses, piers and other infrastructures, such as the 885 km long road 132 around the Gaspé peninsula are affected by these changes. As a reaction, large parts of the coast are now protected by rock armour and other structures, which have adverse ecological and aesthetic consequences.

Several adaptation projects have been undertaken, addressing questions of coastal erosion, flood risks, the impact of climate change on fisheries, etc., using methodologies ranging from "top-down" approaches making use of climate modeling and predictions to more "bottom-up" approaches based on community participation. Among them the Aquatic Climate Change Adaptation Services Program (ACCASP) led by Fisheries and Oceans Canada, the Atlantic Regional Adaptation Collaborative (ACASA), a collaboration between Natural Resources Canada and the four Atlantic provinces, the Atlantic Climate Adaptation Solutions under the Regional Adaptation Collaborative (RAC) Program, also in collaboration between the Atlantic provinces and the federal government. Participative approaches have been adopted for example by the Ecological Action Center in Nova Scotia in a fisheries based program (https://www.ecologyaction.ca/content/adapting-atlantic-canadian-fisheries-climate-change), or the University of Mount Allison in a collaborative mapping project, the Tantramar Community Adaptation Viewer (Lieske 2015). In Quebec, the Ouranos consortium and the Chaire de recherche en geoscience côtière at the University of Quebec in Rimouski have done considerable work on coastal risks associated to climate change (Bernatchez et al. 2008; Savard et al. 2008).

In the face of these difficult challenges, government support, at federal as well as provincial level, for adaptation to climate change is usually considered weak by communities (Chouinard et al. 2006; Chouinard and Martin 2007; Guillemot et al. 2014; Noblet 2015). Most responsibilities for land planning are of provincial resort, although the coastal zone is of divided jurisdiction.[1] The federal government is mainly involved in punctual research projects and in the mitigation of catastrophes through a federal program offering financial assistance to afflicted provinces (Guillemot and Aubé 2015; Lane 2012). The province of New-Brunswick has been working since 1996 on a coastal policy, which was published in 2002, but still has not been translated into legislation and its application is at the discretion of local authorities (Doiron 2012; Fox and Daigle 2012; New Brunswick 2002). In Quebec, the coastal zone has a 10–15 m buffer zone from the high-water line, depending on terrain, that is imposed by the Environment Ministry. Municipalities can however choose to extend this buffer zone. The buffer zone does however allow shore stabilization and protection structures. During the severe storms of 2010, the Quebec Government mainly mobilized the Ministry of Public Security (MSP), which has the mandate to ensure that emergency plans and disaster assistance programs are deployed during extreme events (Decree No. 113-2011 to February 16, 2011). It sought technical advice in order to ensure that the citizens at risk were in security and finally developed guiding plans for risk management for certain municipalities, including Îles-de-la-Madeleine and Ste-Flavie. However, all action oriented towards crisis management is a crisis in the recovery period.

Coastal management and therefore adaptation to climate change thus mainly rests on the shoulders of local authorities. In New-Brunswick, this is particularly problematic because about 70% of the territory and 30% of the population are in local service districts (LSD), without elected officials, legislative capacity or tax revenue. This deficit of local governance has long been recognized (Finn 2008). In order to strengthen local governance, 12 regional service commissions (RSC) were put in place in 2013 to facilitate the management of shared services, including emergency services and land use planning. There are also regional coordinators for emergency measures deputy to the ministry of environment and local governments. Municipalities have produced emergency plans since 2005, and it is only since 2012 that climate change is integrated in those, albeit in a very inconsistent manner (Guillemot and Aubé 2015; Beauchamp and Weissenberger, unpub. res.). In Quebec, the Regional County Municipalities (RCM) devise territorial development plans (French: *schema d'aménagement et de développement*), strategic documents which set the general context for the socio-economic, transport, environmental, agricultural and other development of its territory. Those plans are approved by the Ministry of Municipal Affairs and Land Occupancy (French: *Ministère des Affaires municipales et de l'Occupation du territoire* or MAMOT). Municipalities are responsible for producing local land use plans, which is a guideline for the development of its territory and includes provisions for areas demanding special attention and for Zones to be renovated, restored or protected. Municipalities are

[1]According to the Constitutional Act of 1867 and 1982, the administration of the coastal zone is divides along the mean low-tide mark, the ocean side being under federal and the land side under provincial jurisdiction. Furthermore, the upland part of beaches, above the mean high-tide mark, is private property whereas the foreshore, i.e. the part of the beach between the mean high-tide and low-tide marks remains provincial crown land.

also responsible for putting in place land use by-laws, which are enforceable against residents (MAMROT 2010).

Given the weak provincial governance on climate adaptation, communities (here taken as municipalities and LSD) have to be largely autonomous and develop their own adaptation strategies. In this process, the participation of university or other researchers can act as a catalyst and is usually welcome. In fact, several projects described here were initiated at the request of communities who turn to universities and research centers to help them with the elaboration of climate adaptation plans. The projects described here were undertaken by the University of Moncton, Moncton and Shippagan campuses, the Institut de Recherche sur les Zones Côtières (IRZC) in Shippagan, the University of Québec in Rimouski (UQAR) and the University of Quebec in Montreal (UQAM). As there is no recurrent governmental support for these kind of initiatives, financing has to come from multiple sources, including grants from the New Brunswick Environmental Trust Fund, Environment Canada, Natural Resources Canada (including the Atlantic Climate Adaptation Solutions program), the Social Sciences and Humanities Research Council (in particular the Universities-Communities Research Alliances program on Coastal Communities Challenges CURA-CCC 2010–2016 and Social Economy and Sustainability Research Network 2005–2011). The uncertain and temporary nature of such funding is an inherent problem in community-university collaborations, since communities need to continually work on their adaptation and emergency planning and not just when financing is available.

Study Sites

The studies used in this article stem from several sites in New-Brunswick and Quebec on the Atlantic coast of Canada (Fig. 5.1). These include Bathurst—Pointe Carron (12,275 inh.), Shippagan (2603 inh.), Le Goulet (817 inh.), Bas-Caraquet (1380 inh.), Cocagne (2545 inh.), Grande-Digue (2182 inh.), Pointe-du-Chêne (761 inh.), Sainte-Marie-Saint-Raphaël (SMSR, 955 inh.) in New Brunswick, Maria (2500 inh.), Sainte-Flavie (919 inh.), Bonaventure (2775 inh.), Rivière-au-Tonnerre (307 inh.) in Quebec. Traditionally, most villages on the coast were fishing villages, the other main activities being agriculture and forestry, at some site also shipbuilding. The importance of the fishing industry declined in the 1990s with the collapse of groundfish population stocks. Bathurst used to have an important mining industry. Nowadays, services, including the tourism industry, are the main employer. In the north, in the Acadian Peninsula (Shippagan, Le Goulet, SMSR) as well as in the Côte-Nord in Quebec (Rivière-au-Tonnerre), fishing remains important and the region has experienced an economic and demographic decline. On the southern part of the Acadian coast (Cocagne, Grande-Digue, Pointe-du-Chêne, Richibuctou), as well as in the Gaspé Peninsula (Maria, Bonaventure, Sainte-Flavie) residential use and tourism have led to residential coastal development and an increase in property values. Due to the proximity to the

Fig. 5.1 Map of the study sites

town of Moncton, the largest of the region, professionals, part-time residents and retirees account fro the largest part of the population increase. The Gaspé peninsula as a whole has suffered from a demographic and economic decline, but some towns have grown, transitioning from a natural resources based economy to a service based economy (tourism, education, culture, government agencies, etc.).

New Brunswick and the Gaspé peninsula are faced with accelerating erosion and an increase in extreme meteorological events (Bernatchez et al. 2008; Daigle 2006; Savard et al. 2008). Both territories are inherently vulnerable, due to their geomorphology. The Acadian coast in New Brunswick has little relief and easily

Fig. 5.2 Large rock riff-raff in the region of Pointe-du-Chène

erodible substrates while several segments of the Gaspésie peninsula are characterized by friable cliffs of glacial origin. In New-Brunswick, 70% of all the coast is retreating and 40% of it is protected by dykes (Bérubé n.d.; Daigle 2006). In Quebec, the percentage of coast affected by erosion ranges from 56 to 80% depending on the sites; the part of the coast has doubled over previous decades over certain sectors (Bernatchez et al. 2008). Coastal ecosystems such as dunes and wetlands are, over the long term, threatened by coastal squeeze as the rise in sea level and the retreating shoreline and increasing coastal development leaves increasingly little place for them to retreat (Jolicoeur 2016; Bernatchez and Fraser 2012). Many parts of the coast are now protected by riff-raff and different types of walls or palisades (Figs. 5.2, 5.3 and 5.4).

Methodology

As the projects described here were undertaken by different teams (University of Moncton, Moncton and Shippagan campuses, Institut de Recherche sur les Zones Côtières—IRZC, Université du Québec à Rimouski) and under different grant programs, the methodology is not uniform across sites. The general approach was

Fig. 5.3 Different types of protection structures at Pointe-Carron: *1* heavy shale riff-raff, *2* light rock riff-raff, *3* cement wall, *4* large rock riff-raff. *Source* Chouinard et al. (2009)

however that of participative action research, in that the development of adaptation strategies was built in tandem with the communities, in a process of co-construction. Specific methodological tools employed were: questionnaires, semi-directed interview, focus groups, kitchen assemblies, participative observation, public presentations. Within those broad methods, more specific tools like participative mapping, participative ranking or method of evaluation by group facilitation [MEGF, a structured iterative approach to elicit consensual statements by a group of individuals (Plante and Da Cunha 2015)] were used (Table 5.1). The use of different methodological tools on the same site aims at obtaining complementary information and to cross-validate results through methodological triangulation. As a general approach, projects proceed in three steps: (1) at the onset of the project, negotiate the terms of reference of the project with the communities, including the concerns of the communities, the desired outcome, the timetable, (2) establish a diagnosis of the present situation through questionnaires, interviews, participant observation, bibliographical review, data analysis, cartography, (3) engaging an open dialogue around the implications of climate change and adaptation with the community (incl. focus groups, public meetings, participatory observation and discussion) (Plante et al. 2011). The number of focus groups, public meeting and other tools depends on the progress towards reaching the set objectives. Collective reflexion and decision-making needs to unfold over time.

Fig. 5.4 Wood palisade with filling material in Carleton, Gaspé Peninsula

Details on the number of methodological tools employed and the time-frame at each location are given in Table 5.1.

An important aspect of PAR is to make the research results accessible to and useful for the participants, Therefore, in addition to scientific reports and publications, research was disseminated through public meetings, information brochures, local media, a web-site containing several dozen interviews with academic, governmental and societal actors (http://www8.umoncton.ca/umcm-climat/), the CURA-CCC website and bulletin "Coastal Butterfly/Le Papillon Côtier" (http://www.defisdescommunautescotieres.org/en/home) and the website of the Social Economy and Sustainability Research Network (http://www.msvu.ca/socialeconomyatlantic/English/projectsE.asp) (Novkovic and Brown 2012).

Results—Adaptation Successes Resulting from PAR Projects

Results from PAR projects have to be judged at different levels. The final goal is of course to provoke change and lead to the adoption (and eventual implementation) of adaptation measures. However, the goals of initiating a public dialogue and reorganizing social networks in order to address the problem of climate adaptation

Table 5.1 Field studies in the study sites (non exhaustive list)[a]

	Bathurst/bayshore	Shippagan	Le Goulet	Kent RSC (Cocagne, Grand Digue)	Pointe-du-Chêne	Sout-East RSC (Shediac, Beaubassin-Est, Cap Pelé)	Acadian Peninsula (SMSR, Shippagan)	Acadian Peninsula (Le Goulet, SMSR, Shippagan)
Semi-directed interviews	20	19	6	19	6	12	28	22
Year	2007	2010	2005–2006	2012	2005–2006	2013	2011–2015	2013–2014
Focus groups	1	1 (12 participants)	3 (10–12 participants)	1 (6 participants)	3 (10–15 participants)	1 (7 participants)		1
Year	2007	2011	2006–2007	2012	2006–2007	2013		2013–2014
Public presentations	14	3	3	6	6	3	3	1
Year	2007–2009	2011–2012	2006–2007	2010–2012	2006–2007	2012–2014	2011–2014	2015
Refs.	Plante et al. (2011)	Stervinou et al. (2013)	Chouinard et al. (2006)	Rabeniaina et al. (2014), Chouinard et al. (2012)	Chouinard et al. (2006)	Rabeniaina et al. (2014)	Guillemot and Mayrand (2012)	Guillemot (2014), Guillemot and Aubé (2015)

[a]The IRZC project in Le Goulet, Shippagan, Bas-Caraquet and Sainte-Marie/Saint-Raphaël is ongoing. Interviews and participative research has been lead in other sites of the coast of New-Brunswick by the University of Moncton, Moncton campus. The CURA-CCC project of UQAR extended to three other sites in Quebec (Rivière-au-Tonnerre, Sainte-Flavie, Bonaventure)

Table 5.2 Adaptation priorities at the conclusion of the participatory action-research programs

Study site	Priority actions at the end of the project
Bathurst/bayshore	Protection structures built uniformly according to specifications
Shippagan	(1) Evaluation of the risk of submersion and erosion, identification of risk zones under different sea level rise scenarios (2) Mitigation measures against sea level rise: – Restriction of development in risk zones – Comparison of different adaptation measures (riff-raff, retreat, etc.) – Evaluation of protection options for road 113 (3) Education of the population about effective, environmentally friendly and authorized protection means
Le Goulet	LIDAR measurement of the territory and mapping of flood zones
Cocagne and Grande-Digue	1. Protection: – A system of dykes in accord with IPCC sea level rise predictions – Barriers placed at an angle 2. Accommodation: – Elevate roads and buildings 3. Retreat: – Move the road further inland – Move buildings at risk and provide financial incentives – A government buy-me-out program for properties at risk
Pointe-du-Chêne	Elevation of the bridge linking Pointe-du-Chêne to the mainland

are over the long term equally important. In several cases, adaptation priorities changed over the course of the project (Table 5.2). For example, after the storm surge of 2000, which caused extensive damage, the town of Le Goulet was envisaging to build a flood protection wall and had already mandated a consultant company to write an expertise. However, after the PAR process, residents realized that despite a cost of 3.3 million$ for a 3.8 km long protection wall, in addition to 14 million$ for an aqueduct system, protection would not be complete since the water could circumvent the wall and salt intrusion in the aquifer would continue (Richardson 2010). In consequence, planned retreat of the most exposed properties was increasingly taken into consideration, even though no final decision has been reached yet.

In Pointe-du-Chêne, lying on a peninsula, some options were envisaged, including the dyking-in of the peninsula. At the end of the collective ranking exercise, taking into account cost, efficiency, impacts of the adaptation options and the community's long term development goal, elevating the bridge linking the village to the mainland so that it would not get flooded during storm surges was chosen as the best short-term adaptation goal. With the solid argumentation developed during the research project, the municipality was able to successfully lobby the transport ministry to finance the elevation of the bridge, which already proved successful during the exceptional 2010 flood (Fig. 5.5).

In Le Goulet and Shippagan, the priority identified at the end of the research project from 2005 to 2006 was a better identification of the flood risk zones. As a consequence, funding could be obtained to perform a LIDAR survey of the territory

Fig. 5.5 The newly elevated bridge of Pointe-du-Chêne during the storm of 2010 and under normal conditions

and establish flood risk maps based on the digital elevation models and 2025 and 2100 sea level rise estimates, followed by a risk analysis to infrastructure (roads and buildings). These models allow a collective reflection with citizen around the definition of risk, priority of issues and adaptation strategies, in the shape of public meetings organized by the IRZC (Aubé et al. 2014; Guillemot and Aubé 2015). The mobilization initiated in SMSR in 2011 in the context of the PAR project made it possible to include this study site and erosion scenarios, the principal hazard in this village, will be produced. In following research projects, piloted by the IRZC, those data were used to build risk maps and recommendation maps in a series of work groups with 5–7 participants, assisted by an urbanist from the regional planning commission and a researcher from IRZC (Aubé and Kocyla 2012). The different municipalities preferred different horizons: Bas-Caraquet a hundred-year storm flood in 2011, Shippagan and Le Goulet 50 and 100-year storm surges in 2055, as the 2100 scenarios seemed too catastrophic (Guillemot and Aubé 2015). A preference hierarchy emerged for adaptation options, the options permitting continued occupation of the territory, like protection techniques or architectural measures, being favoured over retreat (Guillemot and Aubé 2015). A longer-term result of the research led in the Acadian peninsula is the groundwork laid towards a regional planification of climate change adaptation. The regional planning commission adopted a resolution in 2013 mandating a committee to think about that question and the IRZC obtained a grant to lead research in that direction.

Bayshore, a district of the Bathurst municipality built on a dunal split, was the only community to hold on to protection as a main mean of adaptation, however in a more planned way with strict guidelines to ensure uniform protection structures. This does not mean that other municipalities had abandoned the idea of coastal protection, but they regressed in order of priority.

In the Acadian Peninsula, it was observed that at the end of the CURA project, certain vulnerability factors were taken into consideration, which had hitherto been ignored, such as problems with water evacuation pumps, contamination of wells, overflow of septic tanks, vulnerability of special infrastructures such as retirement homes, basement floods, etc. There was also a broadened perception of adaptation options. Whereas during the first meetings, participants focused on technical solutions for the short term, they are now considering a portfolio of options, also concerning the long term and future generations, and the necessity to work on the organisational aspect, including a more regional approach. Progress remains however modest.

Not in all cases have tangible results been achieved. The discussion structures put in place during the research project have not all been maintained, although some have, like in Bayshore. However, even though the dialogue does not necessarily result in collective decision-making, it lays the groundwork for it (Guillemot et al. 2014).

Impact on Local Governance

The research projects also had a direct or indirect impact on local governance. In Beaubassin-Est, under the leadership of the regional planning commission, the rural community voted a by-law (CGVD28, by-law 09-1 B, February 2011) that specifically includes future sea level rise in the maximum height of the habitable part of new constructions (Doiron 2012). Similar by-laws have since been adopted by neighbouring communities. In the long-term, this could lead to a general restriction of new residential development in flood-prone zones under future sea-level rise scenarios (an urban planner of RSC7, *pers. comm.*). The municipalitites of Sainte-Marie-Saint-Raphaël and Shippagan in the Acadian Peninsula introduced, within the framework of the provincial policy on coastal zones, a no-build zone of 30 m from any fresh- or saltwater body (Guillemot and Aubé 2015).

In several LSD, the question of incorporation was raised in the course of the PAR. Some participants favoured the creation of a municipality for the territory while others preferred to keep the privileges linked to the LSD status (mainly a lower tax level). In the context of climate change adaptation, the status of a municipality or rural community offers distinct advantages, such as the capacity to vote by-laws and to produce a rural/urban plan, which is not possible for LSD. Thus, three LSD held votes on the creation of a rural community. The referendum in Cocagne in 2013 yielded a majority in favour of the project. The new rural community was officially constituted on May 23rd, 2014. The new rural community of Cocagne has since then produced an emergency measures planning including climate change and is working with the regional service on an urban plan. A consultation in Grande-Digue showed no majority in favour of the project and in consequence, Grande-Digue remains a LSD. In Dundas, a vote held in 2014 resulted in a tie with 420 votes on each side, which meant that the project has for now be rejected.

In the Acadian peninsula, the research projects undertaken at a limited number of pilot sites has resulted in the beginning of a regional planning. In 2013, the board of the regional planning commission (now regional service commission), supported by the forum of mayors, voted a resolution mandating it to work towards a regional planification of climate adaptation. At the same time, the IRZC received a grant from the New Brunswick environmental trust fund to initiate research in that direction (Guillemot and Aubé 2015).

The Importance of Local Knowledge and of the Co-production of Knowledge

One of the essential aspects of PAR is the co-production of knowledge. This allows to create a body of knowledge that is pertinent, responding to the needs of stakeholders, legitimate, being accepted by all, and credible (Lee et al. 2014; Plante et al. 2016; Shaw et al. 2013). This is achieved through a dialogue between actors (including the scientists), which allows to respect the diversity of perceptions and points of view and the complementarity of scientific and local knowledge (Guillemot and Aubé 2015). "Frontier organisms", in this case universities of non-profit organisations can articulate this dialogue (Guillemot and Aubé 2015).

In the case of climate change adaptation, this is particularly pertinent since the scientific knowledge, based on models and scientific observation, and the local knowledge, based on observation and traded knowledge (personal and generational memory) effectively complement each other and can cross-validate one another. Scientific observation is verifiable and follows a rigourous method, whereas local knowledge has the advantage of covering more of the territory and often reaching much further back in time than scientifically measured time series, and can also rely on proxy knowledge. In several instances, participants were able to recall weather events that occurred over a span of 40–50 years.

In New-Brunswick, residents have observed changes in climate and the state of the environment, such as more frequent extreme weather events (floods, storms, freezing rains, etc.), a reduction in ice cover, an increase in erosion of beaches and dunes (Chouinard et al. 2006; Guillemot et al. 2014; Stervinou et al. 2013). A study undertaken in Shippagan showed that there is an excellent correlation between local observations and scientific data regarding environmental changes that occurred over the last 25 years (Stervinou et al. 2013).

Residents observe clear signs of environmental changes, as these examples from Richibuctou, Cocagne, Grande-Digue and Dundas demonstrate (Chouinard et al. 2012) (free translation):

> I have been living here for forty years and it's the first time the water reached my garage
>
> There is a lot of erosion in front of the house, 30 feet since 1975, the grass is all gone…
>
> The tendency on the coast is towards erosion. However, I noticed an advance of dunes on Cocagne Island

Fig. 5.6 Mapping exercise with residents of Le Goulet. *Source* Chouinard and Martin (2007)

> Erosion is more and more frequent. The cliff by my house has lost 2 feet in 6 years. Freezing and thawing are more and more frequent, they accelerate erosion and break the protective structures
>
> I could see the dunes from my property 30 years ago at Pointe-Cormier and the rock wall we built 30 years ago is already not working any more

Taylored tools, such as participative mapping (Fig. 5.6) allow to combine the participant's knowledge and observation and scientifically produced flood risk maps.

Appropriation of Knowledge and Emancipation of Communities

The communities participating in the projects all underlined the importance of co-construction, which allows them to appropriate some of the research tools and products used by scientists in those projects, such as participative mapping and the flood maps produced under different sea level scenarios (Chouinard et al. 2006). A better understanding of the research process and the tools used allows the communities to gain in confidence and be more credible while negotiating with government authorities or grant agencies. Thus, the municipalities of Le Goulet and Pointe-du-Chêne managed to obtain the funding for the adaptation priorities they had established at the end of the project (LIDAR mapping and the elevated bridge). Thus, the PAR project leads to an emancipation of the communities that become better equipped to face current and future challenges.

Municipal councils now find it important to inform the population on the risks linked to the impacts of climate change using such maps. Flood and erosion prevision maps for Shippagan, Pointe-Brûlée and SMSR have been available on the site of the regional service commission of the Acadian Peninsula[2] since 2015. At the beginning, those municipalities were reluctant to share this information, for fear of alarming their citizen and to tarnish the impage of the municipalities (and maybe discourage investors).

The Importance of Facilitators and Local Associations

In its scope and long time scale, climate change represents a new challenge to territorial planning and governance. Besides the traditional governance structures, new actors play an important role in the process of climate adaptation. Collins and Ison (2009) affirm that addressing the impacts of climate change necessitates a participation of all actors and a redefinition of local governance that relies on a capacity to generate collective learning.

Local environmental and development associations have emerged as important actors in the adaptation process and in the mobilization of citizen. They often act as initiators for the collective deliberation process, as intermediates between citizen and authorities and as active participants in research projects. Their role is particularly important in LSD, which don't possess municipal structures (Chouinard et al. 2012). Citizen then often turn to local or regional associations for advice, since they trust them more than provincial representatives (Chouinard et al. 2013, 2015).

During the research projects, researchers play the role of facilitators, as defined by Park (2006) in the context of PAR. They can reinforce ties between decision makers in the public or private sector, citizen, and scientists. They favour de development of purposeful and consensual knowledge. In most instances, the initial facilitators were university groups. However, as in Bayshore, local NGOs can take over this role. The IRZC has played that role in the Acadian Peninsula, federating around it a coalition of communities (municipalities and LSD), provincial authorities, the planning commissions and actors from civil society. Through the access to scientific knowledge, to research funding, and its presumed impartial role, universities or other facilitators are able to stimulate a collective dialogue into

[2]http://www.csrpa.ca/fr/changements-climatiques.

adaptation strategies and the wider perspective of the communities' future development (Aubé and Kocyla 2012; Chouinard and Weissenberger 2015; Guillemot et al. 2014).

Limitations—Hinderance to Adaptation

The PAR projects have managed to initiate a community dialogue around adaptation and put in place a process that in most cases will continue in one form or the other, whether maintaining the structures elaborated during the projects or creating new structures appropriate to changing conditions. Successful adaptation depends on many factors, and some impediments to successful adaptation persist in the study region. Local governance issues are always at the heart of adaptation matters. The material and financial difficulties of communities are partly related to this, but are difficult to overcome, especially in some parts of the study area, which have been economically depressed for a considerable time. Quebec and New-Brunswick have the 3rd and 4th lowest per capita GDP in Canada (ahead of only Nova Scotia and Prince Edward Island, two other Atlantic provinces). In addition, the absence of effective regulatory framework and the fact that most of the coastline is private property leads to a dominance of a pecuniary logic.

At the same time, the coastal zone has been undergoing a change in vocation and perception over the last decades, emphasizing the recreational and esthetic values of the coastal environment (Noblet and Chouinard 2014). This has led to a considerable increase in property prices along the shore (in the territory of the Kent RSC, they have tripled over 10 years), which creates a strong antagonism to stricter coastal regulations (Noblet and Chouinard 2014). The change in demographics, characterized by an influx of new residents from the hinterland also tends to create a rift between new and old residents, full-time and part-time residents and different socio-economic classes which can lead to disagreements on adaptation policies within communities (Chouinard et al. 2013; Rabeniaina et al. 2014).

The PAR process itself contains a number of difficulties or challenges:

- It remains difficult to transition from the co-construction of a vulnerability assessment and possible strategies to collective and individual action and responsibility-taking;
- There is not enough government support, as it focuses on help for reactive measures, usually following a climate hazard;
- The effort in participating in such a time-intensive process must be worth the result The interest of NGO and research organisations in doing so stems from their mandates and financing. For municipalities, it is the mandate to protect its citizen, but short term financial losses and political conflicts sometimes have to be overcome;
- The ability of stakeholders to project themselves into the future, an essential part of the PAR and climate adaptation process, is often insufficient.

These difficulties have also been underlined and discussed in Europe (Vanderlinden 2015) and in three different coastal areas of Canada and the Caribbean (Lane et al. 2013). Nevertheless, the PAR is considered to reinforce the social link and a relation of trust between citizens and decision makers in order to chose the appropriate actions in order to address coastal erosion and inondations in the context of climate change.

Conclusion

Participatory action research has proved over the past decade a successful approach in engaging communities in Quebec and New Brunswick towards efficient and prospective climate adaptation and initiating a community dialogue. It allows communities to appropriate climate change science and tools and to become more autonomous. The projects led over the last decade have resulted in concrete adaptation actions in some communities, and stimulated social change in others that will favour communities' adaptation in the long term. The approach cannot of course overcome all challenges facing communities, including deep-seated governance issues and contemporary development modes of the coastal zone, which are largely inconsistent with sustainable development, especially in the face of climate change. Therefore, long-term adaptation will rely on an inflection of territorial development trajectories, for which the social process initiated through participatory action research can contribute in a significant way. The role of local governments and associations is crucial in tailoring adaptation solutions and in mobolizing the population. In this sense, PAR is successful when it leads to an empowerment of local communities making them eventually able to devise their own adaptation strategies.

Acknowledgements The authors wish to thank Mélanie Aubé, Institut de recherche sur les zones côtières, Université de Moncton, for valuable discussions in the preparation of this manuscript.

References

Aubé MB, Kocyla (2012) Adaptation aux changements climatiques: planification de l'utilisation du territoire à Shippagan, Le Goulet et Bas-Caraquet. Projet ASACCA. Péninsule acadienne. Volet accompagnement des communautés, Institut de recherche sur les zones côtières inc., 62 p

Aubé MC, Hébert A, Doiron (2014) Accompagnement de communautés de la Péninsule acadienne dans la planification de l'adaptation aux changements climatiques: année 3. Exercices de planification stratégique à Shippagan, Bas-Caraquet, Le Goulet et Sainte-Marie-Saint-Raphaël, Cap-Bateau, Pigeon Hill, Institut de recherche sur les zones côtières inc./Fonds en fiducie pour l'environnement du Nouveau-Brunswick, 82 p

Bernatchez P, Fraser C (2012) Evolution of coastal defence structure and consequences for beach width trends, Quebec, Canada. J Coastal Res 28:1550–1566

Bernatchez PC, Fraser S, Friesinger Y, Jolivet S, Dugas S, Drejza A, Morissette (2008) Sensibilité des côtes et vulnérabilité des communautés du golfe du Saint-Laurent aux impacts des changements climatiques. Laboratoire de dynamique et de gestion intégrée des zones côtières, Université du Québec à Rimouski. Rapport de recherche remis au Consortium OURANOS et au FACC, 256 p

Bérubé D (n.d.) L'érosion côtière au Nouveau-Brunswick: tendances et conséquences. Ministère des Ressources naturelles Direction des études géologiques

Bizikova L, Pinter L, Tubiello FN (2014) Recent progress in applying participatory scenario development in climate change adaptation in developing countries. International Institute for Sustainable Develoopment (IISD), Winnipeg, Manitoba

Brown AAH (2009) Learning from stakeholders: some lessons from participatory research and climate change adaptation in practice. Climate change: global risks, challenges and decisions, IOP conference series: earth and environmental science, vol 6, 392014. doi:10.1088/1755-1307/6/9/392014 (P39.03)

Chouinard O, Martin G (2007) Un plan communautaire pour l'adaptation au changement climatique pour Le Goulet, Nouveau-Brunswick: La rencontre de la science, des valeurs et des connaissances locales, Université de Moncton, 28 p

Chouinard O, Weissenberger S (2015) Adaptation aux changements climatiques et à l'augmentation du niveau de la mer en zones côtières—une perspective globale. Online http://www8.umoncton.ca/umcm-climat/uved

Chouinard O, Plante S, Martin G (2006) Engagement des communautés face au changement climatique: une expérience de gestion intégrée à Le Goulet et Pointe-du-Chêne au Nouveau-Brunswick, VertigO, 7. doi:10.4000/vertigo.1912. Online http://vertigo.revues.org/1912

Chouinard OS, Jolicoeur G, Martin S, O'Carroll DB, Kelley B (2009) Pointe Carron: Vivre dans un écosystème côtier. Vers une planification communautaire durable. Comité directeur, l'équipe d'étude sur le problème d'érosion à la Pointe Carron and the Université de Moncton

Chouinard O, Tartibu N, Gauvin J (2012) Adaptation réfléchie en milieu côtier: démarche d'engagement avec les résidents du milieu côtier des DSL de Cocagne et de Grande Digue, Université de Moncton, rapport final pour le Fonds de Fiducie en environnement du Nouveau-Brunswick

Chouinard O, Rabeniaina T, Weissenberger S (2013) Les apprentissages sur l'aménagement côtier dans deux territoires côtiers du littoral acadien du Nouveau-Brunswick vulnérables à l'érosion et aux inondations, Études caribéennes, 26. doi:10.4000/etudescaribeennes.6663. Online http://etudescaribeennes.revues.org/6663

Chouinard O, Weissenberger S, Lane D (2015) L'adaptation au changement climatique en zone côtière selon l'approche communautaire: études de cas de projets de recherche-action participative au Nouveau-Brunswick (Canada), VertigO, Hors-série 23. doi:10.4000/vertigo.16642. Online http://vertigo.revues.org/16642

Collins K, Ison R (2009) Jumping off Arnstein's ladder: social learning as a new policy paradigm for climate change adaptation. Environ Policy Gov 19:358–373

Daigle R (2006) Impacts de l'élévation du niveau de la mer et du changement climatique sur la zone côtière du sud-est du Nouveau-Brunswick, Rapport de recherche, Environnement Canada

Doiron S (2012) From climate change plans to by-laws, it's time to act, Plan Canada. Winter 2012:30–34

Finn JG (2008) Building stronger local governments and regions—an action plan for the future of local governance in New Brunswick. Report of the commissioner on the future of local governance, 196 p

Forbes DL, Parkes GS, Ketch LA (2006) Sea-level rise and regional subsidence. In: Daigle R (ed) Impacts de l'élévation du niveau de la mer et du changement climatique sur la zone côtière du sud-est du Nouveau-Brunswick, Rapport de recherche, Environnement Canada, pp 34–94

Fox M, Daigle M (2012) Planning for sustainability in New Brunswick, Atlantic Climate Adaptation Solutions Association, 36 p

Friesinger S, Bernatchez P (2010) Perceptions of Gulf of St. Lawrence coastal communities confronting environmental change: hazards and adaptation, Québec, Canada. Ocean Coastal Manag 53:669–678

German L, Tiani AM, Daoudi A, Mutimukuru-Maravanyika T, Chuma E, Jum C, Nemarundwe N, Ontita E, Yitamben G (2012) The application of participatory action research to climate change adaptation in Africa: a reference guide. International Development Research Centre (IDRC), Ottawa

Guillemot J (2014) Gestion des risques côtiers et émergence de réseaux d'acteurs dans la Péninsule acadienne (Nord-Est du Nouveau-Brunswick), Actes du colloque international Connaissance et Compréhension des Risques Côtiers: Aléas, Enjeux, Représentations, Gestion 3–4 juillet 2014 IUEM, BREST. 12 p. Online http://cocorisco.sciencesconf.org/conference/cocorisco/pages/cocorisco_2014_actes.pdf

Guillemot J, Aubé M (2015) L'adaptation aux changements climatiques dans la Péninsule acadienne: rôles d'acteurs clés dans l'émergence d'un dialogue articulé à l'échelle régionale, VertigO, Hors-Série 23. doi:10.4000/vertigo.16664. Online http://vertigo.revues.org/16664

Guillemot J, Mayrand E (2012) Décider et agir: quels apports possibles pour une démarche d'accompagnement des collectivités côtiers face aux changements climatiques à Sainte-Marie-Saint-Raphaël et Shippagan (Nouveau-Brunswick), Coastal Zone Canada, Rimouski, 9–14 juin

Guillemot J, Mayrand E, Gillet J, Aubé M (2014) La perception du risque et l'engagement dans des stratégies d'adaptation aux changements climatiques dans deux communautés côtières de la péninsule acadienne, VertigO, 14. doi:10.4000/vertigo.15164. Online http://vertigo.revues.org/15164

Jolicoeur S (2016) Des pressions qui s'opposent: révision des pratiques d'urbanisme, «coastal squeeze» et intégrité de la zone côtière, webinaire, Association of Municipal Administrators of New Brunswick, Jan 14

Jost C, Ferdous N, Spicer TD (2014) Gender and inclusion toolbox: participatory research in climate change and agriculture. CGIAR research program on climate change, agriculture and food security (CCAFS). CARE International and the World Agroforestry Centre (ICRAF), Copenhagen, Denmark

Lane D (2012) C-Change community of practice meeting—Case study: Perth-Andover, New Brunswick. 20 p

Lane D, Mercer Clarke C, Forbes D, Watson P (2013) The gathering storm: managing adaptation to environmental change in coastal communities and small islands. Sustain Sci 8:469–489

Lee E, Jung CS, Lee M-K (2014) The potential role of boundary organizations in the climate regime. Environ Sci Policy 36:2

Leonard JA, Keys CB, Christopher B, Suarez-Balcazar Y, Taylor RR, Davis MI (eds) (2004) Participatory community research: theories and methods in action. Am Psychol Assoc, Washington 271 p

Lieske DJ (2015) Coping with climate change: the role of spatial decision support tools in facilitating community adaptation. Environ Model Softw 68:98e109

Mackinson S, Raicevich S, Kraan M, Magudia R, Borrow K (eds) (2015) Good practice guide: participatory research in fisheries science. Online http://gap2.eu/outputs/pr-handbook/

McClymont Peace D, Myers E (2012) Community-based participatory process—climate change and health adaptation program for northern first nations and inuit in Canada. Int J Circumpolar Health 71:18412. http://dx.doi.org/10.3402/ijch.v71i0.18412

McIntyre A (2008) Participatory action research. Sage Publications, Thousand Oaks 78 p

Ministère des Affaires municipales, des Régions et de l'Occupation du territoire (MAMROT) (2010) Planning and development powers in Quebec, Bibliothèque et Archives nationales du Québec, 18 p

New Brunswick (2002) A coastal areas protection policy for New Brunswick, New Brunswick Environment and Local Government, 36 p

Noblet M (2015) L'adaptation au changement climatique en zone côtière au Canada et au Sénégal, une comparaison Nord/Sud, PhD thesis, Université Picardie Jules Verne, Amiens

Noblet M, Chouinard O (2014) L'adaptation au changement climatique: une question de gouvernance à la lumière de la recherche partenariale: Le cas de la zone côtière du Sud-Est du Nouveau-Brunswick, Canada, In Fontan JM, Buissière D, Klein JL (eds) Le défi de l'Innovation sociale partagée: savoirs croisés, chap 8, PUQ, 189–208

Novkovic S, Brown L (2012) Social economy: communities, economics and solidarity in Atlantic Canada. Cape Breton University Press, Nova Scotia 260 p

Park P (2006) Knowledge and participatory research. In: Bradbury H (ed) Handbook of action research. Sage Publications, New Dehli, pp 83–93

Parkes GS, Manson GK, Chagnon R, Ketch LA (2006) Climatologie—Ondes de tempête, Vents, Vagues et Glaces. In: Daigle R (ed) *Impacts de l'élévation du niveau de la mer et du changement climatique sur la zone côtière du sud-est du Nouveau-Brunswick*, Rapport de recherche, Environnement Canada, pp 95–27462

Plante S, Da Cunha C (2015) La méthode d'évaluation par animation de groupe (MÉAG) comme moyen d'adaptation aux changements climatiques l'exemple de l'ARUC DCC dans la Baie des Chaleurs. In: Chouinard O, Weissenberger S (eds) Adaptation aux changements climatiques en zone côtière: une perspective mondiale, Université de Moncton, part 4.2.4. Online http://www8.umoncton.ca/umcm-climat/

Plante S, Chouinard O, Martin G (2011) Gouvernance participative par l'engagement citoyen à l'heure des changements climatiques: Études de cas à Le Goulet, Pointe-du-Chêne et Bayshore Drive (Nouveau-Brunswick), Territoire en movement, 11. doi:10.4000/tem.1234. Online http://tem.revues.org/1234

Plante S, Vasseur L, Da Cunha C (2016) Adaptation to climate change and participatory action research (PAR): lessons from municipalities in Quebec, Canada. In: Jörg K (ed) Climate adaptation governance. Theory, concepts and praxis in cities and regions (in press)

Rabeniaina TR, Chouinard O, Weissenberger S (2014). L'adaptation en zone côtière: de la parole aux actes: études de cas de deux territoires du Sud-Est du Nouveau-Brunswick (Canada) concernant les dynamiques sociales et les apprentissages mutuels de communautés côtières face aux enjeux climatiques et à la gouvernance locale, Études caribéennes, 27–28. doi:10.4000/etudescaribeennes.6970. Online http://etudescaribeennes.revues.org/6970

Reid H, Alam M, Berger R, Cannon T, Huq S, Milligan A (eds) (2012) PLA 60: Community-based adaptation to climate change, special edition of participatory learning and action. International Institute for Environment and Development (IIED), London

Richardson GRA (2010) S'adapter aux changements climatiques: Une introduction à l'intention des municipalités canadiennes, Ressources naturelles Canada, Ottawa, Ontario, 2010. Online http://www.nrcan.gc.ca/environment/resources/publications/impacts-adaptation/tools-guides/16299

Savard JP, Bernatchez P, Morneau F, Saucier F, Gachon P, Senneville S, Fraser C, Jolivet Y (2008) Étude de la sensibilité des côtes et de la vulnérabilité des communautés du Golfe du Saint-Laurent aux impacts des changements climatiques. Sommaire à l'usage des décideurs. 36 p

Shaw J, Chiara D, Stocker L (2013) Spanning the boundary between climate science and coastal communities: opportunities and challenges. Ocean Coast Manag 86:80–87

Stervinou V, Mayrand E, Chouinard O, Thiombiano AN (2013) La perception des changements environnementaux: le cas de la collectivité côtière de Shippagan (Nouveau-Brunswick, Canada), VertigO, 13. doi:10.4000/vertigo.13482. Online http://vertigo.revues.org/13482

Van Buuren A, Eshuis J, van Vliet M (eds) (2015) Action research for climate change adaptation: developing and applying knowledge for governance. Routledge, Abingdon 198 p

Vanderlinden JP (2015) Prévoir l'imprévu. Ceriscope Environnement et relations internationales. Online http://ceriscope.sciences-po.fr/environnement/content/prevoir-l-imprevu?page=9

Vasseur L, Catto N (2008) Atlantic region (Chapter 4). In: Lemmen D (ed) National climate change assessment. Natural Resources Canada, Ottawa, p 52

Chapter 6
Recent Harm, Problematic Impacts, and Socially Feasible Adaptation Options to Heatwaves and Heavy Rain Events in New York City

Diana Reckien

Abstract The City of New York is highly diverse in socio-economic but also in natural characteristics (proximity to sea, natural areas, building and population density), potentially leading to different impacts, impact perceptions and affectedness of residents by climate change and extreme weather events. Impacts of weather events do not only comprise human fatalities or substantial economic loss, but also less dramatic, daily, small-scale interruptions of daily routines. These may be burdensome for residents and critical, potentially leading to larger impacts in form of knock-on effects. A number of studies and assessments conducted for/on New York City investigate the potential impacts of and adaptation options to climate change, but none of them—to my knowledge—is based on or includes stakeholder knowledge, i.e. the impacts as stated by New York City residents. Furthermore, assessing the impacts of extreme weather events on residents does not necessarily mean that people are strongly affected or feel burdened. However, knowing the impacts and related burden of extreme weather on the city's residents is necessary to effectively plan for adaptation. This study investigates the impact and burden of two form of extreme weather events, heavy rainstorms and heat waves, on New York City residents. The study is based on a questionnaire survey among 762 respondents from the 5 boroughs. To a large extent the study therefore relies on the memory of

---

Most of the work for this article was carried out at the Center for Research on Environmental Decisions, Earth Institute, Columbia University, USA.

D. Reckien (✉)
Department of Urban and Regional Planning and Geo-Information Management, Faculty of Geo-Information Science and Earth Observation (ITC), University of Twente, Enschede, The Netherlands
e-mail: d.reckien@utwente.nl; dianareckien@gmail.com

D. Reckien
Center for Research on Environmental Decisions, Earth Institute, Columbia University in the City of New York, New York, NY, USA

W. Leal Filho and J.M. Keenan (eds.), *Climate Change Adaptation in North America*, Climate Change Management,
DOI 10.1007/978-3-319-53742-9_6

previous events, which is also related to disaster response and management. The study elicits recent harm and damage, strongest impacts and problems, as well as perceived responsibilities for adaptation across sectors and boroughs.

Keywords Fuzzy Cognitive Mapping (FCM) · Climate change impacts · Heat waves · Heavy rain · New York City · Adaptation · Affectedness · Perception · Qualtrics online sample

Introduction

Cities are places of high vulnerability to climate change due to their density of people, assets, and infrastructure and their high diversity of people from different socio-economic backgrounds, gender, ethnicity, and age, determining knowledge and knowledge access, support networks, and other assets for adaptation. New York City (NYC) is a prime example of a large, diverse and complex city. It is the largest urban area in the United States and one of the most populous cities in the world.

As of 2014 the city had an estimated population of 8,491,079 inhabitants (United States Census Bureau 2015a, b), while the total population of the larger urban metropolitan region was 19,746,227 (United States Census Bureau 2015a, b). New York City is located at the east of the United States, directly on the Atlantic coast where the mouth of the Hudson River meets the New York Bay (approximately 40° 42′N 74°00′W). The city is subdivided by five boroughs—the Bronx, Brooklyn, Manhattan, Queens and Staten Island, which serve as identity markers. Four of the boroughs are an island or part of an island; as a consequence, the city has as extensive coastline.

New York City has a temperate, semi-humid, maritime climate (Lauer and Frankenberg 1992), which results in cool and damp winters and hot and humid summers. The city receives on average 1189 mm of precipitation per year (1910–2013) (The Weissman Center for International Business 2015), which is spread relatively evenly across the year. Hurricanes can affect the city in summer (the hurricane season lasts from 1 June to 30 November), with Hurricane Sandy (2012) and Irene (2011) being recent examples. Hurricanes are large storm systems that bring high-speed winds and normally lots of rain from the Gulf of Mexico. New York City can also be affected by Nor'easters in winter. Nor'Easters are North-Easterly winds that are usually accompanied by large amounts of rainfall from the North East Atlantic Ocean. In addition to hurricanes and Nor'easters, the city is prone to heatwaves, usually accompanied by high humidity (Rosenzweig and Solecki 2010).

Climate projections for NYC suggest that heat waves (three or more consecutive days with maximum temperature exceeding 90 °F (~32 °C)) will approximately triple in frequency by the end of the century compared to current conditions (Horton et al. 2011). Already in 2050, NYC is projected to have a climate similar to that of present-day St. Louis (Kalkstein and Greene 1997). Consequently, heat-related deaths are expected to increase significantly in NYC as in many major US cities with over 1 million inhabitants. Knowlton et al. (2007) estimate an increase of 47–95%—with a mean of 70%—in excess heat-related deaths for the NYC area from 1990 to 2050. Precipitation is expected to decrease overall for the North-Eastern region of the US (Blake et al. 2000). However, seasonal increases in winter precipitation may put a burden on areas that are already exposed to flooding and other rain-related hazards (Horton et al. 2010).

Despite a number of climate change, impact and adaptation studies conducted for the city (Solecki 2012) these do not generally include interviews with people. To the knowledge of the author there are no studies that involved assessing the impacts of extreme weather events on NYC residents, let alone on the degree of burden on different social groups and across the city. This study aims to generate knowledge on these issues—important for targeted and effective climate change adaptation planning. However, adaptation might not only be understood as a duty or mandate of policy and planning processes, but also as shared responsibility and concerted efforts among planning, policy and residents. The study elicits the views of the residents on the recent damage, impact distribution, adaptation sectors and assumed adaptation responsibility.

Methodology

The main data base of this study is an online questionnaire survey, which was conducted with New York City residents to investigate the impacts of extreme weather events across the five boroughs using a Fuzzy Cognitive Mapping (FCM) approach. FCM is a semi-quantitative interview and analysis method that aims to develop causal weighted networks with respondents about or of a system under investigation (Reckien 2014). It allows to assess impacts and concepts for which numerical data are absent, difficult to define, objectify or measure, i.e. for many social concepts that are (only) linguistically and culturally shared. FCM does not depend on data sources other than the knowledge of the interviewee. By that it draws on all experience, perception, knowledge, and beliefs, which are important for adaptation as people act on it. The underlying assumption is that significant impacts of climate-related events are remembered and interventions, i.e. in terms of adaptation strategies, can only occur at cause-effect relations that are perceivably understood. In the FCM elicitation phase interviewees were asked to report on impacts of extreme weather events—in this case heat waves and heavy rainstorms, in linguistic terms. After that stage interviewees developed the network of cause-effect relations among the mentioned impacts (however, not reported here).

Additionally, interviewees were asked to answer question on adaptation responsibility, adaptation sectors, and frequently used information channels for disaster-related/adaptation information. Both weather events were addressed in a similar fashion subsequently, controlling for order effects. The questionnaire survey lasted for about 25 min; the respondents were compensated by Qualtrics Survey provider, which executed the survey. This also means that the study targeted the Qualtrics Survey sample population. The interviews were conducted between 05.11.2013 and 08.12.2013. In total 762 completed and valid questionnaires were collected.

The study focuses on New York City and the differences and similarities across the 5 boroughs as well as across weather events. To do so, the elicited impacts will be grouped into impact sectors, such as health, transport, water issues, etc., and their mentioning compared across boroughs, as well as heat waves and heavy rainstorms. Adaptation needs per sectors and responsibilities are also presented. I conclude with implications for adaptation for the NYC planning authorities and legislature.

Results

Interview Sample

The sample comprises of 762 valid, completed questionnaires in total. Most of the interviewees are from Manhattan and Brooklyn, the smallest number of responses were collected on Staten Island, although this number translates into the largest share of interviewees compared with the total borough population (Table 6.1).

Concerning the representativeness of the study sample Fig. 6.1 shows the distribution of respondents in terms of their age and residence in the city. One can see that the distribution is rather even, lowering concerns of the representativeness of online surveys, at least in NYC.

Recent Harm and Damage

Experiencing an extreme event and being harmed is assumed to strongly relate to the beliefs and attitudes towards global warming and climate change, as the recurrent surveys of the Yale Project on Climate Change Communication show

Table 6.1 Sample distribution across NYC boroughs

Borough	Respondents N	Percentage [%] of total respondents	Total population (07/2012 estimate)	Percentage [%] of total population
Bronx	111	15	1,619,090	0.007
Booklyn	194	25	1,408,473	0.014
Manhattan	200	26	2,565,635	0.008
Queens	177	23	2,272,771	0.008
Staten Island	80	10	470,728	0.017
Total respondents in NYC	762	100	8,336,697	0.009

Data Author. *Source* Own draft

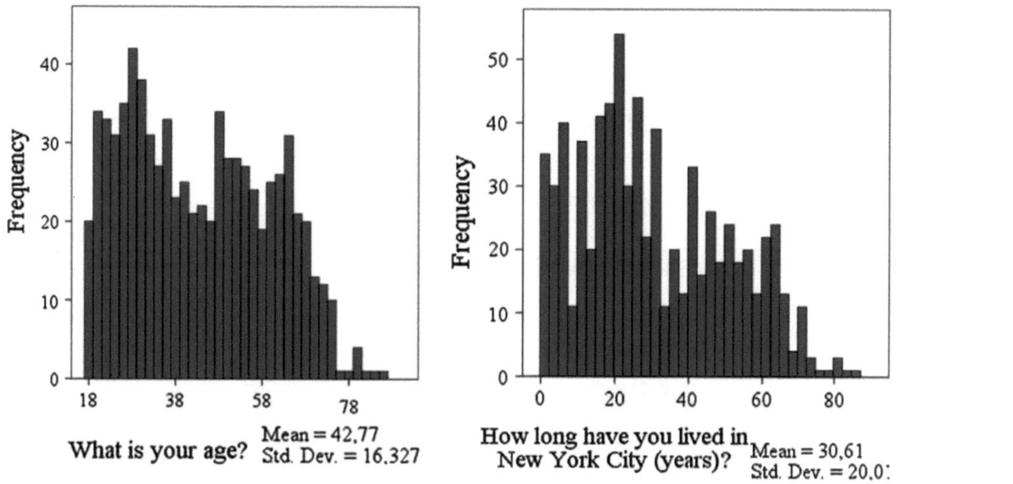

Fig. 6.1 Sample distribution in terms of age of respondents and residence in the city. *Data* Author. *Source* Own draft

(Leiserowitz et al. 2012a, b). Harm from extreme weather events is also assumed to influence individual efforts to adaptation (Slovic 2000). Analogical, this study aims to elicit climate change beliefs and attitudes, including recent damage and harm, to inform climate change adaptation.

Recent was defined as the previous 10 years, i.e. 2003–2013. Respondents were asked: "In the last 10 years, did you personally experience any harm during heat waves/heavy rain events?" The question contains the word "harm", transporting a more general notion, while two of the pre-given answers used "damage", more specifically used for physical structures and visible consequences (Fig. 6.2).

As Fig. 6.2 shows, previous recent harm is more often associated with heavy rain events than with heat waves. There are more respondents that experienced no harm during heat as compared with heavy rain. Rain events mostly cause damage to property and loss of income, whereas heat waves are mostly associated with health-related damages.

The harm/damage question, which particularly speaks to previous, recent events, was raised after a question to impacts of heat waves and heavy rain in general. Text boxes allowing input of free text asked for consequences of heat waves and heavy rain in general. These answers reveal a surprising picture.

Impacts and Problems

To assess perceived general impacts of heat waves and heavy rainstorms in NYC following questions were raised: "Please think about the consequences of heat waves/heavy rain events in New York City, i.e. consequences of a prolonged period

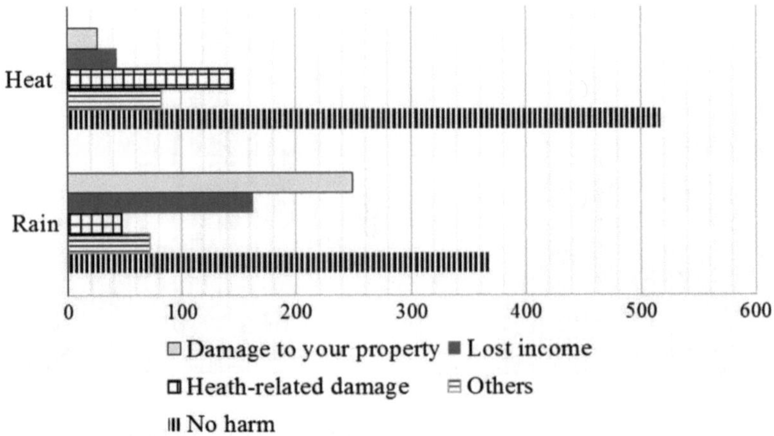

Fig. 6.2 Damage/harm experienced over the last 10 years (multiple entries possible). *Data* Author. *Source* Own draft

of excessive heat. What are aspects of consequences on your family, your neighborhood, the city and how do heat waves/heavy rain events affect you personally? Think about how your life is impacted by heat waves/heavy rain events and why. List key words and/or small groups of words …. Fill in as many as you remember; but, it is not necessary to fill all lines." (in total 8 aspects could be given). This question was raised before asking for recent, personal harm (reported above). These here mentioned aspects of consequences are called "impacts" in Fig. 6.3.

The subsequent question of the questionnaire asked for a rating of the mentioned aspects in terms of personal affectedness, i.e. "How much does each aspect personally affect you?", on a five point scale (affects me very much (1); affects me somewhat (2); affects me not very much (3); affects me not at all (4); don't know (5)). Categories (1) and (2), representing affectedness have been summarized as "problems" in Fig. 6.3. General impacts and personal problems were grouped into sectors or classes, as listed below, left hand side of Fig. 6.3.

The presented results (Fig. 6.3) relate to the answers of 80 respondents in each borough. They were randomly selected in the case of the Bronx, Brooklyn, Manhattan and Queens to reach similarity with Staten Island; which is important for cross-borough comparisons later on in the analyses (see further down).

Figure 6.3 shows that heat waves are associated with more impacts as compared with heavy rainstorms. This also counts for those impacts that are regarded problematic, as they substantially affect respondents personally. In contrast to the mentioned damage and harm during the previous 10 years, which was larger for rain events, respondents report larger personal problems from heat waves. These results point to a larger personal burden caused by heat waves as compared with the damages associated with heavy rainstorms.

Moreover, the nature of the impacts and problems during heat and heavy rainstorms differ. Impact of heat waves are associated with human health and body issue,

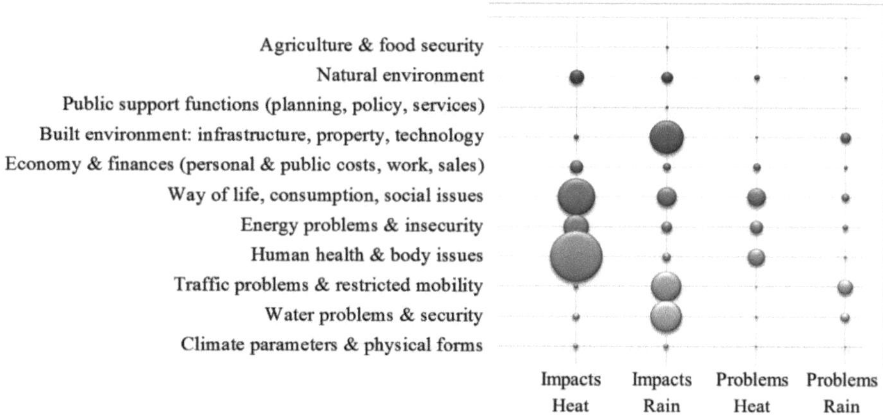

Fig. 6.3 Comparison of number of impacts and problems due to impacts for heat waves and heavy rainstorms across the NYC area. The figure refers to 80 respondents from each borough, i.e. 400 respondents in total. *Data* Author. *Source* Own draft

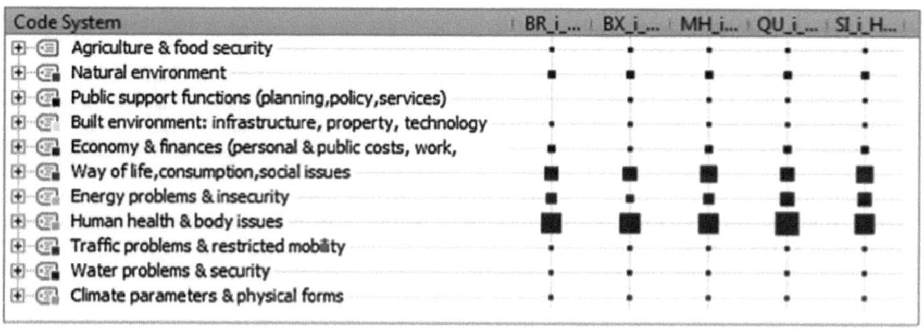

Fig. 6.4 Impacts of heat waves across the 5 boroughs of NYC, assigned to sectors. Key: *Squares* are scaled to matrix sum. *Data* Author. *Source* Own draft

and changes to the way of life. In contrast, the impact of strong rain events is related to the built environment, traffic and water (supply and drainage) problems. Comprehensibly, problems show similar patterns as impacts for both weather events.

Impacts and Problems Across Boroughs

The impacts and problems of the same randomly selected 80 respondents from each borough are now plotted across sectors (Figs. 6.4 and 6.5). The figures show that impacts of heat waves are relatively similar across the boroughs, mainly affecting human health and body issues. This changes when it comes to problems due to heat. Health-related problems are mainly experienced in the Bronx, while in the other

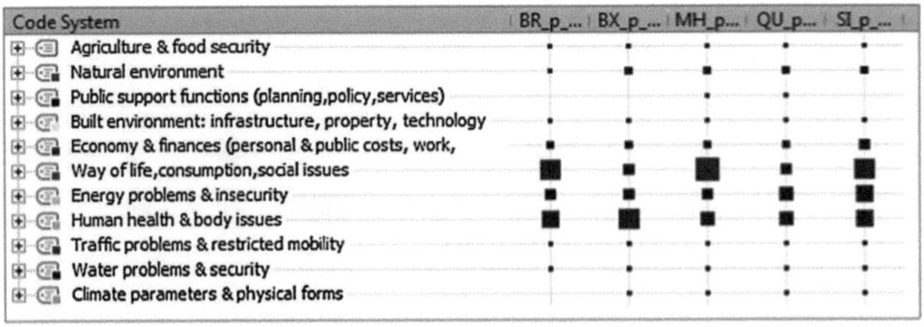

Fig. 6.5 Problems caused by heat waves across the 5 boroughs of NYC, assigned to sectors. Key: *Squares* are scaled to matrix sum. *Data* Author. *Source* Own draft

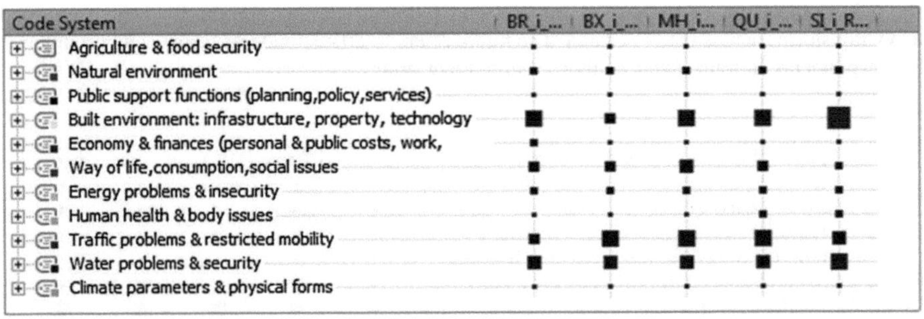

Fig. 6.6 Impacts of heavy rainstorms across the 5 boroughs of NYC, assigned to sectors. *Data* Author. *Source* Own draft

boroughs, such as Manhattan, Brooklyn as well as Staten Island, problems mainly refer to changes in the way of life and social issues. This is important information for adaptation efforts, e.g. those planed by health officials aiming to lower health-related problems during heat.

The same insights can be gained for plotting impacts and problems of heavy rainstorms across the 5 boroughs (Figs. 6.5 and 6.6). Results reveal that impacts of heavy rainstorms differ across the NYC area. Impacts in the Bronx and Staten Island largely relate to the built environment, while respondents from other boroughs mainly name traffic or water related (mostly drainage) issues. The intensity of impacts on the built environment fades when concentrating on problems. In most boroughs, problems of rainstorm are associated with traffic issues, particularly in the Bronx, Manhattan and Queens (Fig. 6.7).

Knowing more about the nature of problematic impacts during heat waves and heavy rainstorms may support decision making for adaptation. It provides general information on potentially efficient adaptation measures and sectors, and curtails adaptation responsibilities. More detailed information on adaptation necessities was elicited by posing additional, dedicated question on adaptation, as described below.

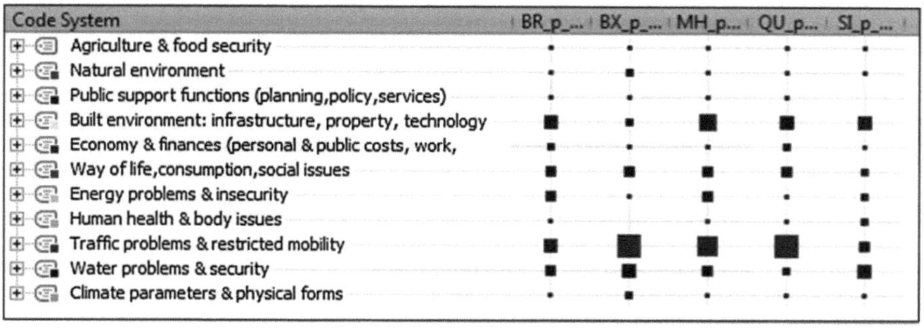

Fig. 6.7 Problems caused by heavy rainstorms across the 5 boroughs of NYC, assigned to sectors. *Data* Author. *Source* Own draft

Adaptation Sectors

Adaptation responsibilities can be shared by national and local governments and planning authorities, and individuals. Often, national governments prepare framework documents and guidelines, which aim to encourage, inform and support local communities to implement adaptation (Reckien et al. 2013; De Gregorio Hurtado et al. 2015; Heidrich et al. 2016). However, the success of national guidelines and its effect on local policies strongly depends on the quality of communication from the national to the local levels (De Gregorio Hurtado et al. 2014). Vice versa, it is important for the local levels to understand their adaptation needs in order to target adaptation measures and effectively liaise with bodies of higher administrative levels. Following question was raised: "The federal government is encouraging communities to prepare for the impacts of extreme weather events so that people and property are protected. How important do you think it is for New York City to take steps to protect the following from heat waves/heavy rainstorm events?"

Figure 6.8 reveals that respondents regard it as particularly important to adapt to heat waves by improvements in the electricity system, the water supply, and public health. It is not perceived as very important to improve the building stock, expand urban greenery and parks, or invest in the subway and rail system—although 'very important' is still the most often ticked category for most of these sectors as well. Only in the building sector the most numerous responses refer to the category 'somewhat important'. This is remarkable, as e.g. heat in homes can be substantially reduced by either more air-conditioning (provided reliant and constant electricity supply—the category assumed most important) or adjustments in the building design, construction, insulation, etc. Apparently, there is little awareness for alternative heat reduction measures, apart from air-conditioning.

In general, adaptation necessities concerning heat waves are assumed to be smaller than for heavy rain events. This is in contrast to the more numerous personal impacts and perceived problems during heat, as mentioned above, but in line

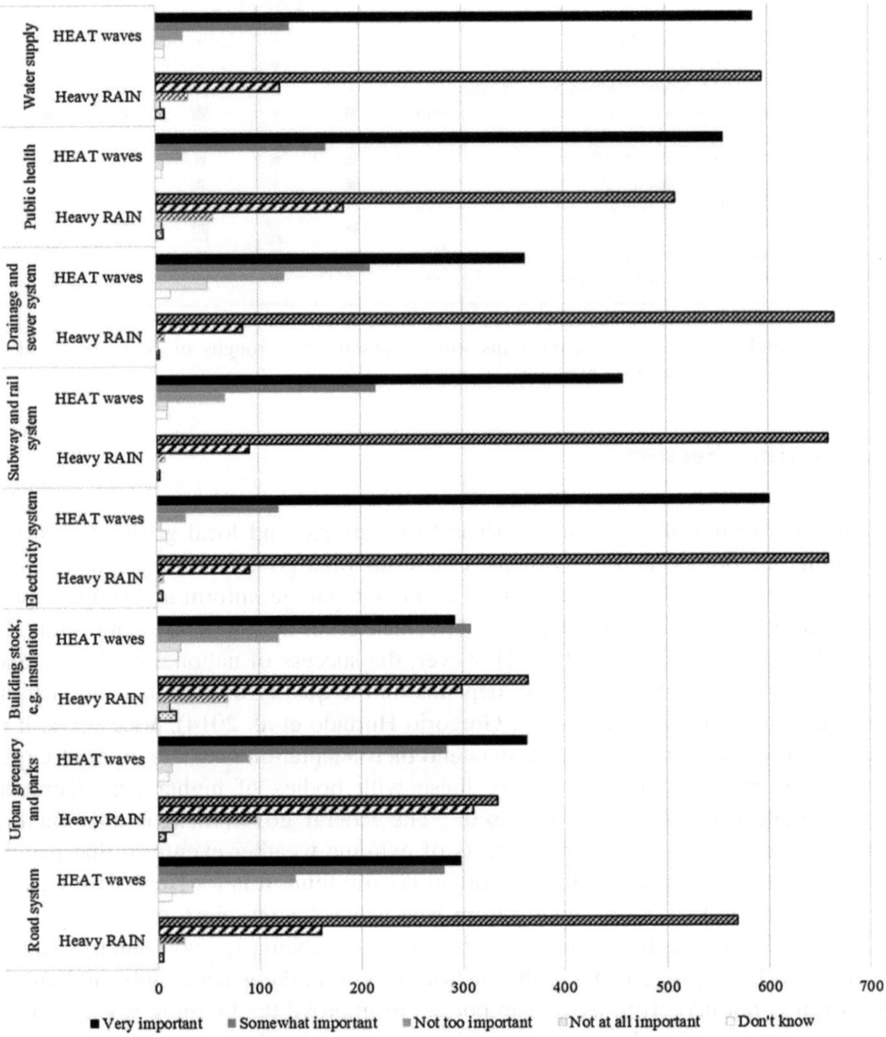

Fig. 6.8 Adaptation necessities per sector and extreme weather event, as indicated by NYC respondents. *Data* Author. *Source* Own draft

with larger recent harm and damage caused by heavy rain events. Apparently, heat waves lead to more numerous impacts and personal problems, but these are perceived as a private burden without many intervention and adaptation possibilities. The result is particularly alarming, since heat waves pose a major climate-related risk. In the US and other countries, more fatalities occur from heat waves than other climate hazards such as floods and hurricanes combined, when adaptation is not taken into account (Klinenberg 2002; Satterthwaite et al. 2007).

Concerning heavy rainstorms, respondents see in particular a need to improve the drainage and sewage, subway and rail, and electricity system. Again, there is seen little need in adaptation through an improvement in urban greenery and parks, and the building stock.

Overall, the electricity, water supply and subway system are seen as prominent adaptation sectors considering both weather events, showing most responses when combined (numbers not shown). The results reveal crucial insights for NYC's planning boards and authorities.

Adaptation Responsibilities

Some form and degree of adaptation responsibility may also lay within the population; a view which was tested by an additional questions. Respondents were asked to think about the adaptation responsibilities of residents in the following way: "Do you think citizens themselves should be doing more or less to protect themselves from the impacts of heat waves/heavy rainstorm events?" Respondents had to choose from a 5-point scale, as shown in Fig. 6.9.

The answers to the latest question indicate that respondents see a substantial adaptation responsibility with the residents, too. Both, during or against heat waves and heavy rain events citizens should do more to protect themselves. However, the second largest category is 'currently doing the right amount', which was selected by more respondents than 'much more'. It indicates that a substantial number of citizens does not see more responsibility with themselves—related to either (1) others being in charge or (2) large-scale adaptation being not necessary. Comparing heat and rain events, more respondents saw citizens in charge of doing more and much more to protect themselves from heavy rainstorms as compared with heat.

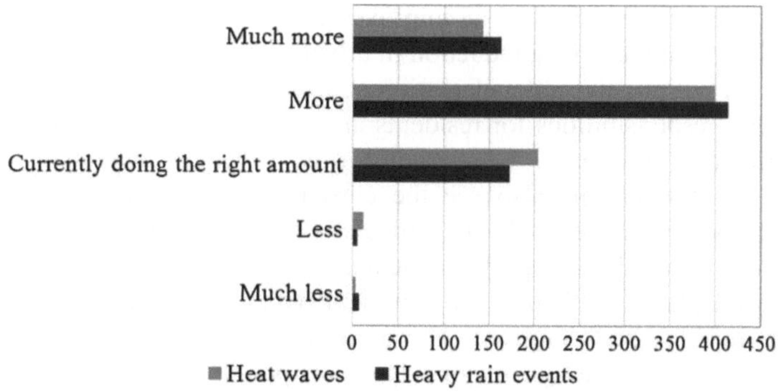

Fig. 6.9 Adaptation responsibility of citizens, as perceived by survey respondents from NYC. *Data* Author. *Source* Own draft

Concerning heat, more people find citizens doing the right amount of protection as compared with heavy rain. The latter underlines an earlier result, i.e. that respondents see few possibilities to alter their situation during heat waves.

Conclusion

This study presented and analyzed impact and adaptation-related information as perceived and viewed by a large sample of NYC residents. In total, 762 residents across the 5 boroughs of NYC reported on recent harm and damage during extreme weather events, strongest impacts and problems of heat waves and heavy rainstorms, as well as perceived responsibilities for adaptation across sectors and boroughs.

The study reveals that recent harm is mostly connected to heavy rainstorms, and in particular to damages to property. In contrast, the largest number of impacts and personal problems are caused by heat waves. Impacts of heat waves are similar across the boroughs, while problems are mostly related to way of life and social issues, particularly on Manhattan, as compared with other sectors and boroughs. Impacts during heavy rainstorms differ across boroughs with the largest effects seen in Staten Island's built environment. This, however, does not correspond with the entailed problems, which are highest in the traffic and mobility sector in the Bronx, Manhattan, and Queens.

Adaptation is largely seen as a municipal responsibility with a need to improve the electricity, water supply and health system during heat, and the electricity, drainage/sewage and subway/traffic system during rain. Overall, the largest need for improvement is seen in the electricity system. Surprisingly, the importance of adaptation in the building sector and urban greenery/parks is seen as low, at least when compared with the importance of the other, above mentioned sectors. The author is deemed to conclude that this implies substantial information and communication needs to residents, as an improvement in urban greenery and parks can substantially contribute to heat reduction in urban areas and air-conditioning is in its current design not an advisable adaptation strategy.

Adaptation responsibilities for residents are seen as higher than currently acted upon, particularly as regards heavy rainstorms. However, overall a substantial portion of the respondents also see the current amount of personal adaptation conducted as being sufficient. This is particularly the case for heat waves, which are seen as more impactful, more problematic and burdensome on a personal level, but also as less adaptable to. This is an important outcome asking for an increase in awareness raising and targeted communication. Heat waves pose major climate-related risks, although major adaptation possibilities exist. Awareness of heat-related risks and adaptation options have to be substantially increased, as more fatalities occur from heat waves than other climate hazards combined, if adaptation is insufficient (Klinenberg 2002; Satterthwaite et al. 2007).

References

Blake R, Khanbilvardi R, Rosenzweig C (2000) Climate change impacts on New York City's water supply systems. J Am Water Resour Assoc 36(2):279–292

De Gregorio Hurtado S, Olazabal M, Salvia M, Pietrapertosa F, Olazabal E, Geneletti D, D'Alonzo V, Di Leo S, Reckien D (2015) Understanding how and why cities engage with climate policy: an analysis of local climate action in Spain and Italy. J Land Use, Mobility Environ 2015(SpI: ECCA):23–46

De Gregorio Hurtado S, Olazabal M, Salvia M, Pietrapertosa F, Olazabal E, Geneletti D, D'Alonzo V, Feliú E, Di Leo S, Reckien D (2014) Implications of governance structures on urban climate action: evidence from Italy and Spain. BC3 Working Paper Series. Bilbao, Basque Centre for Climate Change 02:1–47

Heidrich O, Reckien D, Olazabal M, Foley A, Salvia M, de Gregorio Hurtado S, Orru H, Flacke J, Geneletti D, Pietrapertosa F, Hamann JJP, Tiwary A, Feliu E, Dawson RJ (2016) National climate policies across Europe and their impacts on cities strategies. J Environ Manage 168:36–45

Horton R, Gornitz V, Bowman M, Blake R (2010) Chapter 3: climate observations and projections. Ann New York Acad Sci 1196(1):41–62

Horton RM, Gornitz V, Bader DA, Ruane AC, Goldberg R, Rosenzweig C (2011) Climate hazard assessment for stakeholder adaptation planning in New York City. J Appl Meteorol Climatol 50(11):2247–2266

Kalkstein LS, Greene JS (1997) An evaluation of climate/mortality relationships in large U.S. Cities and the possible impacts of a climate change. Environ Health Perspect 105(1):84–93

Klinenberg E (2002) Heat wave: a social autopsy of disaster in Chicago. The University of Chicago

Knowlton K, Lynn B, Goldberg RA, Rosenzweig C, Hogrefe C, Rosenthal JK, Kinney PL (2007) Projecting heat-related mortality impacts under a changing climate in the New York City region. Am J Publ Health 97(11):2028–2034

Lauer W, Frankenberg P (1992) Welt—Klima. Westermann Schulbuchverlag GmbH, Diercke Weltatlas, p 220

Leiserowitz A, Maibach E, Roser-Renouf C, Feinberg G, Howe P (2012a) Extreme weather and climate change in the American mind. Yale project on climate change communication. Yale University and George Mason University, New Haven, CT, USA

Leiserowitz A, Maibach E, Roser-Renouf C, Hmielowski JD (2012b) Extreme weather, climate & preparedness in the American mind. Yale project on climate change communication. Yale University and George Mason University, New Haven, CT, USA

Reckien D (2014) Weather extremes and street life in India—implications of fuzzy cognitive mapping as a new tool for semi-quantitative impact assessment and ranking of adaptation measures. Global Environ Change-Human Policy Dimensions 26:1–13

Reckien D, Flacke J, Dawson RJ, Heidrich O, Olazabal M, Foley A, Hamann JJ-P, Orru H, Salvia M, Gregorio Hurtado S, Geneletti D, Pietrapertosa F (2013) Climate change response in Europe: what's the reality? Analysis of adaptation and mitigation plans from 200 urban areas in 11 countries. Clim Change 122(1):331–340

Rosenzweig C, Solecki W (eds) (2010) Climate change adaptation in New York City: building a risk management response. Annals of the New York Academy of Sciences, New York City

Satterthwaite D, Huq S, Reid H, Pelling M, Romero Lankao P (2007) Adapting to climate change in urban areas: the possibilities and constraints in low-and middle-income nations. Human Settlements Discussion Paper Series, vol 1. London, UK, IIED, pp 1–124

Slovic P (2000) The perception of risk. Routledge

Solecki W (2012) Urban environmental challenges and climate change action in New York City. Environ Urbanization 24(2):557–573

The Weissman Center for International Business, City University New York (CUNY) (2015) NYCdata TM—It's All here! [online]. https://www.baruch.cuny.edu/nycdata/index.html. Accessed 17 Sept 2015

United States Census Bureau (2015) Factfinder [online]. http://factfinder2.census.gov/faces/nav/jsf/pages/index.xhtml. Accessed 25 July 2015

United States Census Bureau (2015) State and County QuickFacts—New York (city), New York [online]. http://quickfacts.census.gov/qfd/states/36/3651000.html. Accessed 25 July 2015

Part II
Climate Change and the Built Environment

Chapter 7
A Critical Discussion on the Roles of Institutions on Ports' Adaptation to the Impacts Posed by Climate Change

Huiying Zhang, Adolf K.Y. Ng and Austin Becker

Abstract Ports are increasingly vulnerable to the negative impacts posed by climate change, and thus port stakeholders have recently been engaged in different adaptation efforts. However, they find it difficult, if not impossible, to move up to the 'next level'—often staying in the embryonic stage of knowledge sharing and exploration rather than actual planning and implementation of adaptation strategies and measures, and institutions play a significant role on such stagnancy. Understanding such, the article critically reviews the impacts of institutions on the process of climate adaptation planning. It specifically focuses on how institutional embeddedness acts as a significant barrier that hinders ports in progressing through the climate adaptation process. Also, it offers constructive insight on how the institutional structure of planning should be transformed so as to overcome such barriers.

Keywords Climate change · Adaptation · Institutional barriers · Theoretical discussion · Ports

Introduction

Many business activities take advantage of public resources (e.g., water and air) but are also affected by extreme environmental conditions. Many scholars have studied economy-environment interactions and climate change adaptation is at the

---

H. Zhang (✉) · A.K.Y. Ng · A. Becker
Asper School of Business, Transport Institute, University of Manitoba,
181 Freedman Crescent, Winnipeg, MB R3T 5V4, Canada
e-mail: zhangh21@myumanitoba.ca

H. Zhang · A.K.Y. Ng
Department of Supply Chain Management, Asper School of Business,
University of Manitoba, Winnipeg, MB, Canada

A. Becker
Department of Marine Affairs, University of Rhode Island, Kingston, RI, USA

W. Leal Filho and J.M. Keenan (eds.), *Climate Change Adaptation in North America*, Climate Change Management,
DOI 10.1007/978-3-319-53742-9_7

forefront. Changes in climatic conditions have imposed diversified impacts on various business sectors, say, agriculture and forest management (Bouriaud et al. 2015; Chhetri et al. 2010). The impacts posed to human lives and activities can be catastrophic (Keohane and Victor 2010; UNECE 2010; IPCC 2012). Schaeffer et al. (2012) warn that by 2100, sea level may be up to 80 cm higher than today and others put it as high as two meters (Parris et al. 2012; Rahmstorf 2010). It is now probably too late to avoid all the deleterious effects of climate change, in no small part due to uncertainties on how the problem should be addressed (Applegate 2010). Thus, as pointed out by Ng et al. (2016), developing adaptation strategies and measures so as to deal with climate change impacts is not a choice but a necessity.

Located along shorelines, ports are vulnerable to the risks that climate change can pose to their facilities and operations (Becker et al. 2012). Maritime infrastructures and facilities are of critical significance to local, national and international economies. In addition to mitigation efforts, such as the reduction of greenhouse gas emissions (Ng et al. 2013), some port stakeholders have begun to engage in adaptation actions so as to build resilience (the elastic function to the operations of the status quo) (Adger et al. 2005; Osthorst and Mänz 2012). Adaptation to climate change, especially at coastal regions involves internal and external stakeholders, environmental and economic impacts across different scales and sectors, and collective choices from both individual and organizational levels (Brown et al. 2014; Waters et al. 2014). In port research and management, it is widely agreed that adaptation to climate change at ports is in its embryonic stage (Araral 2013; Ekstrom and Moser 2013; Walker et al. 2010). Two global surveys explored stakeholders' perceptions and knowledge towards adaptation to climate change at ports. The findings suggest that port operators have yet taken serious actions (Becker et al. 2012) and indifferent attitude of port operators and managers towards the effectiveness of adaptation measures to climate change impacts (Zhang and Ng 2016). Their views are further supported by a case study at Irish seaports (O'Keeffe et al. 2016). It reveals the failure of port stakeholders to connect their existing strategies and practices with climate change adaptation efforts and thus suggests a lack of knowledge about adaptation to climate change. Indeed, increasing evidence (Ekstrom and Moser 2013; Messner et al. 2016) suggests that stakeholders find it very difficult to move up to the 'next level' of actual planning, funding, and implementation. Even with a better understanding of climatic predictions and potential impacts associated with climate change risks, stakeholders at ports encounter with difficulties to develop implementable adaptation plans (Ford et al. 2011). Nevertheless, the fast changing climatic conditions and their potential hazardous impacts give us no extra time to hesitate or evade this tricky issue of climate change adaptation.

Although barriers to adaptation to climate change can be analyzed from different angles, such as finance and technology (Klein et al. 2001; Xiao et al. 2015), the problems related to stakeholder engagement have grown in importance in light of a considerable amount of literature (Becker and Caldwell 2015; Becker et al. 2015). Ng et al., (forthcoming) point out the significance of 'soft' management such as institutional and financial tools in the adaptation process, instead of 'hard' infrastructures (e.g. engineering solutions like dikes). Past lessons learned from case studies suggest that one key challenge to advance stakeholder collaboration is to include concerns of a wide variety of stakeholders and resolve conflicts of interests (Becker et al. 2015; Messner et al. 2016). To better understand the problems in stakeholder management, the institutional theory has been applied as an analytical perspective (Araral 2013; Barnett et al. 2015; Garrelts and Lange 2011). Taking the full advantage of the existing resources is a more critical issue than tackling the problem of insufficient resources, while institutional capacity is argued as a key enabler to overcome barriers in the process of climate change adaptation (Amundsen et al. 2010; Burch 2010). Despite evidence suggesting that institutional systems will impede climate change adaptation due to ambiguity and competing political interests (Wheeler et al. 2009; Keohane and Victor 2010; Preston et al. 2011; Osthorst and Mänz 2012), with UNCTAD (2012) calling for diversified strategies to develop resilience to climate change impacts, most attention focuses on physical layouts and technical details of capital-intensive engineering projects, e.g., elevation, levee, dykes, etc. (National Research Council of the USA, or NRC 2010). Hitherto, few studies have investigated the institutional aspects of climate change adaptation, especially efforts in reducing uncertainties in decision-making, the development of public policy, and institutional practices. A research gap in the topic clearly exists.

Understanding such deficiency, this article discusses the dilemma of climate adaptation planning from the institutional perspective. It identifies the gap in the literature regarding institutional impacts on adaptation to climate change, queries whether climate change adaptation has catalyzed a transformation of the nature and practice of planning and addresses how institutions play as impediments in the process of climate change adaptation. By doing so, it enriches institutional theory and initiates new thoughts in planning and decision-making in both climate change and other public policy choices. It is a germane reminder to planners and policy-makers that effective climate change adaptation is not limited to physical/engineering technicalities but is an ideological issue that requires a fundamental shift of the existing political, economic and social paradigms.

After this introduction section, the next section reviews the relevant literatures. We highlight the criticality of institutional barriers in climate change adaptation and identify the existing gap within the literature. After then, we offer a theoretical discussion on institutions and climate change adaptation planning in ports. Finally, the conclusion and implementations can be found towards the end of the article.

Institutional Barriers and Gap in Literature

Academia generally agrees on the definition of barriers to climate change adaptation as obstacles and constraints that impede the adaptation process (Eisenack et al. 2014; Moser and Ekstrom 2010; Waters et al. 2014). Barriers and limits are inherently different in terms of whether they can be overcome or surpassed (Eisenack et al. 2014). Moser and Ekstrom (2010, p. 22027) define barriers in climate change adaptation as 'obstacles that can be overcome with concerted effort, creative management, change of thinking, prioritization, and related shifts in resources, land uses, institutions, etc.' With increasing awareness that barriers can be prioritized differently in various sectors by diversified actors (Waters et al. 2014), actor (barrier that is challenging to one actor may be beneficial to another) and context (barriers may vary among different situations) elements were added so as to refine the definition and reflect its complex nature (Eisenack et al. 2014).

Ekstrom and Moser (2014) group the most common barriers into four types: institutional, attitudinal, financial and political, while Waters et al. (2014) review the literature and categorize into groups as institutional, social and cognitive, uncertainty-related and cost-related. Moser and Ekstrom (2010) propose a systemic framework to identify such barriers by following three phases of climate change adaptation: understanding, planning and managing. Afterward, they apply it to the adaptation works in California, US and argue for the pivotal role of institutions and governance-related issues as obstacles in climate change adaptation. Institutional barriers are chosen as the most common ones among 12 categories. Through five case studies in the San Francisco Bay Area, they stress the importance of institutional lens to further investigate barriers in the process of climate change adaptation (Ekstrom and Moser 2013). Their argument is shared by a number of other scholars. Klein (2016) reveals the embeddedness of climate change adaptation into the local context in Finland, while Lawrence et al. (2015) study adaptation works in New Zealand from the institutional perspective. Walker et al. (2015) highlight the crucial governance barriers to deal with climate change risks in transport planning in England.

With increasing evidence suggesting the critical role of institutions in influencing adaptation to climate change, researchers tackle the institutional impacts from a wide variety of theories. Marshall (2013) applies transaction cost theory to build a framework for institutional cost and benefit analysis, with a focus on path dependency that constraints adaptation management. Peñalba et al. (2012) discuss adaptation efforts in the Philippines from the social and institutional perspectives and argue for the pivotal role of cognitive enhancement and adaptive capacity building. A comprehensive model (the 'COW model') is developed by Araral (2013) by synthesizing core views from three fields: Coase's property right, Ostrom's governance principals for the public good and Williamson's transaction cost, and through a case study he verifies its effectiveness and proposes six institutional design principals to reduce transaction cost in adaptation to climate change. Similarly, Brunner and Enting (2014) argue incomplete institutions and incomplete information as obstacles in climate finance from an angle of transaction cost. In particular, unclear property right and vague responsibility allocation are categorized as two major aspects in incomplete institutions.

The second body of literature studying institutional impacts on climate change adaptation mainly is built on a specific theory in institutionalism. By adopting path dependency, Garrelts and Lange (2011) argue the central role of government in adaptation to climate change in German flood management system, and also emphasize functions of other stakeholders (i.e., academia, media, etc.) to offer support and knowledge. Chhetri et al. (2010) examine how path dependency limits farmers' choices to adapt to climate change and show that sub-optimal outcomes are inevitable due to institutional impacts. Barnett et al. (2015) investigate drivers of barriers in climate change adaptation and identified path dependency in institutions as the primary reason.

Nevertheless, the connection between climate change adaptation and institutionalism is incompact (Araral 2013). The first branch of literature covers a wide range of theories and 'borrows' relevant insight to address institutional obstacles in the process of climate change adaptation. This vast reference and discussion have a risk in mismatching theoretical framework with empirical evidence in an inadequate way (Kingston and Caballero 2009). The second part narrows down the theoretical foundation to path dependency but such 'zoom out' may neglect other significant influences that institutions have on climate change adaptation, thus fail to explain institutional barriers from a holistic point of view. In particular, the question of 'in what way' and 'to what extent' institutions matter in climate change adaptation is not fully addressed. Thus, it is necessary to investigate the impacts of institutions on the process of climate adaptation planning to fill the gap and suggest research directions for further study.

Institutions and Port's Adaptation to Climate Change Impacts

Institutions are formal and informal rules that constrain and guide human behaviors (Brunner and Enting 2014; Denzau and North 1994; Roberts and Greenwood 1997). Formal rules, primarily laws and regulations, are explicit and set by legislators to govern strategic choices of actors through various interventions and arrangements, while informal norms, primarily customs, traditions and other social orders, are implicit, emerging and developing through interactions of actors (Tongzon et al. 2015). An institutional system consists of standard practices that structure relationships between agents, both public and private. It can impose preceding constraints on policymaking choices and strategic directions (March and Olsen 1989; Hall and Taylor 1998). It countervails dramatic changes, restricts alternatives and diminishes the rationalities of decision-making to predictable paths according to norms and practices based on culture and hegemonic values of the time (Fuchs and Scharmanski 2009; Glassman 2004) even when they may have become obsolete (North 1990; Hodgson 1993). Institutional systems solidify generally accepted values into predictable practices so as to deter undesirable social outcomes due to individual actions. One inherent character of institutions is change-resistant. From the economic perspective, the transaction cost of switching from the former conventions to the newly established actions may be higher than the benefits generated from the new institutions. On the other hand, institutional embeddedness can bound the pool of alternatives. Decision-makers tend to follow paths that have been created and adapted in the existing institutional system.

However, this does not mean the institutional system cannot evolve. It can 'stretch' (Strambach 2010; Notteboom et al. 2013) to deal with changing circumstances. This usually involves two components, namely the institutional environment and the institutional arrangement. The former refers to informal conventions and norms of which organizations, being parts of a given community, should conform so as to gain legitimacy and general support, and sometimes made compulsory through legally binding rules and regulations (Martin 2000). Also, it includes the mindsets of individuals and political elites. The institutional environment forms the basis for compromise (Gutmann and Thompson 2012), operational characteristics, and receptiveness to new knowledge (Boxer 1991). The institutional arrangement refers to agreements and organizational structures between agents so as to achieve certain objectives or programs governed by the institutional environment, like firms, bureaucracy, policies and cooperative networks. The influence of the institutional system on port planning has been widely studied (e.g., Buitelaar et al. 2007; Ng and Pallis 2010; Notteboom et al. 2013). However, they largely follow a neo-institutional approach that investigates how established institutional environments structure cognition and guide decision-making. Facing new circumstances like climate change, institutional agents may take spontaneous initiatives to re-structure the institutional arrangements, as exemplified by the neo-liberal reforms among ports around the world in recent decades.

Decision-making gets more complicated within an uncertain institutional environment consisting of individual mindsets, ambivalent interests and diversified localities with individualistic and pluralistic traditions (Fishman 2000). Climate adaptation planning possesses such an uncertain environment due to scarce legal standards, direct precedents and readily transferrable scientific knowledge. This causes inadequate understanding, and thus inadequate input, from stakeholders and the general public. With no direct paths to depend on, the institutional environment is a vacuum yet to be filled. Planning should provide clear guidance and practical actions to lead the direction of development, especially in the generation of first plans with many (untried) alternatives to choose from (cf. Wheeler 2008; Preston et al. 2011; Sager 2011). Further problems arise when the new circumstance has yet to reach a critical juncture (Buitelaar et al. 2007) and all parties do not yet deem significant transformation necessary or immediate. A critical juncture is 'when events create visions of institutional change, and divides events into different periods" (Ng and Pallis 2010, p. 2151). The uncertainties (e.g., in the prediction of climate change events, vulnerability assessment, and evaluation of economic losses) are embedded in adaptation to climate change. Without adequate information and knowledge, it can be very challenging for planners to judge whether the critical juncture has come or not and then engage in institutional changes, even when the climatic environment is actually requiring proactive actions to build resilience for the sustainable business in the future. For example, in a study on the ports of Gulfport and Providence, planners and managers widely acknowledge the concerns for resilience from future storms events, but less clear on how these concerns could or should be translated into concrete adaptation actions (Becker and Caldwell 2015). This reveals that climate change has only triggered incremental changes in the institutional system, rather than reforms or major modifications. The widely adopted strategy of 'no regrets' (City of Vancouver 2012; Walker et al. 2010) is a reflection of planners' preference towards an evolutionary approach in climate change adaptation planning. The uncertain institutional environment, uncertainties in climate change adaptation, and high-risk investment in adaptation projects has forced decision-makers to avoid aggressive actions.

The institutional deficiencies caused by the uncertain institutional environment can be investigated using the theory of thick institutionalization. Thick institutionalization is not a new topic in port business (e.g., Tongzon et al. (2015) has used to explain port reform). As mentioned, in recent decades, the major ports have undertaken reforms in ownership structures and operation principles (Cullinane and Song 2002) so as to cope with the new business environment (Ng and Liu 2014; Cullinane and Song 2002). On the first stage, the government as planner builds reform models and rules, while the port authority as implementer carries through the established plans on the second stage. Despite that privatization/public-private partnership/other forms of private participation (the 'formal institution') are quite mature, the unified solution can lead to different reform outcomes (Ng and Pallis 2010) under diversified economic, political, social, and cultural environments.

In climate adaptation, the effect of institutional deficiencies is magnified. The problem is two-folded: uncertainty about stakeholders' roles and loose institutional frameworks. Unlike port reform, the allocation of responsibilities is still fuzzy in climate adaptation. Thus, stakeholders are still not very sure what roles they should be playing. For example, in a study on the port of Providence, the researchers found that stakeholder's perceptions of who should take the lead for adaptation do not always align well with how those 'identified leaders' perceive their own roles and mandates (Kretsch and Becker 2016). Moreover, stakeholders need a robust framework to guide strategic actions. However, who should perform as champion on the first stage of thick institutionalization often remains unclear. Without a clear leader, the institutional arrangements in climate change adaptation are under-developed and result in the lack of a unified solution. Individual participants interpret the formal institution according to their own institutional learning and institutional environment. The government sometimes feels its way out and private agencies often hesitate to take further steps, and other port stakeholders often struggle to agree on a particular framework. As a result, a common effort in adapting to climate change often stagnates on the initial phases of open talk and discussion and, as mentioned, moves adaptation planning to the 'next level'. When the external forces (i.e., the need for climate change adaptation) 'collide' with well-developed institutions, a loose framework is not powerful enough to make progress. Unsurprisingly, such a loose framework often leads to unsystematic implementation and unsatisfactory outcomes.

Even if the government takes the initiative to lead climate adaptation efforts, the synthesized effect of diversified institutional environments among port stakeholders and path-dependency in implementation may determine its 'conventional' planning approach. Indeed, given the insufficient experience in climate adaptation, when designing any adaptation strategies and measures, a port authority may rely on its own traditional planning approach that is tailored to shorter timelines than may be appropriate for climate adaptation. When this happens, it often faces substantial challenges to implement the strategies and measures effectively with different port stakeholders sometimes 'feel being excluded' from the planning process or that their direct interests are being affected. This implies that it becomes extremely difficult to achieve outcomes that are satisfactory or acceptable for each stakeholder involved in ports while the adaptation strategies and measures are implemented effectively. In this case, the failed experience of the port of San Diego (Messner et al. 2016) offers strong indication that a complete reliance on the government (or public sector) to initiate and lead climate adaptation planning may not work particularly well.

Undertaking technical responses to adapt to climate change may appear straightforward (Walker et al. 2010) but in practice can require an intricate action system: they often require heterogeneous resources (e.g., time, money and people), inputs from different scales, and collaboration from various sectors. Even with a better understanding of the 'paradigm shift' in adaptation planning that requires stakeholders to work together more holistically and on a longer time horizon (Becker and Caldwell 2015), knowledge on how institutions impede the process of climate adaptation is critically important to address the institutional barriers and subsequently provide feasible suggestions. Without any doubt, adaptation to climate change risks is a regional or even national issue, for which a port authority should not be the sole decision-maker. The governance mechanism that is currently deployed to tackle regional adaptation efforts heavily relies on voluntary, network-based processes with stakeholders from public and private sectors (Becker 2016). The lack of coherent guidelines and regulations makes it even more difficult to establish a hierarchical governance structure. Nevertheless, few examples yet exist that show how these types of networks successfully allocate responsibility and resolve conflicts in a way that suits many different stakeholders. The institutional deficiency determines the ineffectiveness of the current informal networks (e.g., climate change councils) to advance the process of adaptation to climate change. Without direct power over the stakeholders, these networks only provide a platform for stakeholders to share and discuss the adaptation work, instead of being the leader to direct further actions. Hence, in future research, researchers must put more efforts to develop organizational structures where leadership can be made more effective, particularly in the mediation of limited resource allocation. The focus should be on finding a more suitably empowered agency that can address the stagnating process of adaptation planning in ports, and provide the 'breakthrough'.

Conclusions and Implications for Future Research

The article clearly illustrates that institutions play significant roles in affecting the process of climate adaptation planning and its effective implementation. It is a far-reaching attempt for any port to address the impacts of climate change alone. Some re-structuring of a port's institutional arrangements should be done and port planners must be aware on the necessity of paradigm shift from previous planning practices.

However, the uncertainty in the institutional environment and the likely speculative attitude of major participants can strengthen the perception that political controversies might hinder future implementation. Without resolving these challenges, climate adaptation plans may become more of visionary guidance rather than real action plans, and planners may be forced to 'muddle through' the planning process in an evolutionary approach. This is not surprising, as the objective of the institutional system is to deter undesired shocks to society due to individual actions (Weber 1922). Under such an uncertain institutional environment, the neo-liberal ideology that emphasizes on minimal public intervention (Harvey 2005) may continue to dominate (and dissipate the quality of) planning decisions. Under such influences, planners may be forced to adopt an evolutionary approach in climate adaptation planning, even if they favor a more revolutionary one.

Also, the government (or a single public agency) should not act as the sole decision-maker in adaptation to climate change, as the issue of climate change poses impacts to the whole region surrounding the ports (Becker et al. 2013). The regional nature of climate change adaptation fosters the presence of the current networks (e.g., climate change councils) to facilitate the process of adaptation. However, problems arise when the institutional deficiencies of the current network are not appropriately addressed. The vague allocation of responsibility and conflicts of interest make it difficult, if not possible, for port stakeholders to move from the stage of knowledge sharing and open discussion to the 'next level' of planning and implementation. A less hierarchical governance structure that reserves certain autonomy may contribute to a smoother process of adaptation to climate change.

Last but not least, the article is an early attempt to theoretically dissect climate adaptation planning from the institutional perspective. We believe that institutional systems strongly shape the planning process, and institutional deficiency may hinder the effective tackling of climate adaptation. It highlights various important structural principles of climate adaptation planning and existing loopholes that require paradigm shift solutions. Needless to say, our suggestion needs further verification and development, and a broader spectrum and a greater volume of relevant literatures should be reviewed so as to get a more comprehensive conclusion. Apart from ports, we need to further examine our suggestions from other sectors along the supply chains. Even for ports, as many ports nowadays are joint ventures between local and national government entities, further research is required to examine more closely the link between such federalism and climate adaptation planning. It will be particularly interesting to investigate how federal institutions can be involved to mobilize the process of climate adaptation to a broader range of stakeholders (not just port stakeholders). While the impacts posed by climate change to the world are likely to become even more explicit in the future, continuous research germane to reduce uncertainty in decision-making dedicated to climate adaptation is an absolute necessity. Further research is needed to investigate how institutional systems affect climate adaptation planning. The impacts of the findings can be substantial for the well being of our future generations.

Acknowledgements The University of Manitoba Transport Institute (UMTI) funds this study. The usual disclaimers apply.

References

Adger WN, Arnell NW, Tompkins EL (2005) Successful adaptation to climate change across scales. Glob Environ Change 15(2):77–86

Amundsen H, Berglund F, Westskog H (2010) Overcoming barriers to climate change adaptation—a question of multilevel governance? Environ Plan C Gov Policy 28(2):276–289

Applegate JS (2010) Embracing a precautionary approach to climate change. In: Driesen DM (ed) Economic thought and US climate change policy. MIT Press, Cambridge, pp 171–196

Araral E (2013) A transaction cost approach to climate adaptation: insights from Coase, Ostrom and Williamson and evidence from the 400-year old zangjeras. Environ Sci Policy 25:147–156

Barnett J, Evans L, Gross C, Kiem AS, Kingsford RT, Palutikof JP, Pickering CM, Smithers SG (2015) From barriers to limits to climate change adaptation: path dependency and the speed of change. Ecol Soc 20(3):5–15

Becker A (2016) The state of climate adaptation for ports and the way forward. In: Ng AKY, Becker A, Cahoon S, Chen SL, Earl P, Yang Z (eds) Climate change and adaptation planning for ports. Routledge, Abingdon, pp 265–274

Becker A, Caldwell M (2015) Stakeholder perceptions of seaport resilience strategies: a case study of Gulfport (Mississippi) and providence (Rhode Island). Coast Manag 43(1):1–34

Becker A, Inoue S, Fischer M, Schwegler B (2012) Climate change impacts on international seaports: knowledge, perceptions, and planning efforts among port administrators. Clim Change 110(1–2):5–29

Becker A, Acciaro M, Asariotis R, Carera E, Cretegny L, Crist P, Esteban M, Mather A, Messner S, Naruse S, Ng AKY, Rahmstorf S, Savonis M, Song D-W, Stenek V, Velegrakis AF (2013) A Note on Climate change adaptation for seaports: a challenge for global ports, a challenge for global society. Clim Change 120(4):683–695

Becker AH, Matson P, Fischer M, Mastrandrea MD (2015) Towards seaport resilience for climate change adaptation: stakeholder perceptions of hurricane impacts in Gulfport (MS) and providence (RI). Prog Plan 99:1–49

Bouriaud L, Marzano M, Lexer M, Nichiforel L, Reyer C, Temperli C, Peltola H, Elkin C, Duduman G, Taylor P, Bathgate S, Borges JG, Clerkx S, Garcia-Gonzalo J, Gracia C, Hengeveld G, Kellomäki S, Kostov G, Maroschek M, Muys B, Nabuurs G-J, Nicoll B, Palahí M, Rammer W, Ray D, Schelhaas M-J, Sing L, Tomé M, Zell J, Hanewinkel M (2015) Institutional factors and opportunities for adapting European forest management to climate change. Reg Environ Change 15(8):1595–1609

Boxer B (1991) Societal contexts of ocean pollution science: cross-national comparisons. Glob Environ Change 1(2):139–156

Brown S, Nicholls RJ, Hanson S, Brundrit G, Dearing JA, Dickson ME, Gallop SL, Gao S, Haigh ID, Hinkel J, Jiménez JA, Klein RJT, Kron W, Lázár AN, Neves CF, Newton A, Pattiaratachi C, Payo A, Pye K, Sánchez-Arcilla A, Siddall M, Shareef A, Tompkins EL, Vafeidis AT, van Maanen B, Ward PJ, Woodroffe CD (2014) Shifting perspectives on coastal impacts and adaptation. Nat Clim Change 4(9):752–755

Brunner S, Enting K (2014) Climate finance: a transaction cost perspective on the structure of state-to-state transfers. Glob Environ Change 27:138–143

Buitelaar E, Lagendijk A, Jacobs W (2007) A theory of institutional change: illustrated by Dutch City-Provinces and Dutch Land Policy. Environ Plan A 39:891–908

Burch S (2010) Transforming barriers into enablers of action on climate change: insights from three municipal case studies in British Columbia, Canada. Glob Environ Change 20(2): 287–297

Chhetri NB, Easterling WE, Terando A, Mearns L (2010) Modeling path dependence in agricultural adaptation to climate variability and change. Ann Assoc Am Geogr 100(4): 894–907

City of Vancouver (2012) Climate change adaptation strategy. City of Vancouver, Vancouver, Canada

Cullinane K, Song D-W (2002) Port privatization policy and practice. Transp Rev 22(1):55–75

Denzau AT, North DC (1994) Shared mental models: ideologies and institutions. Kyklos 47(1): 3–31

Eisenack K, Moser SC, Hoffmann E, Klein RJT, Oberlack C, Pechan A, Rotter M, Termeer CJAM (2014) Explaining and overcoming barriers to climate change adaptation. Nat Clim Change 4 (10):867–872

Ekstrom JA, Moser SC (2013) Institutions as key element to successful climate adaptation processes: results from the San Francisco Bay Area. In: Moser SC, Boykoff MT (eds) Successful adaptation to climate change—linking science and policy in a rapidly changing world. Routledge, Abingdon, pp 97–113

Ekstrom JA, Moser SC (2014) Identifying and overcoming barriers in urban climate adaptation: case study findings from the San Francisco Bay Area, California, USA. Urban Clim 9:54–74

Fishman R (ed) (2000) The American planning tradition: culture and policy. The Woodrow Wilson Center Press, Washington, DC

Ford JD, Berrang-Ford L, Paterson J (2011) A systematic review of observed climate change adaptation in developed nations. Clim Change 106(2):327–336

Fuchs M, Scharmanski A (2009) Counteracting path dependencies: rational investment decisions in the globalising commercial property market. Environ PlanA 41(11):2724–2740

Garrelts H, Lange H (2011) Path dependencies and path change in complex fields of action: climate adaptation policies in Germany in the realm of flood risk management. Ambio 40 (2):200–209

Glassman J (2004) Transnational hegemony and US labor foreign policy: towards a Gramscian international labour geography. Environ Plan D 22(4):573–593

Gutmann A, Thompson D (2012) The spirit of compromise. Princeton University Press, Princeton, NJ

Hall PA, Taylor RCR (1998) Political science and the three new Institutionalisms. In: Soltan K, Soltan E, Uslaner EM (eds) Institutions and social order. University of Michigan Press, Ann Arbor, pp 14–44

Harvey D (2005) A brief history of neoliberalism. Oxford University Press, Oxford

Hodgson G (1993) Economics and evolution: bringing life back in economics. Polity, Cambridge

Intergovernmental Panel on Climate Change (IPCC) (2012) IPCC special report on managing the risks of extreme events and disasters to advance climate change adaptation. IPCC, Geneva, Switzerland

Keohane RO, Victor DG (2010) The regime complex for climate change. Discussion paper, Harvard project on international climate agreements, Cambridge, MA (January 2010) (No. 2010-33)

Kingston C, Caballero G (2009) Comparing theories of institutional change. J Inst Econ 5 (02):151–180
Klein J (2016) Embeddedness of climate change adaptation: established procedures and contending discourses for flood protection in Espoo, Finland. Local Environ 21(2):1–18
Klein RJ, Nicholls RJ, Ragoonaden S, Capobianco M, Aston J, Buckley EN (2001) Technological options for adaptation to climate change in coastal zones. J Coast Res 17(3):531–543
Kretsch E, Becker A (2016) Leadership and responsibility for long-term hurricane resilience: port of providence, RI. Paper presented at the transportation research board conference for committee on maritime transportation system (CMTS), Washington, DC
Lawrence J, Sullivan F, Lash A, Ide G, Cameron C, McGlinchey L (2015) Adapting to changing climate risk by local government in New Zealand: institutional practice barriers and enablers. Local Environ 20(3):298–320
March JG, Olsen JP (1989) Rediscovering institutions: the organizational basis of politics. The Free Press, New York
Marshall GR (2013) Transaction costs, collective action and adaptation in managing complex social-ecological systems. Ecol Econ 88:185–194
Martin R (2000) Institutional approaches in economic geography. In: Sheppard E, Barnes T (eds) A companion to economic geography. Blackwell, Oxford, pp 77–94
Messner S, Becker A, Ng AKY (2016) Port adaptation for climate change: the role of stakeholders and the planning process. In: Ng AKY, Becker A, Cahoon S, Chen SL, Earl P, Yang Z (eds) Climate change and adaptation planning for ports. Routledge, Abingdon, pp 9–23
Moser SC, Ekstrom JA (2010) A framework to diagnose barriers to climate change adaptation. Proc Natl Acad Sci 107(51):22026–22031
Ng AKY, Pallis AA (2010) Port governance reforms in diversified institutional frameworks: generic solutions, implementation asymmetries. Environ Plan A 42(9):2147–2167
Ng AKY, Liu J (2014) Port-focal logistics and global supply chains. Palgrave Macmillan, Basingstoke
Ng AKY, Chen SL, Cahoon S, Brooks B, Yang Z (2013) Climate change and the adaptation strategies of ports: the Australian experiences. Res Transp Bus Manag 8:186–194
Ng AKY, Becker A, Cahoon S, Chen SL, Earl P, Yang Z (2016) Time to act: the criticality of ports in adapting to the impacts posed by climate change. In: IDEM (eds) Climate change and adaptation planning for ports. Routledge, Abingdon, pp 1–8
Ng AKY, Wang T, Yang Z, Li KX, Jiang C (forthcoming) How is business adapting to climate change impacts appropriately? Some insight from the commercial port sector. J Bus Ethics (in press). doi:10.1007/s10551-016-3179-6
North DC (1990) Institutions, institutional change and economic performance. Cambridge University Press, Cambridge
Notteboom T, de Langen P, Jacobs W (2013) Institutional plasticity and path dependence in seaports: interactions between institutions, port governance reforms and port authority routines. J Transp Geogr 27:26–35
NRC (2010) Adapting to the impacts of climate change. National Academies Press, Washington
O'Keeffe JM, Cummins V, Devoy RJ, Lyons D, Gault J (2016) Stakeholder awareness of climate adaptation in the commercial seaport sector: a case study from Ireland. Mar Policy 2:27
Osthorst W, Mänz C (2012) Types of cluster adaptation to climate change. lessons from the port and logistics sector of Northwest Germany. Marit Policy Manag 39(2):227–248
Parris A, Bromirsji P, Burkett V, Cayan D, Culver M, Hall J, Horton R, Knuuti K, Moss R, Obeysekera J, Sallenger AH, Weiss J (2012) Global sealevel rise scenarios for the US national climate assessment. NOAA Technical Report, National Oceanic and Atmospheric Administration 37
Peñalba LM, Elazegui DD, Pulhin JM, Victor O, Cruz R (2012) Social and institutional dimensions of climate change adaptation. Int J Clim Change Strat Manag 4(3):308–322
Preston BL, Westaway RM, Yuen EJ (2011) Climate adaptation planning in practice: an evaluation of adaptation plans from three developed nations. Mitig Adapt Strat Glob Change 4:407–438

Rahmstorf S (2010) A new view on sea level rise. Nature Rep Clim Change, 44–45

Roberts PW, Greenwood R (1997) Integrating transaction cost and institutional theories: toward a constrained-efficiency framework for understanding organizational design adoption. Acad Manag Rev 22(2):346–373

Sager T (2011) Neo-liberal urban planning policies: a literature survey 1990–2010. Prog Plan 76:147–199

Schaeffer M, Hare W, Rahmstorf S, Vermeerm M (2012) Long-term sea-level rise implied by 1.5 °C and 2 °C warming levels. Nat Clim Change 2:867–870

Strambach S (2010) Path dependency, path plasticity and the co-evolution of institutions and innovation: the German business software industry. In: Boschma RA, Martin R (eds) Handbook for evolutionary economic geography. Edward Elgar, Cheltenham, pp 406–431

Tongzon JL, Ng AKY, Shou EC (2015) Institutions, transport infrastructure governance, and planning: lessons from the corporatization of port authorities in East Asia. Environ Plan C Gov Policy 33(6):1467–1483

UNCTAD (2012) Ad hoc expert meeting on climate change impacts and adaptation: a challenge for global ports. Geneva, Palais des Nations, 29–30 September 2011: main outcomes and summary of discussions, Geneva, Switzerland (September 2011)

UNECE (2010) Climate change impacts on international transport networks. Working paper, working party on transport trends and economics, UNECE, Geneva, Switzerland

Walker WD, Liebl DS, Gilbert LE, LaGro J, Nowak P, Sullivan J (2010) Adapting to climate change: why adaptation policy is more difficult than we think (and what to do about it). Wisconsin Initiative on Climate Change Impacts

Walker BJ, Adger WN, Russel D (2015) Institutional barriers to climate change adaptation in decentralised governance structures: transport planning in England. Urban Stud 52(12): 2250–2266

Waters E, Barnett J, Puleston A (2014) Contrasting perspectives on barriers to adaptation in Australian climate change policy. Clim Change 124(4):691–702

Weber M (1922) Economy and society. In: Roth G, Wittich C (eds) (edited version in 1978). University of California Press, Berkeley

Wheeler SM (2008) State and municipal climate change plans: the first generation. J Am Plan Assoc 74(4):481–496

Wheeler SM, Randolph J, London JB (2009) Planning and climate change: an emerging research agenda. Prog Plan 72:210–222

Xiao Y-B, Fu X, Ng AKY, Zhang A (2015) Port investments on coastal and marine disasters prevention: economic modeling and implications. Transp Res B Methodological 78:202–221

Zhang H, Ng AKY (2016) Climate change and adaptation planning for ports: a global study. In: Proceedings of the world conference on transport research (WCTR) 2016, Shanghai, China, 10–15 July

Chapter 8
Beyond Restoration: Planting Coastal Infrastructure

Rosetta S. Elkin

Abstract The following research project proposes a model whereby the biological arrangement of plants in space and over time can lead to a new paradigm using the modification of coastal ecologies, to move beyond restoration, ultimately dissolving the limiting dichotomy between green and grey infrastructure.

Keywords Resilience · Green infrastructure · Ecology · Botany · Coastal landscape · Planning · Design

The future of coastal resilience may rest on the adaptive capacities and changing dynamics of plants. Yet plant matter rarely serves as the basis of resilient strategies. In place of vain notions of permanence, the following model of coastal living for this century begins with plants and ultimately challenges the defining characteristics of restoration and green infrastructure. The plant is used as a scalar device, amassing relationships and providing the foundation for a portfolio of interventions. Plant life enables new programmatic incentives, increased terrestrial resources, and flexible attenuation measures, without resorting to restrictive or outmoded procedures associated with reconstruction. Once acknowledged for its contribution to shaping coastlines, the plant establishes a prominent position within the ecological sequence, an adaptive model for an uncertain future. Such a proposition also necessitates a reexamination of the word resilience, as its application has been recently mainstreamed following major storm events. In particular, this project proposes that plant dynamics are the raw matter of any valid model of resilience, suggesting that the trajectory of resilience planning as a metric of defense may be quite different than the one that is currently being imagined.

The ideas offered as part of this research synthesize around the necessity for new paradigms in practice, in order to move away from the commonly held idea that the materials of grey infrastructure (concrete, steel) are resilient to storm events and that

R.S. Elkin (✉)
Department of Landscape Architecture, Harvard University Graduate School of Design, Cambridge, MA, USA
e-mail: rselkin@gsd.harvard.edu

W. Leal Filho and J.M. Keenan (eds.), *Climate Change Adaptation in North America*, Climate Change Management,
DOI 10.1007/978-3-319-53742-9_8

the role of green infrastructure is not one of protection or defense, but of ecosystem services. By offering a consideration of plants in which woody material becomes the primary wave attenuating agent, the proposal exploits disturbance regimes to cultivate rhizomes, plant species with extended and interconnected root mass, as structural components. The influence of the root zone can be measured alongside other built features, yet has the capacity to adapt uniquely over time. Faced with uncertainty as a result of a changing climate, self-organizing processes can be harnessed to generate ecological transitions and alternative scenarios that do not replicate a past equilibrium or offer a false promise of stability. The project is therefore both a construction detail and a methodology that studies not just the isolated behavior of certain plants and their ability to sprout back, but also their contribution to the entire ecological fabric of a shifting coastal, marine, and estuarine landscape. Accepting the ongoing transitions of plant life as a design feature releases the anxiety that surrounds restorative safeguarding and changes the course of resilient measures.

The Urbanized Coast

Resilient strategies often find themselves in a deeply contradictory position when determining the balance between security and vulnerability during stress periods. When applied to coastal development along the North Atlantic, the suggestion of resilience cannot overcome the reality of an extremely susceptible and exposed built condition (Fig. 8.1. The Urbanized Coast). Development is typically centered on the affluent coastline; the thin threshold between land and water where the impacts of sea-level rise are most severe and the possibility of retreat, buyouts, and relocation mobilize policy (Adger et al. 2011). Retreat does not align with contemporary practices that prioritize the capacity to build back and unite around a return to life as usual. Retreat tends to signify defeat. As a result, event-based reconstruction has emphasized rebuilding armored structures, largely by increasing elevation and installing berms, albeit with a growing emphasis on 'natural systems' that seemingly indicate more flexible typologies in the face of indeterminate conditions. Termed 'green infrastructure' projects rely heavily on status quo standards of valuation based engineering, reasserting the enduring distinction between artificial and natural systems. Most green infrastructure is either a conservative restoration project or green veil obscuring a grey infrastructure project—a vegetal embellishment draped across a higher seawall elevation. The grey performs, the green distracts, and together they are called resilient. Following the devastation of superstorm Sandy, resilience has also come to define the repair and reconstruction of such features that are none the less called natural; the restoration of swales, dunes, wetlands and marshes that are expected to offer a more nuanced and adaptable type of coastal security.

The terms restoration and resilience have evolved over the past decade to describe both the diagnosis and the cure of climate based coastal management

Fig. 8.1 The North Atlantic can be read as a series of fortified shoreline conditions, constructed borders and armoring projects. The urbanized coastline disregards its oceanic position, securing its terrestrial claim instead. Tracing the influence of major events such as Hurricane Sandy reveal extended thresholds between land and water. *Credit* Ocean State project

practices. More precisely, they threaten to become codependent as resilience takes on greener intentions and the role of restoration is empowered by state and federal regulation (Hollnagel 2009). This combined authority has stymied more experimental approaches. A false sense of security surfaces within restored communities or adjacent reconstructed seawalls, expressed by the proposition of designing for stable and predictable outcomes. This tendency not only rehearses current practices but also limits the potency of future or alternative outcomes. The environmental narrative of decline services the nostalgic considerations of conservative environmentalism, and has become emblematic of current 'resilience' funding and resilient strategies (Bennett 2001). Yet the term holds great promise and as such, deserves to be problematized in order to become a catalyst for thinking about a very different model of resource management.

From Systems to Individuals

Plants emerge and 'spring back' based on regimes of disturbance. This attribute is a result of their fundamental competitiveness, a primary function of reproduction and development that warrants their use along the coastline, regardless of native or restorative ideologies. Ecology is composed of layers of interconnected and interrelated organisms; the animation of this ecology is driven by an individual species' ability to flourish. When new species appear within a former ecology, the environment is presenting an alternative vision of the future, a reorganization leading to the growth of a new ecological cycle. Within this system, the more energy a plant devotes to recovery—or defense, in a broader context—as opposed to growth, the longer it will live (Del Tredici 1999). Using the concept of rejuvenation and working within the morphological characteristic of sprouting in temperate trees, a plant's natural capacity towards recovery can be operationalized to respond to noticeable, or uncharacteristic stresses in the environment. This project proposes that the temporality of plant formation be applied to the shifting coast in order to develop green infrastructure that actually performs. As of yet, individual plants and their impact on the micro-conditions that amalgamate through biological evidence have been overlooked in the development of coastal resilience.

A hurricane is an example of swift destruction whether or not it is elevated through climate narratives. As sea levels rise, hurricanes are having greater impact along coastal environments, due to a host of phenomena including higher water tables, erratic surge, and stronger winds (Emanuel 2008). Further evidence has proven that its influence is aggravated by an increase in surface water temperatures, which creates more vapor and stronger lift (Emanuel 2005). In every case, the storm is also disturbance to terrestrial ecosystems. In ecological terms, this external force instigates an adaptive cycle on the land as energy is allocated to a systems recovery, rather than on producing new structure or material (Chapin et al. 2009; Holling 1986; Walker et al. 2004). Holling termed this phase 'creative destruction' (2001). Therefore, disturbance instigates a reorganization, and creates time and space for an altered system to emerge. More resilient systems recover rapidly, while more mature, rigid systems are slower to recover (and thus less resilient) due to closed cycles of accumulation and storage.

Derived from its lineage in systems ecology, resilience is a value typically associated with the scale of an ecosystem rather the scale of an individual organism. As a result, resilience is likened to the ambitions of conservation and landscape planning in a way that lends it nostalgic currency.[1] Building back conflates past and

[1]I borrow the term nostalgia from Svetlana Boym in *The Future of Nostalgia*, where she differentiates between nostalgia as a state-building initiative as opposed to a personal project. With this in mind, I propose that restoration is used nostalgically as means to restore the greatness of former ecologies, rather than focus on its transitions or the negative impact of human influence, sanctioning the funds for a construction project rather than initiating a change in behavior: "*What is crucial is that nostalgia was not merely an expression of local longing, but a result of a new understanding of time and space that made the division into "local" and "universal" possible.*

present, heightening the longing for a previous state, when homes were not lost, species did not invade and hurricanes were tame. In much the same way that conservation narratives generated protection rather than management, the influence of restoration when it is linked to resilience only refines a prefix—rebuilding, recuperation, restoration, reconstruction—cashing in on the green and grey infrastructures that are repaired and restored following an event. This backward momentum inhibits the future and trivializes the role of resilience.

The allure of stability and predictability underlies the success of ecological restoration within coastal development projects, as regional manifestations of climate change remain uncertain. Accordingly, local strategies prioritize known management practices and known ecologies, which estimate visual rather than functional attributes (Hilderbrand et al. 2005). For instance, along the evolving North Atlantic coast, as beaches erode and recede, pressure is put on the adjacent land that once profited from coastal low lying conditions and wide-open vistas. As fringing beaches and salt marshes attenuate or shrink with the dynamics of a changing climate, the higher, more solid ground and upland ecology emerges as the future coastline (Fig. 8.2).

The land that once served as a backdrop for coastal lifestyles is slowly becoming an edge condition. The only actions prohibiting this ecological shift to high ground are the restoration projects that promote dredge as a means to elevate salt marshes and the re-nourishment projects that raise beach profiles. These changing conditions are the systems that help create space for vegetal migration, so that the plants of the vertical high ground can gradually creep into the lower areas, sending seeds and spouts closer to the shore as conditions change. In many ways, upland plants are trying to secure bottomland habits. This is evidenced by marsh migration— the ongoing subsidence that is evidenced as *Phragmites* sp. replaces *Spartina* sp., disturbed by fluctuating hydrologic cycles. Local ecologists tend to be aware of these shifts, but their tools are limited to documentation of extents and the measure of increased fluctuations.[2] Despite their immediate and visual impact, these changes invite a productive exchange of plant life, as a different communities emerge when conditions shift.

Rhode Island: The Ocean State

Rhode Island is a small state with a long shoreline. Residents are quick to claim that no one lives further than half an hour from the shore. Also known as the 'Ocean State,' its shoreline extends inland along multiple waterways including large bays,

(Footnote 1 continued)

The nostalgic creature has internalized this division, but instead of aspiring for the universal and the progressive he looks backward and yearns for the particular."

[2]Ongoing conversations with director of restoration at 'Save the Bay,' Wenley Ferguson, reinforce this local frustration. See also SAMP reports such as: http://www.beachsamp.org/wp-content/uploads/2015/06/Rhode-Island-Sea-Level-Affecting-Marshes-Model-Technical-Report-11.pdf (accessed March 12, 2016).

Fig. 8.2 The Upland and Lowland condition is most visible from within the low-lying salt marshes, which are generally attenuating due to hydrologic disturbance. Current restoration practices tend to favor elevating the salt marsh with dredge, rather than appreciate its transformation towards open water. *Credit* Author (2014)

rivers, coves and streams. Consequently, communities are sensitive to the immediacy of littoral shifts and aware of their vulnerability within the context of coastal dynamics and seasonal hurricanes. This is especially the case in Narragansett Bay, a mixed estuary that bisects Rhode Island in a north-south direction. Providence lies at the northern reaches of the bay and Newport sits on the southern periphery. Shoreline types include fringing and meadow salt marshes, bulkhead and other modified perimeters, which comprise 25% of Narragansett Bay's perimeter.[3] Its

[3]The Bay extends approximately 45 km along this north-south axis, reaches 18 km at its widest point, and covers an area of 342 km^2. For more details on Bay dynamics: http://www.savebay.org/bayfacts.

coastal areas feature a combination of considerably disturbed sites, preserved marshland or post-industrial fill. Narragansett Bay's estuarine condition makes it unique along the highly developed Atlantic coast, as it has resisted coastal engineering such as seawalls and surge barriers.

Relative to other North Atlantic states, coastal development is limited in Rhode Island. The low density, coupled with its wealth of natural features and a strong network of environmental stewardship, positions the State to become a leader in coastal resilience, offering an exemplary model to other coastal and estuarine populations along the coast that are facing similar risks. The range of eco-tones at the land-sea (aquatic-terrestrial) interface includes numerous inland ponds, salt marshes, and coastal dunes (Fig. 8.3. Rhode Island State Map). The mucky threshold between land and water is both sandy and flat: a slim horizontal field that is perpetually shifting to open water, inch by inch. Local ecologists have been tracking and monitoring this trend, yet current environmental authority only sanctions its restoration, leaving little opportunity for a change in practice for local organizations.[4] As the shore subsides, the firmer, higher metamorphic rock of the upland materializes as the potential future waterfront: a steep, rocky coastline substitutes a flat, granular shoreline.[5] The feedback between ecology and geomorphology forms this critical eco-tone, which is transforming due the increased frequency of periodic inundation.

Most of Rhode Island's oceanfront property is privately owned. In theory, this leaves proprietors facing the prospect of lost land with only two options for their parcel: sell their lot at a net loss or elevate vertically. The municipality is also left with few actionable options: defend the beach for its economic value, protect the road for the community, or appeal for a planning strategy that acknowledges retreat. According to the Rhode Island Coastal Resources Management Council, retreat is quickly becoming a feasible proposition.[6] Retreat as a strategy promotes resettlement to higher ground or a permanent relocation, simultaneously acknowledging the authority of sea-level and its breaching of human jurisdiction. As retreat becomes more and more feasible, the question of how to use, manage, or zone the abandoned plots and lands deserves deliberation. Coupling the subsidence of the intertidal zone with the creation of increased state land holdings through resettlement and relocation reveals the opportunity to reshape the coastline as both a novel resilient ecology and a remarkable new public space.

The ecological transitions initiated by disturbance and altered hydrology can be perceived most directly in local plant life, as seed dispersal, colonization, and

---

[4]With particular appreciation to Wenley Ferguson (Ecologist, Save the Bay) and Janet Friedman (Coastal geologist), at CRMC for leading numerous site visits.

[5]In the case of Rhode Island, that scale of subsidence would result in a significant area of land loss, up to 20% of its total land area.

[6]http://www.crmc.ri.gov.

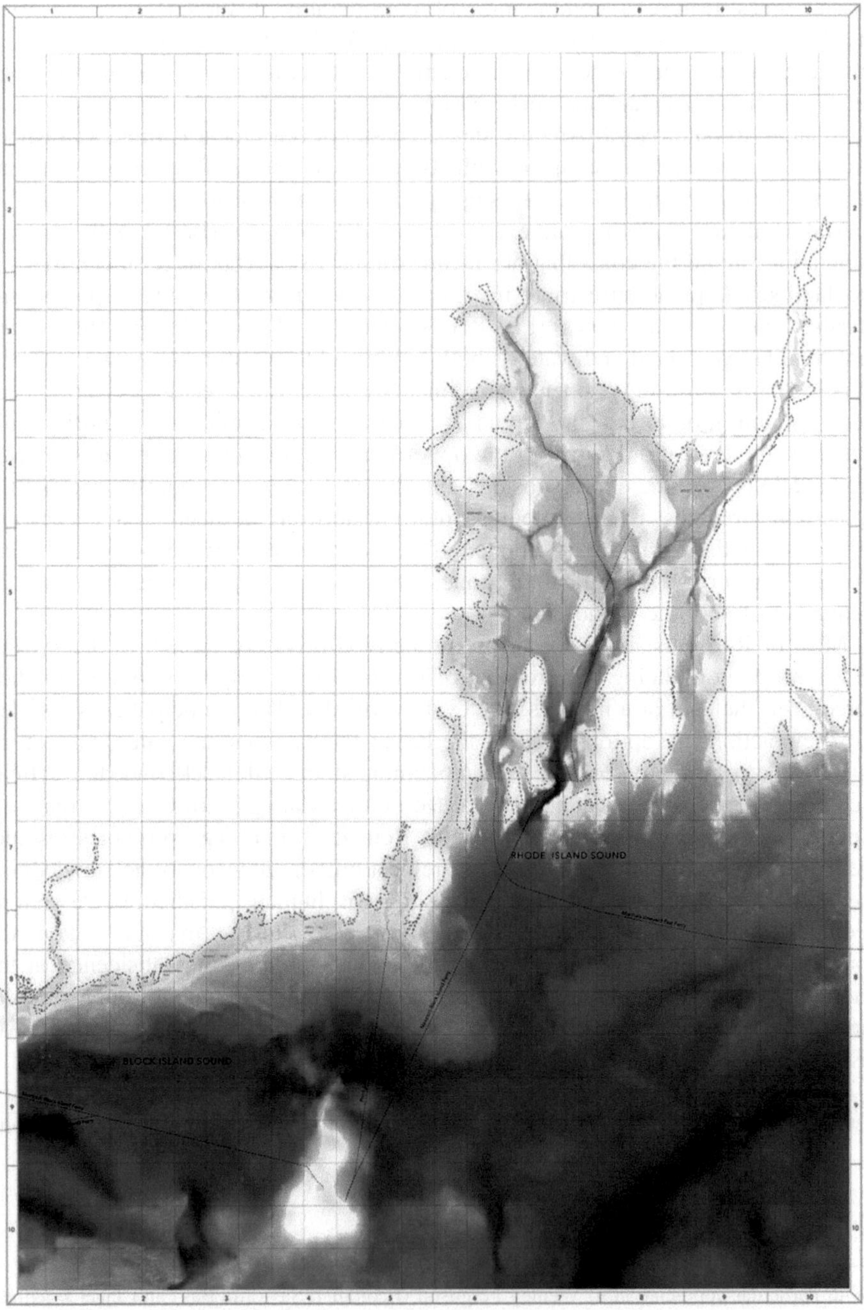

Fig. 8.3 Rhode Island sound is a rich coastal estuary where shoreline movement will undoubtedly impact its extended shores. This projection traces the margins between lowland and upland ecology. *Credit* Ocean State project

reproduction are instigated. Some plants wilt and dieback (the gradual decay of shoots), while others develop the means to prosper gaining competitive advantage. The new challenge is not how to control or halt their spread, but how to praise and encourage it. If plant life is adapting most rapidly to a changed condition, then the model for resilience cannot be reduced to restoration, but lies in the embedded intelligence of the plants that survive and flourish with shifting conditions. Simply stated, as the coastline recedes, it creeps closer to a shelf of higher ground, which is comprised of dense, mixed woodlands. If the intertidal lowland and bottomland is no longer artificially elevated with dredge and the beaches were no longer nourished by imported sand, then open water would dominate and the upland would likely become the new coast.[7] Examining the transitional species that have appeared and flourished post-hurricane Sandy suggests that given the appropriate conditions, upland species can thrive in bottomland sites (Fig. 8.4, Sassfrass sp. in a tidal marsh). Plants are adapting quickly, and, while they might move slowly, they are nonetheless always trying to gain advantage, to compete with other species and advance. In this case, they may even be outcompeting humans in their appropriation of the coast.

Operation Resilience

The word resilience did not enter contemporary usage until well after its first acceptance into ecology in 1973.[8] The word was cleverly appropriated from its application to elasticity gradients in material sciences by the prominent ecologist C. S. Holling in order to help him explain complex systems dynamics. The term has a well-documented foundation that has been heavily debated, but in every case the current definition embraces the notion of a return to stability, implicit in the objectives of recovery (McAslan 2010; Folke et al. 2010). Resilience thus tends to be juxtaposed against disturbance. It does not evoke low stability, volatile states or non-equilibrium. Instead, it has come to suggest a kind of inherent stability or toughness. When we think of resilience in 2016, we think of strength, perhaps nowhere captured more obviously that in the title of New York City's post-Sandy reconstruction plan, "A stronger, more resilient New York" (Bloomberg 2013). When Holling introduced the term to systems ecology, the world had not yet invented a 'superstorm' and climate change narratives had not touched ground. The misuse and over-appropriation of the term resilience over the past decade is a direct

[7]This speculation is based on conversations between the author, Dr. Del Tredici and Grover Fugate at a meeting in June 2014 at 'Save the Bay,' Providence, Rhode Island.

[8]W. Skeat in A Concise Etymological Dictionary of the English Language first defined the term. (Oxford: Claredon 1882) The origin is traced by the Latin root *resilire*, which translates to 'leap back'. Subsequent to this, it is defined by The Oxford Advanced Dictionary (2016) as the "ability (of a person or animal) to withstand or recover quickly from difficult conditions." A clear modification from its original application and principle etymology.

Fig. 8.4 This stand of Sassafras trees is located at the edge of a salt marsh near open water, thriving only 2 years after Hurricane Sandy made landfall. *Credit* Author (2014)

result of the funding surge that followed Hurricanes Sandy and Katrina. In other words, resilience has slowly evolved from a noun into a concept that encompasses all procedures, policies, and operations that are symptomatic of adversity. As such, we risk it going the wayside with other dated neologisms (Károly 2011; Friend and Moench 2013; Benson and Craig 2014). In particular, resilience has become an expression of recovery that implies a level of stability that is both local and feasible, fostering fantasies of a 'return' to a state of normalcy—the literal bouncing back that currently leads the discourse. The liberal application of the term invites a passionate nostalgia for a time prior to the disturbance, a moral authority that fosters the reinstating of former ecologies and advances the paradigm of restoration.

Yet, when Holling appropriated the term resilience, he applied it suggestively to time rather than space, using evidence collected through observation of behavior, amplitude, and frequency of oscillations: "*Individuals die, populations disappear, and species become extinct*" (Holling 1973). By relating it to duration, he could justify monitoring periods of dormancy and decline in an ecosystem, charting the intervals of recovery. Holling proposed that when considering the performance of systems, attention ought to emphasize the conditions and time-scales that allow organisms to persist, rather than targeting momentary equilibrium states which overlook dynamic values. Therefore, he revealed the association or amount of impact from a disturbance could either result in the absorbance of a shock or in an alternative configuration. Prior to this suggestion, single equilibrium assumptions had dominated the global stability of ecological thought. Holling's 'multi-stable state reality' revealed instability as a predictable outcome, an important attribute of the behavior of ecological systems (Gunderson et al. 2010). The term resilience was applied as a means to designate this novel form of instability, one that was predicated on high variability and behavior expressed far from equilibrium. Time became a generative feature of ecological theory as Holling ascertained that the most resilient systems are simultaneously those that display low stability.

The notion that resilience references a future condition, rather than retrospective one, offers expanded opportunities to the discourse that surrounds coastal design strategies. Further, the modifications that disturbance-based events create ought be respected and evaluated as a measure of ecological resilience (Gunderson 2000). This line of inquiry also profoundly denies the image of a hopeful past, as evidenced through the success of 'before and after' images that propagate following coastal storm events. Such popular documentation invokes past states by vividly indexing what has been lost or destroyed. Each photograph signifies an effort to record the hard times of the present, by expressing a convincing fiction of prosperity and stability about the past. Images of this kind suggest that restoration to former states is a tradition of progress and a virtue that can enable humans to persevere in a time of crisis: before and after and then before again. Such restorative practices negate the value of duration and time based events, sanctioning static images and creation myths: the marsh, the dune, the pond are reinstated as non-native species are eradicated and water levels are controlled by the pipes, backflow preventers, and culverts that offer long term ecological life support.

The Planted Coast

What if the loss of beachfront can be balanced by a significant gain in forest cover? It is not difficult to imagine that elevated boardwalks and tents can replace umbrellas and bikinis. Accepting a transitional coast means that in increasingly important ways, the questions that the living environment presents are about how to shape it not how to preserve it.[9] Designing a series of forests in lieu of restoring existing ecology forces a consideration of how the environment is valued and managed. In this case, deliberate species selection realizes the potential of plants to shape the environment, providing a foundation to develop a design that can be manipulated and measured alongside typical construction materials (Fig. 8.5. Selected species charts). The research proposes a design strategy that proposes to plant a considerable area of coastal forests along the intertidal zone, creating a biological fabric that is both adaptable and resilient to persistent saltwater inundation, accelerating the shifting coastal morphology. An assortment of disturbance-adapted woody trees and shrubs instigates accumulation and land accretion on hummock-like formations.[10] Construction occurs as a process. As species are selected for their ability to sprout through burls, roots, and stems, their root zone thickens and their adaptive capacity to withstand disturbance multiplies (Del Tredici 2001). Each forest can benefit from the current particulate condition, reinforced when required and planted by clumps of new species (Fig. 8.6. Design drawing, construction detail). The proposal of a novel ecosystem is equally a redesign of the coast, as grasses and sedges are replaced with shrub thickets and

[9]See Purdy, J. *After Nature: A Politics for the Anthropocene.* Cambridge: Harvard, 2015 (11).
[10]Plant lists developed in collaboration with Dr. Peter Del Tredici (see appendix).

multi-stem tree species. Coastal forests will provide essential infrastructural protection for adjacent roads and evacuation routes, including wind mitigation, debris capture, expanded recreational opportunities, and more importantly a critical setback for development that prevents more open water, as bottomlands are appropriated by rising sea levels.

Sprouts, Cuttings, and Reiteration

A tree is characteristically imagined as having one single stem, rising up to a wide and dense canopy. This particular form, while culturally valuable, makes the tree susceptible to environmental disturbances. However, if a central role for disturbance is accepted as an attribute of resilient design, then the form of the tree becomes pliable. Trees can be expected to break or fall, sprouting secondary trunks —an induced response to injury common to temperate angiosperm trees (Del Tredici 2001). This behavior can be classified into four different morphologies: root sprouts, collar sprouts, sprouts from underground stems, and layered, opportunistic sprouts (Fig. 8.7. Root morphology). Root Sprouts—more commonly known as root suckers—produce genetically identical clones that emerge from an injured or non-injured root, a feature of a many trees and shrubs of the Northeast. Collar sprouts emerge from the junction between roots and sprout, very close to the ground, and have great potential to develop secondary trunks from this point. Sprouts from underground stems dominate the rhizosphere as they emerge from underground and form adventitious roots that can often appear far from the parent tree. Finally, opportunistic sprouts are the least common, spreading through layering or reiteration. Reiteration occurs when a low-hanging branch reaches the soil, produces roots and eventually sprouts vertical shoots.

Sprouting in trees is not a biological attribute that is generally considered desirable by designers who, with the curious exception of white birches, seem to favor single stemmed specimens. This preference leads to an overreliance on stock material and planting seedlings with great labor budgets, denying the true potential of designing with plant behavior. Ultimately, such generic practices reduce the selection of plants to a desire for above-ground form, silhouette or attractive seasonal qualities, features of planning that ignore the root system. Planting with behavior in mind necessitates a profound acceptance of an environment in constant transition, for example through fluctuations in salinity levels and altered hydrologic flows. Behavior is predictive, not reflective, and equally cannot help to restore former plant communities or associations. In order to establish a planted forest so close to subsidence conditions, the design relies on creating hummocks by re-distributing fill (Fig. 8.8. Design drawing plan). Plants are introduced on higher elevations, eventually holding the ground and spreading laterally as the soil accumulates. The ability of plants to colonize the ground and build soil becomes a critical consideration in the design proposal, especially relevant now that the challenges have been scaled irreversibly by the complexity of climate change.

	Root Suckering Species	Stump Sprouting Species	Burl Sprouting Species	Conifers (non-sprouting)
Bottomland Species	Acer negundo* Asimina triloba# Liquidambar styraciflua *# Nyssa sylvatica	Acer rubrum Acer pseudoplatanus*+ Acer saccharinum Alnus glutinosa*+ Alnus meritima*# Betula nigra# Catalpa speciosa *# Celtis laevigata# Gleditsia triacanthos* Ilex opaca Magnolia virginiana# Platanus occidentalis* Platanus x acerifolia *+ Populus deltoides* Quercus bicolor* Quercus palustris* Quercus phellos*# Salix nigra* Salix fragilis+		Chamaecyparis thyoides* Larix decidua+ Larix laracina Picea mariana Taxodium distichum*# Thuja occidentalis Thuja plicata+
Upland Species	Diospiros virginiana# Fagus grandifolia Gymnocladus dioicus*# Maclura pomifera*# Populus tremuloides* Prunus virginiana% Robinia pseudoacacia* Sassafras albidum *	Aesculus hippocastanum*+ Amelanchier arborea Betula lenta Carpinus caroliniana Carya cordiformis Carya tomentosa* Celtis occidentalis Juglans nigra* Koelreuteria paniculata*+ Prunus serotina * Quecus acutissima+ Quercus coccinea Quercus rubra* Sorbus alnifolia*+	Ginkgo biloba+ Magnolia acuminata# Quercus alba Tilia americana* Tilia cordata*+	Juniperus virginiana* Picea abies+ Picea glauca* Picea pungens* Pinus heldreichii*+ Pinus mugo*+ Pinus parviflora*+ Pinus rigida* Pinus strobus Pinus thunbergii*+

*Species with some degree of salt-tolerance
#Southern species
+Non-native species
%Rhizomatous species

Note: Bottomland species grow well in upland conditions, but not vice versa

Fig. 8.5 Selected resilient and sprouting tree species chart. *Credit* Ocean State project and Dr. Peter Del Tredici

Exploiting the sprouting and regrowth of the living tree stump, or stool, is a common to coppice forestry, a practice that produces significant economic value. Yet, the biological benefits are less culturally notable since the root zone lies out of sight and makes no significant capital contribution. Characteristically, these forests are valued for the poles or whips that provide firewood, fence material, and such. But, the ability of seedlings or roots to resprout following damage actually enhances their survival through disturbance is less known. Further, individuals that

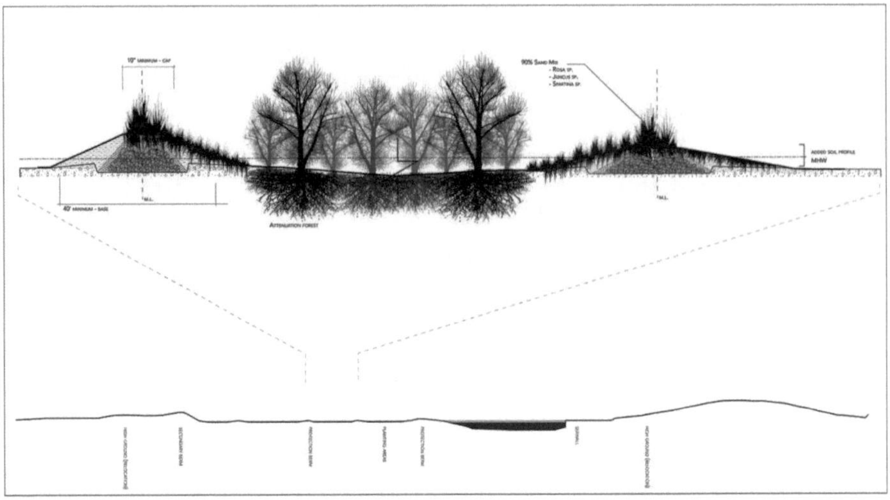

Fig. 8.6 The interaction between hardened and softened infrastructure is drawn in section to describe how the landform helps to attenuate, protect and control inundation. *Credit* Ocean State project

Fig. 8.7 Three types of tree sprouting morphologies from *left*: Root Suckering, Stump Suckering and Burl Sprouting. *Credit* Dr. Peter Del Tredici

grow and adapt to disturbed sites are likely to grow more vigorously and retain their sprouting ability longer than those who undergo less stress (Del Tredici 2001). This is evidenced by the number of ancient stools around Europe, the physical result of hundreds of years of coppicing. In much the same way, the trees selected for this project simply adapt to disturbance by living longer and getting stronger, as their

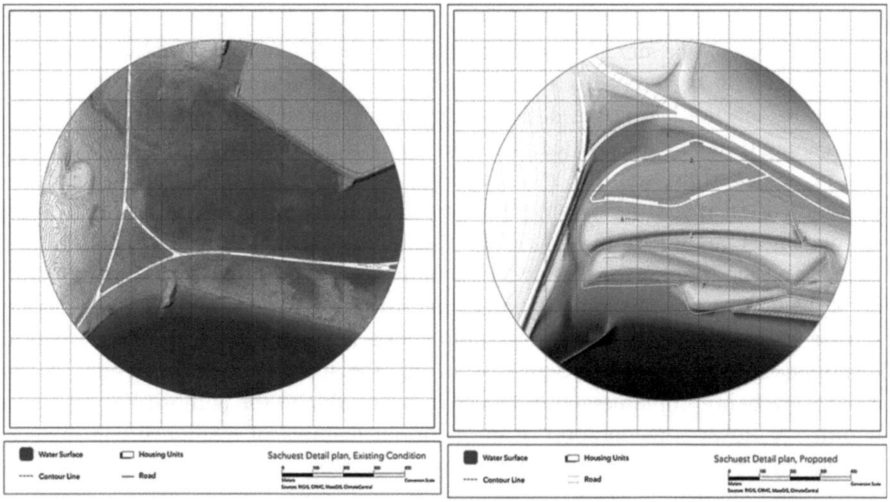

Fig. 8.8 Typical design plan. The typical attenuation backfill from the former condition (*left*) to the future, indicating programmatic potential before planting (*right*). *1* visitor parking, *2* picnic areas, *3* multi-use trail, *4* re-aligned river, *5* maidform river jetty, *6* transverse dune, *7* fore-dune system. *Credit* Ocean State project

structures below ground thicken to promote survival. Such morphological features actually exploit disturbance in order to build and confirm resilience.

The research relies on known morphological behaviors, rather than unpredictable horticultural features. By relying on disturbance and integrating maintenance into the initial concept, the design is necessarily time based, as upkeep is imagined though rigorous cycles of pruning and mowing so that the entire tree can become more durable in the face of strong winds and erosive waves. These forests of multi-stemmed, disturbance-adapted specimens are predicted to reproduce, creating restructured woodlands that can be both productive and resilient in the face of change. Imagining stands of clonal, fragrant *Sassafras* sp., against tall, green mounds of *Rhus* sp. and groves of multi-stemmed *Tilia* sp., or dense suckering thickets of *Cornus* sp., implies a different image of the temperate forest, one that forces a reevaluation of resilience. Reconsidering plant species and the role of plants provides novel ways to harness disturbance as a mechanism for building and strengthening.

Future Resilience

Modern resilience is paradoxical in the sense that the universality of its application in sanctioning the restoration of landscape features. This universal acceptance reinforces the notion that recovery is predictable and familiar, an acknowledgment that engenders a false sense of stability. At the scale of the community and within the regulatory environment, reconstruction efforts support restorative practices. While the intentions themselves are progressive, this research argues that the design professions are not taking the notion of instability and disturbance seriously, since the meaning of resilience is often misunderstood or ill defined. If high resilience can only be achieved through the modifications embedded in low stability, and if the outcomes often reveal new configurations, then renovating a seawall, restoring a salt marsh, or renourishing a dune are the lowest possible forms of building resilience and the highest form of securing failures. The promise to rebuild or restore lies at the core of this misapplication, so that resilience as a term loses meaning, and is reduced to a label that can only help make sense of the fear of destruction. Misapplication is especially explosive along the urbanized coast, where design projects are being implemented in record numbers due to the authority of resilience funding. A closer consideration of the defining features of resilience can increase the motivation for a change in status quo, endorsing novel ecologies, adaptable species, and alternative programs in order to draft a new image of the coast. The proposition deflates the unproductive dichotomy between green and grey systems and elevates the role of plants in the paradigm of resilience.

Acknowledgements This project was originally developed as part of SCR: Structures of Coastal Resilience, with support from the Rockefeller Foundation. With particular thanks to Michael J. Luegering for his collaboration. For team assembly, consultants and more information, see: Structures of Coastal Resilience: www.structuresofcoastalresilience.org.

References

Adger WN, Brown K, Nelson DR, Berkes F, Eakin H, Folke C, Galvin K, Gunderson L, Goulden M, O'Brien K, Ruitenbeek J, Tompkins EL (2011) Wiley Interdisc Rev: Climate Change 2(5):757–766

Bennett O (2001) Cultural pessimism: narratives of decline in the postmodern world. Edinburgh University Press, Edinburgh

Benson MH, Craig RK (2014) The end of sustainability. Soc Nat Res: Int J 27(7):777–782

Bloomberg M (2013) A stronger, more resilient New York. City of New York, New York

Chapin FS III, Gary P Kofinas, and Carl Folke (2009) A framework for understanding change. In: Chapin FS III, Kofinas GP, Folke C (eds) Principles of ecosystem Stewardship: resilience-based natural resource management in a changing world. Springer, New York, New York, USA

Del Tredici P (1999) Aging and rejuvenation in trees. Arnoldia. Winter:11–16

Del Tredici P (2001) Sprouting in temperate trees: a morphological and ecological review. Bot Rev 67(2):121–140

Emanuel KA (2005) Increasing destructiveness of tropical cyclones over the past 30 years. Nature 436:686–688

Emanuel KA (2008) The hurricane-climate connection. Bull Am Meteorol Soc 89

Friend R, Moench M (2013) What is the purpose of urban climate resilience? Implications for addressing poverty and vulnerability. Urban Climate 6:98–113

Folke C, Carpenter SR, Walker B, Scheffer M, Chapin T, Rockström J (2010) Resilience thinking: integrating resilience, adaptability and transformability. Ecol Soc 15(4)

Gunderson LH (2000) Ecological resilience—in theory and application. Ann Rev Ecol, Evol, Syst 31:425–439
Gunderson LH, Allen CR, Holling CS (2010) The evolution of an idea—the past, present, and future of ecological resilience. In Gunderson LH, Allen CR, Holling CS (eds) Foundations of ecological resilience, pp 1–13
Hilderbrand RH, Watts AC, Randle AM (2005) The myths of restoration ecology. Ecol Soc 10(1)
Holling CS (1973) Resilience and stability of ecological systems. Ann Rev Ecol Syst 4:1–23
Holling CS (1986) Resilience of ecosystems: local surprise and global change. In: Clark WC, Munn RE (eds) Sustainable development and the biosphere. Cambridge University Press, Cambridge, UK, pp 292–317
Holling CS (2001) Understanding the complexity of economic, ecological, and social systems. Ecosystems 4(5):390–405
Hollnagel E (2009) The four cornerstones of resilience engineering. In Nemeth Christopher P, Hollnagel E, Dekker S (eds) Resilience engineering perspectives: preparation and restoration. Ashgate, Farnham, UK
Károly K (2011) Rise and fall of the concept sustainability. J Environ Sustain 1(1):1
McAslan A (2010) The concept of resilience: understanding its origins, meaning and utility. Torrens Resilience Institute, Adelaide
Purdy J (2015) After nature: a politics for the anthropocene. Harvard, Cambridge
Walker B, Holling CS, Carpenter SR, Kinzig A (2004) Resilience, adaptability and transformability in social–ecological systems. Ecol Soc 9(2):5

Chapter 9
COREDAR for Cities: Developing a Capacity Building Tool for Sea-Level Rise Risk Communication and Urban Community-Based Adaptation

A. Saleem Khan

Abstract Sea-level rise (SLR) is one of the greatest threats for densely populated low-lying coastal cities. Building capacities at community level to address the challenges of SLR is an important first step towards adaptation planning. However, efforts and attention for community based adaptation (CBA) in urban coastal cities are often ignored and not given much importance. Thus, building capacity through SLR risk communication and involving communities in framing urban CBA is a high priority for cities. Nevertheless, it is a difficult task for climate scientists to communicate complex SLR science and build capacity at local level. To address these challenges, this study has put forth three research questions through the lens of SLR risk communication and urban CBA, as (1) What, if any, community engagement in risk communication in addressing SLR risk occurring in urban areas; (2) What information does communities need and (3) How does it need to be communicated, in order to be better prepared and have a greater sense of agency? To answer these questions, by following the framework on SLR risk communication and urban CBA, this study has resulted in evolving "COREDAR" (COmmunicating Risk of sea-level rise and Engaging stakeholDers in framing community-based Adaptation stRategies), a capacity building tool for SLR risk communication and urban CBA. Thus, this study seeks to provide insights on communicating the risk of SLR and to evolve a robust picture of urban CBA through effective decision-making that is grounded in pressing community priorities by developing a capacity building tool for urban coastal cities.

Keywords Climate change · Sea level rise · Capacity building · Tools · Risk communication · Urban CBA

A.S. Khan (✉)
Center for International Earth Science Information Network (CIESIN), The Earth Institute, Columbia University, PO Box 1000 Palisades, New York 10964, NY, USA
e-mail: asaleemkhan.cc@gmail.com; skhan@ciesin.columbia.edu

W. Leal Filho and J.M. Keenan (eds.), *Climate Change Adaptation in North America*, Climate Change Management,
DOI 10.1007/978-3-319-53742-9_9

Introduction

The scientific literature has documented the growing risks of flooding posed for coastal cities by the combination of climate change, as reflected in sea level rise (SLR) and intensified storms and storm surges, and ongoing urban growth in low-lying coastal zones (Fuchs et al. 2011). Settlements in coastal low lands are especially vulnerable to risks resulting from climate change, yet these low lands are densely settled and growing rapidly (McGranahan et al. 2007). Thus, accelerated SLR is a major long-term outcome of climate change leading to increased inundation of low-lying areas and put global cities and communities at greater risk. Particularly, urban poor communities will bear the brunt of its effects since they live in informal settlements that are more exposed to hazards like SLR. Therefore, building capacity for resilience and adapting to climate change threats like SLR are increasingly high priorities for urban coastal cities. There has been considerable focus on the potential of cities to contribute to mitigation, yet there is growing recognition of the need for urban adaptation (Romero Lankao and Dodman 2011).

Capacity building and improved knowledge management towards increasing the resilience and adaptive capacity of the coastal communities to current and future climate risks (Sales 2009) is an important dimension to view the challenges of SLR for low-lying urban coastal cities. However, community-based adaptation (CBA) experiences emphasize that it is important to understand a community's unique perception of their adaptive capacities in order to identify useful solutions and that scientific and technical information on anticipated coastal climate impacts need to be translated into a suitable language and format that allows people to be able to participate in adaptation planning (IPCC 2013). Thus, the ability to act collectively to develop and implement adaptation responses based on sound knowledge of climate science, intervention of policies and the role of societies are prime factors in building adaptive capacity (Bulkeley et al. 2009; CAP 2007; Tanner et al. 2009; Wilson 2006). Unfortunately, the planning community has had a limited role building and implementing adaptation response to climate change in urban areas (Sanchez-Rodriguez 2009). Thus, capacity building through urban CBA is a key issue for climate adaptation measures in urban areas (Hartmann and Spit 2014) and it is the urgent requirements for addressing climate risk like SLR, particularly at local level (IFRC 2009). Nevertheless, capacity building has been defined as the practice of enhancing the strengths and attributes of, and resources available to, an individual, community, society, or organization to respond to change (IPCC 2013). In other words, it is defined as an effort aimed to develop human skills or societal infrastructures within a community or organization needed to reduce the level of risk (UNISDR 2009). Importantly, capacity building through risk communication strategies would be the first step towards addressing the challenges of SLR and CBA for urban coastal cities. To meet these challenges and requirements, this study puts forth three research questions (1) What, if any, community engagement in addressing SLR is occurring in urban areas; (2) What information does communities need and (3) How does it need to be communicated, in order to be better prepared and have a greater sense of agency?

Importantly, Article 6 and Article 11 (capacity building for climate action on Paris Agreement) of the UNFCCC addresses the importance of climate change communication and engaging stakeholders in the decision-making process. It highlights the responsibility of participating countries to develop and implement educational and public awareness programmes and to ensure public access to information and to promote public participation. Nevertheless, this is one of the most difficult challenges for scientists to communicate complex climate science and to build capacity at local level (Khan et al. 2012). Similar to climate change communication, SLR risk communication presents challenges involving complex science, uncertainty, and invisibility (Covi and Kain 2015). Messages about SLR risks may fail to adequately inform audiences or persuade them to respond to consequences (Harvatt et al. 2011) because SLR information may be too general for people to relate to; scientific and technical information is too specialized; and messages are not sufficiently tested with representative audiences to gauge potential reactions (Covi and Kain 2015). Given its critical importance, public understanding of climate science (SLR) deserve the strongest possible communication science to convey the practical implications of large, complex, uncertain physical, biological and social processes (Pidgeon and Fischhoff 2011). Thus, to communicate the risk of SLR and to engage communities in urban CBA planning, capacity building tools are considered as a powerful vehicle to accomplish this task. Emphasis has been placed on participatory tools and methods for urban communities to plan strategies (Dodman and Mitlin 2013). Particularly, designing a capacity building tool for cities with the objective of developing the capacity of cities has promoted a sense of ownership over city-level resilience-building initiatives, and the evidence emphasizes that there needs to be demand from the city itself in order for a capacity building process to be successfully taken up (Archer and Dodman 2015). It is important to engage in community mobilization and awareness rising through designing activities that are tailored to local practices and establish strong relationships with the communities to enable sustainable actions to involve the key stakeholders in adaptation action and enhance capacity building (Khan 2013) particularly in the urban context. In this purview, the main aim of this article is to develop and design a capacity building tool for SLR risk communication and urban CBA for coastal cities, to communicate complex SLR science in a simple way, to better understand knowledge needs of SLR, to educate and create awareness about SLR adaptation and to emphasize the importance of community engagement in the SLR decision-making process.

Need of Capacity Building Tool for SLR Risk Communication and Urban CBA

The lack of effective communication about climate risk like SLR has contributed to improper interpretation of scientific findings pertaining to climate change and to a poor mobilization of vulnerable groups for developing appropriate response actions

(MRC 2010). Therefore, improving public awareness and developing overall communication strategies to make climate change science accessible to the average citizen and could reduce their vulnerability (UNFCCC 2007). This, in turn, will enhance capacities of various stakeholders in the community and improve sustainability at the local level. However, communicating climate risks like SLR, especially the unfamiliar, non-voluntary, scientifically complex, and politically controversial may be one of the most difficult tasks scientists, city planners, managers, and policy-makers face in dealing with adaptation to climate change and SLR (Covi and Kain 2015). Nevertheless, many scientists recognize the importance of communicating scientific findings to citizens to help them become aware of the urgent need to act and to enable them to carry out anticipatory actions. However, this is one of the most difficult and sensitive objectives of SLR communication and education (Khan 2013). Hence, effective risk communication requires more than conveying accurate information (Fischhoff 1995) it should raise awareness, increase understanding, and move audiences to action (Boholm 2009). To be effective in presenting information and moving audiences to action, communication about SLR risk must use appropriate framing, compelling visuals, and accessible language, preferably tailored to the audience for the information (Covi and Kain 2015). Furthermore, experience with adaptive learning and capacity building activities has shown that despite the existence of many of the tools, decision-makers still struggle with adaptation. Clearly, this has implications for how the approaches are applied (MRC 2010). Therefore, there is a pressing need for developing a "capacity building tool" that address and integrates SLR risk communication and urban CBA. However, the tool is defined as "a means or instrument by which a specific task is accomplished". The process of developing climate risk communication tools and methods revealed a number of key insights that can inform efforts to engage vulnerable communities, policy-makers, and other stakeholders (MRC 2010). In this context, this article advocates for urgent need of SLR risk communication and urban CBA for coastal cities. It is believed that despite of availability various other tools, the development of simple tools that holistic synthesize all perspective of SLR information is an urgent need of the hour.

Methodology

SLR and urban CBA capacity building tool have been constructed based on the rationale and guiding principle of COREDAR framework evolved by Khan et al. (2015). A broad understanding of the trends in thought on tools for adaptation planning through capacity building as a first step was considered a necessary foundation for this framework. This understanding emerged from examining scholarly discussions on the subject of SLR and urban CBA. The analysis was also guided by a set of criteria that determine an approach to adaptation planning can be considered through capacity building. However, the framework was developed following the guidelines and recommendations of IPCC Assessment Report 5,

Working Group I Report on Climate Change 2013: The Physical Science Basis; Working Group II Report on Climate Change 2014: Impacts, Adaptation and Vulnerability; Climate Change 2014: Synthesis Report is followed with an emphasize on SLR and urban CBA. The main objective of this framework is to communicate complex climate science to different stakeholders to build capacity at all levels of SLR adaptation decision-making. It provides a platform to integrate "science-policy-society" nexus of SLR and urban CBA with a systematic step-wise approach to address different dimensions of SLR risk communication and urban CBA (Fig. 9.1).

Systematic step-wise approach is constructed based on complex SLR science, such as delineation of the climate and city profile of the urban coastal city; documentation of past sea-level trend, based on tide gauge data; future SLR projection based on IPCC-AR5 dependent climate (SLR) model under different Representative Concentration Pathways (RCPs) scenarios; prediction of SLR impact based on Geographic Information System (GIS) impact assessment methods; identification of vulnerable human communities and prioritization of stakeholders based on vulnerability assessments such as stakeholder analysis; communication of SLR risk based on risk communication strategies; framing urban CBA strategies to SLR by engaging stakeholders through participatory approaches and techniques; and mainstreaming urban CBA strategies through policy intervention. Detailed descriptions of a wide range of methodologies involved in developing this tool such as analysis of tide gauge data, descriptions of climate models, impact and vulnerability assessment methods, risk communications strategies, participatory approaches and techniques etc., underlines the limitations of this paper. However, by bringing together different methods and analytical frames, the assessment sought to provide decision-makers and stakeholders in participating cities with information and tools necessary to better adapt to climate change while also recognizing the current successes and strengths of each city (Barclay et al. 2013).

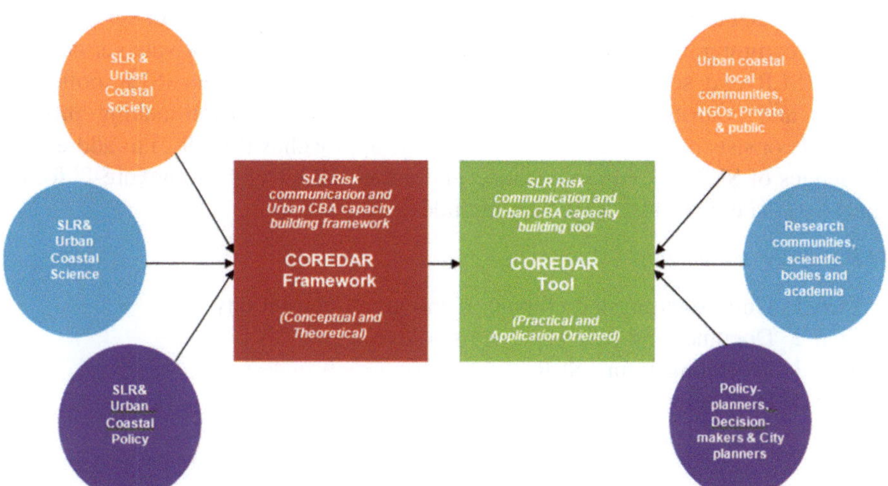

Fig. 9.1 SLR science, society and policy nexus of COREDAR framework and tool

Results

Design of the COREDAR Tool

SLR and an urban capacity building tool named COREDAR have been developed and designed based on the rationale and the guiding principle of COREDAR framework. The term COREDAR is an acronym and it stands for COmmunicating Risk of climate change and Engaging stakeholDers in framing community-based Adaptation stRategies (Khan et al. 2015; USIEF 2015; UNAI 2016; State of the planet 2016). The tool has been designed in the form "checklist" (a comprehensive list of important or relevant information and steps to be taken in a specific order) in a "check sheet" (a simple data recording sheet, custom designed to enable a user to readily interpret the results of an activity or process). In this case, this tool gathers holistic information on SLR risk communication and urban CBA in a systematic step-wise approach with information ranging from climate and the profile of the urban coastal city, past sea-level trends, future SLR projections, predicted SLR impacts and vulnerabilities, framing SLR and urban CBA strategies and mainstreaming SLR adaptation policies. Importantly, it gives a space for all stakeholders who have stake on this issue such as scientists, researchers, academicians, policy-planners, decision-makers, NGOs, the general public and anyone who is interested in the risks of SLR and how to develop adaptation strategies that are grounded in community priorities.

Structure of the COREDAR Tool

Structuring the tool with checklists may also be used as an initial step to provide an improved information basis for selecting SLR risk screening, assessment (Trærup and Olhoff 2011) and communication. Each step offers a checklist for scientific and research communities (SLR science); local communities, NGO, private partners and public (SLR and society); policy-planners and decision-makers (SLR policy) to involve and contribute to the SLR and urban CBA decision-making process. The tool consists of eight systematic step-wise approaches (Fig. 9.2) to address the challenges of SLR risk communication and urban CBA (refer "Annexure-I" for the checklist). The eight step approach includes:

SLR science dimension:

Step 1 Understanding the climate of the urban coastal city
Step 2 Documenting past observed sea-level trend
Step 3 Projecting future SLR
Step 4 Predicting SLR impact

Fig. 9.2 Outline of systematic step-wise approach of COREDAR tool

SLR society dimension:

Step 5 Identifying vulnerable communities and stakeholders
Step 6 Communicating SLR risk

SLR policy dimension:

Step 7 Framing urban CBA strategies to SLR
Step 8 Mainstreaming urban CBA strategies

Step 1 *Understanding the climate of the urban coastal city*

This step provides a checklist to fill the information related to geographical nature, climate and administrative structure of the urban coastal city. It includes geographical location, total area, and total population, administrative structure of the urban coastal city, governing body of the city such as a municipality, corporation and other relevant department involves in city governance, and importantly, overall climate pattern of the urban coastal city such as temperature, precipitation, wind speed, humidity, elevation from mean-sea level, information on coastal disasters etc.

Step 2 *Documenting past observed sea-level trend*
Tidal datum is used as references to measure local sea levels. The estimates of twentieth century SLR are primarily based on the historical tide gauge data maintained by various services such as Permanent Service for Mean Sea Level (PSMSL) (Woodworth and Player 2003). This tool provides a checklist for documenting tide gauge station names, duration of data availability and the time span of annual mean sea-level trend of the urban coastal city. Importantly, it provides space for documenting a list of recent coastal extreme events that the urban coastal city has experienced in the recent past.

Step 3 *Projecting future SLR*
Future SLR projections, based on climate models, can provide valuable information for the robust decision-making in adaptation policy-planning. The checklist provides column for future SLR projections for different RCP scenarios of IPCC AR 5 such as RCP 2.0; RCP 4.5; RCP 6.0; RCP 8.5 (IPCC 2013) for different time scales from 2020 to 2100 with the focus on short-term and long-term projections. Importantly, to meet the uncertainties in different SLR projections, it provides opportunity for users to take effective decision-making by selecting appropriate scenario of their choice, based on the city planning objectives and goals.

Step 4 *Predicting SLR impact*
There are number of impact models using various methodologies are available to study the impact of SLR at the city level (Titus and Richman 2001). One such example is GIS based inundation models and it can be used to predict the area of inundation to rising sea level. The checklist provides space to quantify the area of inundation to a projected SLR for e.g. 0.5 m SLR and ranking vulnerable regions as high vulnerable, medium vulnerable and low vulnerable regions to SLR. In order to meet the uncertainties, it provides opportunity for users to take effective decision-making by selecting a suitable region of their interest based on short and long term planning process in the city.

Step 5 *Identifying vulnerable communities and stakeholders*
Identifying vulnerable communities to rising sea-level in the urban coastal cities are one of the most important prerequisites to plan for city adaptation strategies to rising sea-level. Importantly, prioritizing stakeholders that include local communities, city governments, NGOs, the public and others are an important first step. Stakeholder analysis is one such technique frequently used to identify and investigate any group or individuals who will be or are affected by a change and whether they are equipped to deal with it. The checklist classifies stakeholders as key stakeholders, primary stakeholders, secondary stakeholders and tertiary

stakeholders based on their exposure to risk of SLR and importance in decision-making.

Step 6 *Communicating SLR risk*
The checklist provides a platform for SLR risk communication to concerned stakeholders by two interventions through participatory approach such as (1) Convey information: one-way transmission of information to provide basic understanding of SLR science and its impacts on the study area to the concerned key stakeholders, provide data on the local SLR e.g. information campaigns, news article, exhibits, posters, brochures etc. and (2) Build understanding: a two-way transmission of information that aims to facilitate opportunities for key stakeholders to develop their own methods to understand the concept of SLR e.g. vulnerability mapping, issue investigation, focus group interviews, citizen science programs etc. (Khan et al. 2012).

Step 7 *Framing urban CBA to SLR*
CBA draws on participatory approaches and innovative participatory methods to help communities analyze the causes and effects of climate change, to integrate science and community knowledge of climate change, and to plan adaptation measures (Reid et al. 2009). Thus, identifying appropriate adaptation options should then follow, building on information about existing community capacity, knowledge and practices used to cope with climate hazards (Huq 2008). The checklist provides columns for framing short-term and long-term adaptation strategies with its descriptions, relevance, benefits and constraints.

Step 8 *Mainstreaming urban CBA strategies*
Mainstreaming means integrating climate concerns (SLR) and adaptation responses into relevant policies, plans, program and projects at the national, sub-national and local scales it includes city planning as well (USAID 2009). A more realistic approach is needed to use existing methods and strategies of coastal adaptation that inform and meet new challenges of climate change induced vulnerabilities (Cheong 2010) like SLR and particularly for urban coastal cities. This step offers a checklist to list all existing coastal, urban and national policies at one end on the other hand to list all identified SLR adaptation policies. Importantly, it provides opportunity to make an intervention to mainstream SLR adaptation strategies with existing policies.

Application of the COREDAR Tool

COREDAR tool has been designed to help users in two ways (1) *Capacity building*: it provides a systematic step-wise approach to understand the science behind SLR and risk of SLR (step 1 to step 4) and plays a valuable awareness-raising and educational role. Communities can use this tool to build capacity to limit their exposure to hazards, save lives, limit public expenditures on armoring and emergency response, and protect valuable natural resources that provide natural flood protections and other environmental services (Grannis 2011); (2) *Decision making*: it provides a platform for all stakeholders to play a role in decision-making, ranging from researchers, policy-planners, decision-makers, local communities and public and whoever take stake on this issue of SLR and particularly it emphasizes the importance of urban CBA as an integral element of adaptation plan (step 5–step 8). This tool is designed to help policymakers manage the complexity of adoption by identifying and organizing adaptation strategies to SLR. This tool will help policymakers identify and weigh different solutions so that they can plan for and begin to adapt to SLR before impacts occur (Grannis 2011).

SWOT Analysis of the COREDAR Tool

One way to approach an evaluation of a tool like this could be through a SWOT (Strengths, Weaknesses, Opportunities and Threats) analysis. SWOT is used to encourage awareness of factors, positive and negative, that may affect planning and decision-making (Goodrich 2013). It is used to provide a descriptive assessment of tool, and provide a comparison of the factors affecting the effectiveness of the tool. There are several interesting observations that can be useful for anyone considers the application of tools or the development of new approaches, particularly when it comes to weaknesses, opportunities and threats (MRC 2010). The following are some findings of SWOT analysis of the COREDAR tool. *Strength*: (i) synthesizes different dimensions of SLR information holistically for urban coastal cities; (ii) offers a platform for SLR capacity building for different stakeholders of cities; (iii) integrates SLR science, society and policy for effective decision-making; (iv) provides simple checklist tools for easy access; (v) applicable to any coastal cities at any geographic location that are under the threat of SLR. *Weakness*: (i) difficulties in gathering too much scientific information on SLR; (ii) mistrust of climate information; (iii) challenges in transferring scientific knowledge of SLR to non-specialized audiences; (iv) lack of technical expertise of multidisciplinary perspective of SLR risk communication and urban CBA to handle the tool. *Opportunities*: (i) developing SLR risk communication and urban CBA capacity building training modules for capacity building training programs and workshops for different end users and stakeholders; (ii) developing web-based online capacity building tool for SLR risk communication and urban CBA; (iii) developing the capacity building smart phone application as a citizen science application or decision-making application for SLR risk communication and urban CBA. *Threats*: (i) threat of using wrong or incorrect SLR information; (ii) tool may be very complex for integrating wide-range of SLR information; (iii) gaps in expertise as it related to the SLR. The above mentioned are some major findings that captured within the SWOT analysis of these tools so that they may be taken up during the future course of action for developing tool deliverables and evolving strategies and modules (NIDM 2013).

Discussion

The SLR is a significant challenge that requires a coordinated response from all members of the community particularly in the urban context. While SLR can be a sensitive topic for public engagement, research has shown that approaches with an emphasis on learning, collaboration and openness can help the City to achieve wide public support for action (Barisky 2015). Thus, the aim of (SLR) science-policy effort is to establish a capacity to improve understanding about how climate change will affect cities and to provide more detailed regional impact assessments (OECD 2008) and to facilitate effective urban adaptation decision making to changing climate.

However, building capacity for adaptation in urban areas is one of the major challenges, yet an essential requirement to address climate change, and capacity building tools are used to educate and empower stakeholders and involve them in the decision-making process. There are a number of climate change tools are available with various contexts, for various sectors, and for various locations in general. However, some specific tools focusing on CBA and SLR are Climate Vulnerability and Capacity Analysis—CVCA(CARE 2009); Climate Adaptation Knowledge Exchange–CAKE (Eco adapt 2010); Designing Climate Change Adaptation Initiatives: A Toolkit for Practitioners (UNDP 2010); Adaptation tool kit: SLR and coastal land use (Georgetown Climate Centre 2011); The Community-based Risk Screening tool: Adaptation and Livelihoods—CRiSTAL (IISD 2011); Opportunities and Risks from Climate Change and Disasters-ORCHID (IDS 2011); Regional Adaptation Collaborative Toolkit (ARCCA 2014) etc.

However, application of these tools, as a more generic or separate effort, can be an appropriate means of increasing the awareness among key stakeholders and decision-makers on climate- related risks and their potential implications for vulnerability and development. Awareness is a prerequisite for integrating climate change risks and responses into planning and decision-making at all levels, so particularly in cases where the integration of climate change adaptation into development is in very early stages, generic or separate efforts may be an important first step. In addition, the tools may be used to build capacity among key stakeholders for integrating climate change issues (Trærup and Olhoff 2011). Thus, the emphasis on participatory development of risk communication tools and methods created greater stakeholder ownership over the process and is thus more likely to trigger learning and engender capacity for action (bottom-up approach) as compared with risk communication strategies that rely on top-down approaches for information dissemination about climate change (Padgham et al. 2013). Subsequently, there is a need for user friendly climate risk screening tools and their potentials for application that targets stakeholders. This need is amplified by the sheer volume of climate change mainstreaming guidance documents and risk screening and assessment tools available and currently under development (Trærup and Olhoff 2011)

Nevertheless, communicating the risk of SLR is one of the greatest challenges of adaptation planning because climate change associate SLR acceleration is a global concern, but SLR impacts are experienced locally (Covi and Kain 2015) and poses a serious threat particularly to densely populated low-lying coastal cities. On the other hand, SLR science is complex from observing a past SLR trend to future SLR projection, its predicted impact, identifying vulnerable communities, evolving place based urban CBA and mainstreaming adaptation in city planning.

Importantly, the practitioners and various stakeholders in the urban space not only need an understanding of climate science but also the linkages to climate impacts to city systems, such that this information could be logically applied to decision-making processes and could be inbuilt into the development paradigm (APAN 2014). Hence, there is a pressing need to develop a capacity building tool that address the challenges of communicating the risk of SLR science and methods to evolve urban CBA strategies, despite of the availability of a number of other climate change tools. The multidimensional approach of developing a new tool such as COREDAR provides promising opportunities and scope of coastal adaptation policy-planners and decision-makers, urban city planners, research and academia, urban coastal communities and other stakeholders who take a stake on this issue can better understand, prepare and adapt to rising sea-level.

Conclusion

Accelerated SLR is a major long-term outcome of climate change leading to increased inundation of low-lying areas and put global cities and urban coastal communities at greater risk. Building community resilience and adapting to SLR are increasingly a high priority for cities. Capacity building is a key issue for climate adaptation measures in urban areas (Hartmann and Spit 2014). It is one of the most urgent requirements for addressing climate risk, particularly at the local level (IFRC 2009). However, one fundamental condition of adaptation planning in urban contexts appears to be the availability of a good planning capacity, including the knowledge and participation base needed for informing decision making for the urban future (Johnson and Beril 2012). Thus, the ability to act collectively to develop and implement adaptation responses based on sound knowledge of climate science, an intervention of policies and the role of societies are prime factors in building adaptive capacity. Communicating risk of SLR and engaging stakeholders in framing CBA is one such an attempt of capacity building. However, risk communication is now a major bottleneck preventing science from playing an appropriate role in climate policy (Sterman 2008). Capacity building tools are considered as a powerful vehicle to address this challenge and effectively communicate complex SLR science in a simple way and to educate, and engage stakeholders in the decision-making process. In this context, this article seeks to provide insights on communicating risk of SLR and to develop a robust picture of urban CBA through effective decision making that is grounded in pressing community priorities. The article introduced a capacity building tool "COREDAR" to assist users in addressing the challenges of SLR risk communication and urban CBA holistically by systematic step-wise approach that addresses the nexus of SLR science, society and policy. It provides a platform to integrate both top-down and bottom-up approaches to address the challenges and requirements of SLR and urban CBA. Thus, the main finding of this study is introducing a new capacity building tool "COREDAR" for SLR risk communication and urban CBA for coastal cities. One of the main strengths of this framework is, it is relevant to any urban coastal cities across the world and for global audiences (UNAI 2016). However, this tool requires testing and enhancement before it aid for capacity-building and decision-making. Nevertheless, the tool could be used as a foundation to systematically gather information on different perspectives of SLR and holistically synergies SLR information for potential capacity building and effective decision-making such as framing urban CBA to SLR.

Acknowledgements This work was funded by the Department of State, Government of the United States of America and Government of India through the Fulbright-Nehru Postdoctoral Research Program 2015–2016. I place my sincere record of appreciation to Council for International Exchange of Scholars (CIES) and Institute of International Education (IIE), Washington, DC, USA and United States Indian Educational Foundation (USIEF), New Delhi, India. I am indeed grateful to Dr. Robert S. Chen (Director) and Dr. Alex de Sherbinin (Associate Director) of CIESIN, Columbia University; Prof. A. Ramachandran and Prof. K. Palanivelu of Centre for Climate Change and Adaptation Research (CCC&AR), Anna University, Chennai, India for their support. I am very much thankful to other potential partners of COREDAR, Prof. Cynthia E. Rosenzweig (NASA Goddard Institute for Space Studies and UCCCRN); Prof. Radley Horton (Centre for Climate Systems Research, Columbia University); Prof. Adam Paris (Science and Resilience Institute at Jamaica Bay, Brooklyn College); Prof. Laxmi Ramasubramanian (The Institute for Sustainable Cities, Hunter College); Dr. Ebru A. Gencer (Centre for Urban Disaster Risk Reduction and Resilience); Mr. Kytt MacManus, Ms. Elisabeth Sydor, Dr. Malgosia Madajewicz and Dr. Sandra Baptista of Columbia University for their generous support.

Annexure-I

Checklist outline of the COREDAR capacity building tool for SLR risk communication and urban CBA for coastal cities.

Name of the COREDAR Project	
Project Duration Period	
Institution/Organization Name	
Name of the Funding Agency	

Step 1: Understanding the climate of the urban coastal city

(a) About the city

Name of the coastal City	City	District/County	State/Province	Country
Geographical location	Latitude		Longitude	
Total area in ha/sq.km etc				
City Size Classification				
Population	Total population			
	Male		Female	
GNI Classification (Gross National Income)				
Human Index Classification(HDI)				

(b) Administrative structure of the city

Governing body of the City	
Name of the Municipality/Corporation	
Other relevant department involved in the City Governance	

(c) Climate of the city

Average Temperature	
Average Precipitation/Rainfall	
Average Wind Speed	
Average Humidity	
Elevation from Mean-sea level	

Step 2-Documenting past observed sea-level trend	
Name of the tide gauge station near to the city	
Station ID/ Reference Number	
Latitude	
Longitude	
Duration	
Time Span	
Annual mean Sea Level Trend (mm/yr or inches/yr)	
List of recent coastal extreme events that the city experienced in the recent past	1.
	2.
	3.
	4.
	5.

Step 3- Projecting future SLR				
(a) Local SLR Projections in cm/inch/ft :				
	Short-term projections		**Long-term projections**	
Year / IPCC AR5 RCPs	2020	2040	2070	2100
RCP 2.6				
RCP 4.5				
RCP 6.0				
RCP 8.5				
(or) SLR Projections in cm/inch/ft based in IPCC AR5 by 2100:				
(b) To meet the uncertainty: Select scenario of choice				

Decide RCP scenario of interest *(e.g. RCP 4.5)*		
Decide its scale for short-term and long-term projections: *(e.g 0.5m and 1m SLR)*	Scale for short-term projection	Scale for long-term projection

Step 4-Predicting SLR impact						
(a) List and quantify regions of the city that at risk to the predicted impact of SLR:						
			Short-term predictions		Long-term predictions	
S.No	Name of the regions of the urban coastal city	Total area of the regions of the urban coastal city ha/sq.km	Total area of the region are at risk of inundation to the predicted impact of ___ m SLR	Ranking* vulnerable regions to the predicted impact of ___ m SLR	Total area of the region are at risk of inundation to the predicted impact of ___ m SLR	Ranking* vulnerable regions to the predicted impact of ___ m SLR
1.						
2.						
3.						
4.						
			* Ranking vulnerability: 1= High; 2 = Medium; 3= Low			
(b) To meet the uncertainty: Select vulnerable regions of interests and justify the reasons						
Most vulnerable region to short-term prediction of the impact of SLR						
Most vulnerable region to lone-term prediction of the impact of SLR						

Step 5- Identifying vulnerable communities and stakeholders					
Name of the chosen vulnerable region of the urban coastal city to short-term SLR impact	Identifying and prioritizing stakeholders to the predicted impact ___m SLR	Name of the stakeholders and quantity numbers	Name of the chosen vulnerable region of the urban coastal city to long-term SLR impact	Identifying and prioritizing stakeholders to the predicted impact ___m SLR	Name of the stakeholders and quantity numbers
	Key Stakeholders			*Key stakeholders*	
	Primary stakeholders			*Primary stakeholders*	
	Secondary stakeholders			*Secondary stakeholders*	
	Tertiary stakeholders			*Tertiary stakeholders*	

Step 6- Communicating SLR Risk					
Focal Point	Climate Change (SLR) Context	Purpose	Some non-formal climate change (SLR) communication, education and learning strategies (below are only examples)	Checklist of activities done	Short description about the activity
To convey information	Science of SLR and its threats	To disseminate information on SLR risk and to raise awareness	*Information campaigns*		
			News articles, posters, brochures		
			Community radios		
			Exhibits		
			Others		
To build understanding	Impacts and vulnerability of SLR and building sense of urgency	To exchange ideas and provide dialogue, to build a sense of place, to clarify and enhance understanding of information and issues and to generate concern	*Vulnerability Mapping*		
			Issue investigation		
			Focus group interviews		
			Citizen science program		
			Structured surveys		
			Others		

Step7-Framing urban CBA to SLR						
Urban CBA in the locality within the region are at risk to the predicted impact of SLR		Adaptation options to SLR	Description	Benefits	Constraints	Specific Relevance
	Short-term adaptation strategies					
	Long-term adaptation strategies					

Step 8- Mainstreaming urban CBA strategies		
List of existing national/state/city coastal policies (with an emphasize on communities)	Checklist for possible interventions	List of prioritized urban CBA strategies to SLR
1.		1.
2.		2.
3.		3.
4.		4.
5.		5.
6.		6.
7.		7.

References

APAN (2014) Asia Pacific climate change adaptation forum, Kula Lampur, Malaysia. Asia Pacific adaptation network (APAN). Online http://www.asiapacificadapt.net/adaptationforum/2014/about%20. Accessed 25 Apr 2016

ARCCA (2014) Regional adaptation collaborative toolkit, California, USA. Online http://www.arccacalifornia.org/toolkit/. Accessed 25 Apr 2016

Archer D, Dodman D (2015) Making capacity building critical: power and justice in building urban climate resilience in Indonesia and Thailand. Urban Clim 14:68–78

Barclay P, Bastoni C, Senhauer D, Hassan M, Lopez M, Mekias L, Ramachandran S, Stock R (2013) Climate change adaptation in Dayton, OH. Great lakes adaptation assessments for cities, Dayton, Ohio, USA. Online http://graham.umich.edu/media/files/Climate%20Change%20Adaptation%20in%20Dayton%20OH.pdf. Accessed 25 Apr 2016

Barisky T (2015) A public engagement toolkit for sea-level rise. Prepared for the Sustainability Group, City of Vancouver. Vancouver, British Columbia, Canada, p 72

Boholm A (2009) Speaking of risk: matters of context. Environ Commun 3(3):335–354

Bulkeley H, Schroeder H, Janda K, Zhao J, Armstrong A, Yi Chu S (2009) Cities and climate change: the role of institutions, governance and urban planning. Report prepared for the World Bank urban symposium on climate change

CAP (2007) Cities preparing for climate change: a study of six urban regions. Clean air partnership Toronto. Online http://www.cleanairpartnership.org/pdf/cities_climate_change.pdf. Accessed 25 Apr 2016

CARE (2009) Climate vulnerability and capacity analysis. CARE International. Online http://www.careclimatechange.org/files/adaptation/CARE_CVCAHandbook.pdf. Accessed 25 Apr 2016

Cheong SM (2010) Guest editorial on coastal adaptation. Clim Change. doi:10.1007/s10584-010-9999-y

Covi MP, Kain DJ (2015) Sea-level rise risk communication: public understanding, risk perception and attitudes about information. Environ Commun. Online http://dx.doi.org/10.1080/17524032.2015.1056541. Accessed 25 Apr 2016

Dodman D, Mitlin D (2013) Challenges for community-based adaptation: discovering the potential for transformation. J Int Dev 25(5):640–659

Eco Adapt (2010) Climate adaptation knowledge exchange—CAKE. Online http://www.cakex.org/. Accessed 25 Apr 2016

Fischhoff B (1995) Risk perception and communication unplugged: twenty years of process. Risk Anal 15(2):137–145

Fuchs R, Conran M, Louis E (2011) Climate change and Asia's coastal urban cities can they meet the challenge. Environ Urban Asia 2(1):13–28

Georgetown Climate Centre (2011) Adaptation tool kit: sea-level rise and coastal land use. Georgetown, Washington DC, USA. Online http://www.georgetownclimate.org/files/Adaptation_Tool_Kit_SLR.pdf. Accessed 25 Apr 2016

Goodrich R (2013) SWOT analysis: examples, templates and definition. Business News Daily. Online http://www.businessnewsdaily.com/4245-swot-analysis.html. Accessed 25 Apr 2016

Grannis J (2011) Adaptation tool kit: sea-level rise and coastal land use how governments can use land-use practices to adapt to sea-level rise. Georgetown Climate Center, Georgetown, Washington, DC. Online http://research.fit.edu/sealevelriselibrary/documents/doc_mgr/456/SLR_Adaptation_Tool_Kit_-_Grannis_2011.pdf. Accessed 25 Apr 2016

Hartmann T, Spit T (2014) Capacity building for the integration of climate adaptation into urban planning processes: The Dutch experience. Am J Clim Change 3:245–252

Harvatt J, Petts J, Chilvers J (2011) Understanding householder responses to natural hazards: flooding and sea-level rise comparisons. J Risk Res 14(1):63–83

Huq S (2008) Community based adaptation. Tiempo 86(7)

IDS (2011) Opportunities and risks for climate change and disasters (ORCHID). Institute of Developmental Studies, Brighton, United Kingdom. Online http://www.ids.ac.uk/climatechange/orchid. Accessed 25 Apr 2016

IFRC (2009) Climate change adaptation strategies for local impact key messages for UNFCCC negotiators. Prepared by the, technical paper for the Inter Agency Steering Committee. Task force on climate change, international federation of red cross and red crescent societies, p 11. Online http://unfccc.int/resource/docs/2009/smsn/igo/054.pdf. Accessed 25 Apr 2016

IISD (2011) CRiSTAL (community-based risk screening tool—adaptation and livelihoods) International Institute for Sustainable Development. Online https://www.iisd.org/cristaltool/. Accessed 25 Apr 2016

IPCC (2013) Summary for policymakers. In: Stocker TF, Qin D, Plattner GK, Tignor M, Allen SK, Boschung J, Nauels A, Xia Y, Bex V, Midgley PM (eds) Climate change 2013: the physical science basis. Contribution of Working Group I to the fifth assessment report of the intergovernmental panel on climate change, Cambridge University Press, Cambridge, United Kingdom and New York, NY, USA

Johnson K, Beril M (2012) Conceptualizing urban adaptation to climate change-findings from an applied adaptation assessment framework. SSRN Electron J. doi:10.2139/ssrn.2077476

Khan AS (2013) Climate change induces sea-level rise and its impacts on the Tamil Nadu coast, India: framing ecosystem and community—based adaptation strategies. Ph.D Thesis. Anna University, Chnenai, Tamil Nadu, India, p 226

Khan AS, Ramachandran A, Usha N, Arul Aram I, Selvam V (2012) Rising sea and threatened mangroves: a case study on stakeholders' engagement in climate change communication and non-formal education. Int J Sustain Dev World Ecol 19(4):330–338

Khan AS, Chen RS, De Sherbinin A, Ramachandran A, Palanivelu K (2015) COREDAR framework-communicating risk of sea-level rise and engaging stakeholders in framing community-based adaptation strategies. American Geophysical Union (AGU), fall meeting, 16 Dec 2015, San Francisco, California, USA. Online https://agu.confex.com/agu/fm15/meetingapp.cgi/Paper/78074. Accessed 25 Apr 2016

McGranahan G, Balk D, Anderson B (2007) The rising tide: assessing the risks of climate change and human settlements in low elevation coastal zones. Environ Urban 19(1):17–37

MRC (2010) Review of climate change adaptation methods and tools. MRC technical paper 34. Mekong River Commission, Office of the Secretariat in Vientiane (OSV), Vientiane, Lao PDR. Online http://www.mrcmekong.org/assets/Publications/technical/Tech-No34-Review-of-climate-change.pdf. Accessed 25 Apr 2016

NIDM (2013) Preparing long-term training and capacity building strategy for disaster risk reduction in India, under NCRMP. National Institute of Disaster Management, India, p 209. Online http://nidm.gov.in/pdf/ncrmp/Deliverable%205.pdf. Accessed 25 Apr 2016

OECD (2008). "Competitive cities and climate change. In: Proceedings of conference on the organization for economic co-operation and development, 9–10 Oct 2008, Milan, Italy

Padgham J, Devisscher T, Chuluun T, Mtilatila L, Kaimila E, Mansingh I, Agyemang-Yeboah F, Obeng FK (2013) Building shared understanding and capacity for action: insights on climate risk communication from India, Ghana, Malawi, and Mongolia. Int J Commun, 7:970–983. Online http://ijoc.org/index.php/ijoc/article/viewFile/1475/899. Accessed 25 Apr 2016

Pidgeon N, Fischhoff B (2011) The role of social and decision sciences in communicating uncertain climate risks. Nat Clim Change 1:35–41

Reid H, Alam M, Berger R, Cannon T, Huq S, Milligan A (2009) Community-based adaptation to climate change: an overview. In: Theme section: community-based adaptation to climate change, participatory learning and action 60, International Institute for Environment and Development, Russell Press, Nottingham, United Kingdom, p 210. Online http://pubs.iied.org/pdfs/14573IIED.pdf. Accessed 25 Apr 2016

Romero Lankao P, Dodman D (2011) Cities in transition: transforming urban centers from hotbeds of GHG emissions and vulnerability to seedbeds of sustainability and resilience. Curr Opin Environ Sustain 3(3):113–120

Sales RFM (2009) Vulnerability and adaptation of coastal communities to climate variability and sea-level rise: their implications for integrated coastal management in Cavite City, Philippines. Ocean Coast Manag 52(7):395–404

Sanchez-Rodriguez R (2009) Learning to adapt to climate change in urban areas. A review of recent contribution. Curr Opin Environ Sustain 1(2):201–206

State of the Planet (2016) Helping communities respond to climate change. Earth Institute, Columbia University, New York, USA. Online http://blogs.ei.columbia.edu/2016/02/23/new-communications-tool-to-help-communities-respond-to-climate-change/. Accessed 25 Apr 2016

Sterman J (2008) Risk communication on climate: mental models and mass balance. Science 322:532–533

Tanner T, Mitchell E, Polack B, Guenther (2009) Urban governance for adaptation: assessing climate change resilience in ten Asian cities. Institute of Development Studies, University of Sussex, Brighton, UK. Online http://www.preventionweb.net/files/7849_Wp31520web1.pdf. Accessed 25 Apr 2016

Titus JG, Richman C (2001) Maps of lands vulnerable to sea level rise: modeled elevations along the US Atlantic and Gulf coasts. Clim Res 18(3):205–228

Trærup SLM, Olhoff A (2011) Climate Risk screening tools and their application: a guide to the guidance. UNEP Risoe Centre, p 49. Online http://orbit.dtu.dk/files/6238714/Climate%20Risk%20Screening%20Tools.pdf. Accessed 25 Apr 2016

UNAI (2016) New tool helps communities better understand climate change. United Nations Academic Impact, United Nations Headquarters, New York, USA. Online https://academicimpact.un.org/content/new-tool-helps-communities-better-understand-climate-change. Accessed 25 Apr 2016

UNDP (2010) Designing climate change adaptation initiatives: a UNDP toolkit for practitioner. UNDP Bureau for Development Policy, Environment and Energy Group, New York, USA. Online https://sustainabledevelopment.un.org/content/documents/951013_Toolkit%20for%20Designing%20Climate%20Change%20Adaptation%20Initiatives.pdf. Accessed 25 Apr 2016

UNFCCC (2007) Climate change: impacts, vulnerabilities and adaptation in developing countries. United Nations Framework Convention on Climate Change Secretariat, Bonn

UNISDR (2009) Terminology on disaster risk reduction. United Nations international strategy for disaster reduction (UNISDR). Online http://www.unisdr.org/eng/terminology/terminology-2009-eng.html. Accessed 25 Apr 2016
USAID (2009) Adapting to coastal climate change: a guidebook for development planners. U.S. Agency for International Development, Washington DC 162
USIEF (2015) COREDAR Framework-Communicating risk of sea-level rise and engaging stakeholders in framing community-based adaptation strategies. New-Delhi, India
Wilson E (2006) Adapting to Climate change at the local level: the spatial planning response. Local Environ Int J Justice Sustain 11(6):609–625
Woodworth PL, Player R (2003) The permanent service for mean sea level: an update to the 21st century. J Coastal Res 19(2):287–295

Part III
Conflicts and Synergies: Adaptation, Resilience and Multi-hazard Mitigation

Chapter 10
A Region Under Threat? Climate Change Impacts, Institutional Change and Response of Local Communities in Coastal Yucatán

Lysann Schneider and Tobias Haller

Abstract The Yucatán peninsula is the region in Mexico most at risk in the face of climate change, and it is a focal point of sociopolitical and institutional changes that have led to serious environmental degradation. The northern part is particularly affected by irregular rainfall, droughts, and tropical storms, as well as institutional changes (mainly in private property land rights) leading to different uses of forests that are undermining reforestation efforts. The area also experiences catastrophic wildfires that often follow tree-toppling hurricanes exacerbated by climate change. These conditions affect the livelihoods of local people, forcing them to adjust. In addition, inequality in development between the wealthy Caribbean tourist coast and its rural hinterland hinder the sustainability and improvement of livelihoods inland, discouraging rural people from investing in their land and leading them to migrate for work. However, a stable population and long-term economic investment in forests and forestry institutions could be the basis for a more climate-resilient landscape. To understand these processes, this chapter focuses on historical and institutional transformations of economic exploitation, social and political marginalization, and specific adaptation strategies like circular labor migration of the local population in northeastern Yucatán. This institutional history has made the population and the landscape more vulnerable to climate events and climate change, and has made local collective action for reforestation more difficult. The only major area that remains forested is communally owned. The chapter discusses the implications of the commons for sustainable development and increased resilience under conditions of peripheralization and climate change.

The original version of this chapter was revised: Author name has been corrected. The erratum to this chapter is available at https://doi.org/10.1007/978-3-319-53742-9_22

L. Schneider (✉)
Centre for Development and Environment, Institute of Social Anthropology, University of Bern, Bern, Switzerland
e-mail: schneiderlysann@gmail.com

T. Haller
Institute of Social Anthropology, University of Bern, Bern, Switzerland
e-mail: haller@anthro.unibe.ch

W. Leal Filho and J.M. Keenan (eds.), *Climate Change Adaptation in North America*, Climate Change Management,
DOI 10.1007/978-3-319-53742-9_10

Keywords Climate change adaptation · Institutional change · Reforestation · Commons · Mexico

Introduction

Mexico is particularly vulnerable to damage in the face of climate variability and change because of its geographical location, topography, and socioeconomic conditions. Future scenarios developed in 2006 (see Regional Climate Change Index of Giorgi) show that it is a hotspot for climate change, in particular due to a reduction in precipitation and increase in precipitation variability. Karmalkar et al. (2011), using regional climate models, showed that projected warming for Mexico and Central America is greater than the global temperature increase. The Yucatán Peninsula experiences the highest warming in both the wet and dry seasons (Karmalkar et al. 2011: 626). The already noticeable weather and climate anomalies as well as future warming and precipitation variability may lead to a loss of biodiversity and water resources and to stress on socioeconomic sectors such as subsistence farming (Karmalkar et al. 2011: 627).

The likelihood that most of Mexico, especially Yucatán, will experience drier conditions is a great threat, given that a third of Mexico's population depends on agriculture and forestry and a quarter of the population lives in rural areas (Ballesteros-Barrera et al. 2011: 3). Because crops are mainly grown and mature under rain-fed and seasonal conditions, they are particularly sensitive to extreme climatic variations such as droughts and floods. Especially for the Yucatán Peninsula, a disappearance of the right conditions for cultivation and a decrease in agricultural production are predicted (Ballesteros-Barrera et al. 2011: 8). Climate change researchers in Mexico have also confirmed the high vulnerability of the Yucatán Peninsula to climate change and have criticized the lack of effective climate change actions by the government, such as protection of mangrove forests (Mexico Daily 2015), which would protect coastlines from storms.

Research and analysis of climate change–related impacts often fail to take into account their social-anthropological, institutional, and political-ecological aspects. Climate change is thus perceived as having fallen from the sky, with no attention paid to historical transformations that made populations vulnerable to it when it began to occur (see Ribot 2010; Haller 2015). This chapter explains why local private owners of land are not interested in investing in collective strategies to save, restore, and expand forests, which are so crucial to buffering people from climate-related stressors such as heavy rainfall, drought, and storms that uproot trees and increase fire risks. The chapter shows that despite the government's provision of economic incentives to extend forest cover, deforestation continues.

We argue that this climate insecurity does not result from the "tragedy of the commons" so frequently described in the literature. Land and related forest resources were in the hands of a timber investor who had a sawmill and employed people in the area. After timber production was exhausted, the land was redistributed to local peasants in small parcels as part of a pro-peasant land reform program. However, these private plots are too small for investment in forests to be practical—given the insecure conditions in agriculture due to changing climate conditions—despite government funding. The only forest cover that remains is held in common. It could serve as a model of coping with climate change and with tenure insecurity of size, as transactions costs can be reduced for actors and as resilience can be established. It can also form the basis for restoring Maya agricultural systems, which would provide climate- and market-specific benefits and could help reduce problems related to out-migration.

The ancient Maya had a highly efficient, productive, and sustainable system of resource management. Humans have influenced this ecosystem for many millennia (Ford and Nigh 2009: 214). Ford and Nigh (2009: pp. 216) suggest that this highly managed, anthropogenic landscape evolved as an adaptation of the Maya to a changing climate during the preceding millennia. One of the features of this landscape, called Maya forest gardens, was that they included non-timber forest resources (e.g. food, raw materials, and animals) as a conservation strategy. The complex *milpa* swidden cycle was the axis of this agroforestry system. Researchers have suggested that the remnants of this ancient system are still practiced (e.g. Gómez-Pompa 1987; Ford and Nigh 2009; Diemont et al. 2011). However, the area under study has undergone major changes in land and related common-pool resource management since colonial times, including the transformation from common to state ownership and then back to common ownership in the *ejido* system (discussed in more detail below) and to today's private property regime.

Moreover, as prices for timber and beef were relatively high in the early 20th century, deforestation and cattle ranching have threatened the viability of this cultural landscape and tropical ecosystem, which would be much more resilient to fires, storms, and soil degradation than the current land use system. This in turn is also related to increasing damage caused in the region by environmental disasters such as fires, hurricanes, and long droughts. It would be appropriate for the government to invest more in the rural hinterland of the Yucatán. Statements by inhabitants and research studies have shown that when catastrophic events such as hurricanes have hit the research area, state aid has flowed quickly and efficiently to the touristic region of the Riviera Maya, while residents in the hinterland have waited a long time for support (woman, Colonia Yucatán, March 2013; Alscher 2008). This trend reflects, and likely exacerbates, the inequality of economic development in the region. It also demonstrates that environmental problems exist that are caused by climate change, and clearly shows the relationship between institutional change and climate change.

The New Institutionalism as a Conceptual Framework for Climate Change Studies

The chapter makes use of the center–periphery relationship in the study region to apply the "new institutionalism" approach (see North 1990; Ostrom 1990) from a social anthropology perspective (Ensminger 1992; Haller 2010). This approach can explain historical environmental changes and current regional development, both in the hinterland and on the Caribbean coast. It provides an analysis of background issues like changes in ecological, economic, and political conditions, which can be linked to climate change issues and climate change adaptations by the local population.

Institutions are central for understanding human and social action and economic, social, and cultural exchange. Institutions are understood here as human-made and widely shared rules that help people coordinate interactions with others. They include formal (written) rules, such as laws, regulations, and policies, as well as informal rules such as habits or norms. Effective institutions reduce transaction costs (for gathering information on other actors and situations, monitoring, and sanctioning) and thus increase the predictability of behavior. In this way, institutions can provide a basis for the sustainable use and fair distribution of resources, which for example would hinder overuse via selfish and exploiting behavior, such as described in discussions of the tragedy of the commons (Ostrom 1990; Haller 2007, 2010).

Ostrom (1990) in *Governing the Commons* proposed eight general design principles for an efficient and robust reduction of transaction costs by institutions to manage common-pool resources like forests or fisheries. Cox et al. (2010) analyzed these principles, supported most of them, and reformulated only a few. The principles explain what conditions are necessary to build trust and reciprocity, both of which are needed for collective action, which in turn is needed for good management of common-pool resources (Cox et al. 2010). Pagdee et al. (2006) conducted a worldwide metastudy to identify factors that influence the success of community forest management and defined three principles for robust self-governance and consequently for successful community action: well-defined property rights, effective institutional arrangements, and community interests and incentives (Pagdee et al. 2006: 49).

Our approach differs from Ostrom's in that we address the issue of power, as institutions are often formed or changed by powerful actors. This in turn has implications for local settings and affects the bargaining power of actors, their forms of organization, their ideologies, and the way they choose, transform, and justify formal and informal institutions. This raises questions of how formal and informal institutions affect decision-making by resource users, who often are confronted with trade-offs between economic and ecological and other goals, and how these institutions change. According to the model proposed by Ensminger (1992, modified in Haller 2010, 2013), changes can be explained in terms of endogenous and exogenous factors. Exogenous factors that encourage outmigration include new infrastructure that enables migration as well as new job opportunities (in this case, mostly on the Riviera Maya). Exogenous changes lead to local price changes, for example by opening up the market and the connection to the world economic system, but also to an increase in outsiders' interest in buying local land. This in turn leads to further migration of local people and changes in their constellation and their need to find work in other regions. An essential endogenous factor is the alteration of land rights through agrarian reform.

Local actors rarely influence external factors such as the political and economic environment, population dynamics (in and out-migration), and technological change (for example, road networks and logging). More importantly, external institutional and policy factors shape what are called relative prices for goods and services. External factors can change the relative value that an area is given, and this has influences on the internal level. Here, power relation issues are important, because this changes the power structures and the bargaining power of different users such as local, state, and commercially oriented actors as well as the interests of internal actors. The way that institutions are chosen depends on the interests and bargaining power of actors, the way they organize, and the way they are able to legitimate the institutional choice to be made. These transformations lead to the issue of who has the power and the interest to devise institutions.

Thus we argue that climate change occurs in settings that have already been institutionally transformed—for example, by changes in property and use rights that exclude previous users, thus making them even more vulnerable to changing climate conditions by depriving them of access to resources that could have enabled them to buffer shocks or crisis. Loss or change of collective choice arrangements can have a similar effect. This is why we adopt the concepts of political ecology and new institutionalism presented by Ribot (2010) and Haller (2015), which retrace historical institutional changes in order to understand how local actors became vulnerable and how they try to adapt to climate change today. This mixed approach can explain how institutional change affecting the governance of resources leads to a situation in which the change from common to private property does not support badly needed reforestation on private lands, while the small portion of forest common property remains.

Study Area and Research Methodology

Physical Geographic Characteristics

Research for this chapter took place on the Mexican peninsula of Yucatán, which includes the states of Yucatán, Quintana Roo, and Campeche, bordering the Gulf of Mexico to the west and north and the Caribbean Sea to the east. Data were gathered mainly in the community of Colonia Yucatán and its environs but also along the Caribbean coast. Colonia Yucatán is located in the northeastern part of the state of Yucatán, while the Riviera Maya lies in the state of Quintana Roo.

The Yucatán Peninsula is one of the most prominent climate change hotspots in the tropics (Giorgi 2006: 1). It lies in two Köppen-Geiger climate zones: tropical wet and dry over most of the peninsula, with a pronounced dry season in winter and rain during the summer months, and hot semiarid in the north, with hot summers, mild to warm winters, and a longer dry period. The peninsula lies within the Intertropical Convergence Zone, which determines the precipitation pattern. In the humid tropical areas, it rains throughout the year, while in the drier parts, rain is relatively rare from November to April; March and April are characterized by particularly high temperatures and low rainfall. About 80% of the annual precipitation occurs in the rainy season, from May to October. The northeastern trade winds are dominant, and between November and February cold winds from the north (*Nortes*) often lead to temperature drops to below 10 °C for about three days.

The peninsula is located in the Atlantic hurricane belt, rendering its flat terrain vulnerable to hurricanes and tropical storms coming from the southeast. The east and southeast areas are especially frequently exposed to heavy rain and flooding during the hurricane season (August to November). The strong tropical storms that develop into hurricanes in the Caribbean Sea can cause widespread destruction of forests, fields, and houses. This has an important impact on the study area, which is situated near the northeastern coast of Yucatán, within the tropical wet and dry climate zone but close to the hot semiarid zone, making climate conditions more extreme and unpredictable.

Wildfires are a frequent danger in this region, made more severe by the stormy climate, and both fires and hurricanes may become more intense in the future (Myers and van Lear 1998). Disastrous fires are mostly caused by traditional swidden farming methods, which include burning the fields. The consequences of large fires are the loss of peasants' belongings and more difficult regeneration of the natural vegetation.

The predominant natural vegetation types of Yucatán are short and tall tropical jungle plants. Like the climate zones, the vegetation zones are in transition in the study area, from evergreen forest in the east to semideciduous forest in the west. The natural vegetation in the dryer climate consists of tropical dry forest, which can reach a height of 18–25 m. In the mid 20th century, precious woods (such as cedar and mahogany) and hardwoods dominated in that area (Villarino 1984: 8), but these are now hard to find. Many useful tree species, such as fruit and breadnut trees, can be found in the remaining forest areas, which were part of the Maya forest ecosystem and were once cultivated by them.

History of Livelihoods and Sociopolitical Structures

The study focused on the history of Colonia Yucatán, which is not an indigenous community, during the first half of the 20th century, when immigration and local settlements played a major role in the development of northeastern Yucatán. Colonia Yucatán was founded in the late 1930s in the heart of the largest and last continuous forest on the northern peninsula to serve the emerging timber industry. However, to understand the changes which took place then, it is necessary to go back to the precolonial era, during which local forests were not completely wild but a cultural landscape of forest ecosystems created by the Maya people. They used the forest in an extensive way via hunting and gathering as well as some horticulture. Institutionally, it can be assumed that the ecosystem and its resources were held and accessed under a local common property tenure system, as it was influenced by the Maya economic system based on local community production and access to common-pool resources (Quezada 2010; Bracamonte y Sosa 2003).

Extensive use of the forest landscape ecosystem was partially maintained during the 300 years of Spanish colonization. The major institutional change in resource management in this area came in the mid 20th century when rising prices for forest products attracted commercial interest, mainly in hardwood and precious wood for export in log form to the United States, Cuba, and Puerto Rico (Ruiz Silva 2011: 62), followed by privatization as the institutional choice of the actors with the most bargaining power. They transformed the landscape via commodification of the forests and established a timber business with a sawmill and wood-products factory attracting wage labor. Additionally, in the late 1970s the agrarian reform—a political decision taken after the Mexican revolution in 1917 (Pérez Castañeda 2002: 8)—was implemented in this area and greatly reduced the remaining forest (Secretaria de la Reforma Agraria 1978: 46; Villarino 1984: 8; Ruiz Silva 2011: 144). Thus, it became an environment with only a sparse distribution of trees, which has remained until today. This has implications for local peoples' livelihoods and resilience in the context of climate change.

Colonia Yucatán at the time of its establishment was a center around which small locations or camps were strategically set up, especially on the routes used to transport logs, supplies for human needs, and construction materials. Most of these sites later grew into small communities. Job seekers, above all Maya from the Yucatán Peninsula, immigrated in the newly founded colony (Ruiz Silva 2013). Residents obtained their income mainly through labor in the timber industry and in addition they received subsidized products, which were provided to them by the factory (statements by various interlocutors, Colonia Yucatán, 2013 and 2014; Villarino 1984: 14). The extraction of raw materials had a great impact on the environment. Tropical tree species from the forest, which had formed part of the precolonial landscape ecosystem and a common-pool resource, became marketable commodities. With the rise of the timber industry, not only the environment but also local socioeconomic and cultural conditions were greatly transformed. Today the majority of the inhabitants of Colonia Yucatán are mestizos descended from Yucatec Maya, while in the surrounding communities, Yucatec Maya are the predominant ethnic group.

Methodology

Qualitative research embedded in a mixed-methods approach was necessary to explore deeper cultural and social structures, their processes and problems, and their visible structures. It was also necessary for understanding institutions, behavior, and responses to changes and problems, because neither institutions nor behavior are independent of their context. Qualitative ethnographic data were necessary to work out the emic perception and valuation concepts. Because qualitative data had limitations for verifying the hypothesis, they were combined with quantitative data, such as statistics on fires and tropical storms. The authors prefer to use an interdisciplinary approach combining social and cultural anthropology with natural sciences, especially with regard to climate data such as precipitation trends.

This chapter relies on qualitative data generated and statistical data obtained by L. Schneider during the field research, using ethnographical research and empirical anthropological methods in the Mexican states of Yucatán and Quintana Roo. Field research took place in 2013 and 2014 for periods of four and five months. The main methods applied for data generation were continuous participant observation and qualitative interviews (open, guided, and problem-oriented interviews and thematic biographies).

To gain greater insight into adaptation strategies like migration and into emic perceptions, L. Schneider conducted an ethnographic study in several places and compared and contrasted the results. She conducted interviews with three groups of rural people connected to the hinterland who are affected by climate and environmental change: those who live in the hinterland, those who commute between the hinterland and the coastal Riviera Maya or the Yucatán capital, Mérida, and emigrants who have left the rural area completely. She conducted expert interviews with members of governmental organizations, including the National Forest Commission and the Secretariat of Agriculture, Livestock, Rural Development, Fisheries and Food, as well as with nongovernmental organizations such as Pronatura Yucatán and national institutions such as the Autonomous University of Yucatán.

Some of the data used in this chapter stem from interviews, group discussions, and participation in everyday activities, during which audio and video recordings and photos were taken. The sources of statistical data, legal documents, reports, and maps were mainly official institutions, universities, ministries, and administrative organizations at the state and national level in Mexico, including the National Forest Commission, National Institute of Statistics and Geography, and Ministry of the Environment and Natural Resources.

The study was conducted to generate findings that would be useful for national and international decision-making organizations, nongovernmental organizations, and development organizations, in matters concerning local climate change adaptation and sustainable regional development.

Results

This section discusses in more detail the geographical and historical context of the study area. It describes a cultural landscape from precolonial times in which common-pool resources were managed by common-property institutions, and describes local strategies from the new-institutionalist and political-ecology perspectives.

History of Institutional Change in Land Use and Property Rights

In pre-Hispanic times, the research areas were part of the Maya chiefdom Chikinchel, which was located on the north coast of Yucatán (Roys 1957: 103; Góngora Salas 2001: 17). Beside fishing and the production and trading of salt and other trade goods, forests served as an important livelihood source. Chikinchel is described mostly as a wooded area (Roys 1957: 103). The forests were communal, but we do not know exactly how land and forest use were governed (Quezada 1998: 476). The Maya had a different valuation of forest use and management. In their conception of the universe, the land was the giver of life, growing the forests and providing sustenance for animals and humans. Land was not likely to be privatized, much less an object of trade or sale. The governing lineages did not control access to the forests; this was the prerogative of the deity Yumbalam (Quezada 2010: 26). In the political administrative system of the ancient Maya, zones were defined in order to maintain control over the land, but more from a social than a territorial perspective (Quezada 2010).

This kind of organization was linked to the land, resource use, and resource management, and therefore it was also closely connected with the settlement structures. As it was an extensive resource management system, the settlement patterns were dispersed in order to be close to the productive areas (Ford and Nigh 2009). The milpa farmers, who belonged to the lower class of society, had to supply the elite with tributes in the form of food products and thus served as a labor force. They also served in wartime as members of the army (Bracamonte y Sosa 2003: pp 18).

In some regions of the peninsula, the Maya still practice agroforestry, though somewhat modified from the ancient Maya practice. The forest gardens are managed in a rotation system and require careful human management. The land use cycle consists of closed-canopy forest, followed by open fields dominated by annual crops (the milpa), fruit tree gardens, and again closed-canopy forest. The gardens include vegetables, fruit trees, herbs, and medicines. In addition, they provide areas for raising domesticated animals such as pigs, turkeys, and ducks, as well as bee colonies for honey production (Flores-Delgadillo et al. 2011). The milpa is the most labor-intensive part and produces vegetables, tubers, herbs, and plants that repel insects and promote soil formation and nutrients (man, Colonia Yucatán, March 2014). Through additional planting of certain tree species during the incipient phases of the milpa, the future secondary forest is shaped to the needs of the cultivator.

During the conquest of Yucatán in the first half of the 16th century, the ownership and control of land and forests were not defined. Through the process of conquest, the population was substantially diminished. The Maya were moved by the missionaries several times and their numbers continuously dwindled (Roys 1957: 105), in part because of diseases brought by the Europeans to which the natives were not resistant. During its long colonial era, between 1535 and 1821, the Yucatán was legally considered to be the property of the Spanish Kingdom, but the de facto use of the forest and land is unclear. Spain had relatively little interest in this area, and its population density remained low. Throughout the colonial period, the demographic and economic center was the northwestern part of the peninsula (Moßbrucker 1994: 7). The Spaniards only had control of that area, while the remaining territory was hardly accessible to them. Thus, it served partially as a refuge for the Maya (Moßbrucker 1994: 7), and it is likely that they used the forests for hunting and gathering, among other things.

After Mexico's independence from Spain in 1821, the area remained under state control as territory of the new and often form-changing Mexican state. Northeastern Yucatán remained sparsely populated. There was no uniform regime for the area as a whole or for the use of its forests. It is likely that they were used by small farmers and hunter-gatherers. Between 1847 and 1901, the Caste War of Yucatán took place. The exact impacts on the research area are unknown, but generally there were large population losses in the east and south of the peninsula and the area remained unpacified until the late 19th century (Moßbrucker 1994: 9). The eastern part of the peninsula remained dominated by the rebellious Maya, who did not resort to subsistence farming. Rather, they traded with the English, who were encroaching over Belize to Quintana Roo, and who delivered, among other things, weapons and munitions in exchange for precious wood, honey, and other foods (Moßbrucker 1994: 9).

In the first third of the 20th century, the Mexican Revolution took place, a political and social upheaval whose main achievement was land reform, which was implemented at different times and in different ways depending on the region. The state enforced the redistribution of land that had been part of large estates (*haciendas*) to smallholders, leading to the establishment of *ejidos* (communal land) (Moßbrucker 1994: 16). The development of the study area remains unclear, but the land around Colonia Yucatán must have belonged to a private owner, as a timber factory installed in the late 1930s received the land title from a bank that must have received the land from a former owner (Benjamin 1951: 22). According to older inhabitants of Colonia Yucatán, the land was uninhabited before the factory and the surrounding settlements were built. The common-pool resources of the forests, such as rubber, were used by *chicleros* (rubber gatherers) (man, Colonia Yucatán, April 2014). From an institutional perspective, it is of central importance that private property led to the commercial use of timber throughout the forest after that. The timber industry began to decline in the 1970s.

Balancing Between Institutional and Climate Change: Adaptation Strategies on the Periphery

To analyze local processes of adaptation, we begin after the decline of the timber companies in Colonia Yucatán in the mid 1970s and the start of an in- and out-migration process. The majority of the now unemployed residents who worked in the timber factory emigrated—some to Yucatán's capital, Merida, most to Cancún (Ruiz Silva 2013: 95). In 1978, the agrarian reform was implemented in Colonia Yucatán and the land was divided into parcels given to the inhabitants as private land. In addition, a small common land was established (Secretaria de la Reforma Agraria 1978). This led to the immigration of Yucatec peasants from rural regions, who cut down the remaining forests to cultivate the land (Ruiz Silva 2011). Thus, the economic regime changed from timber extraction to agriculture and livestock raising. These changes in the economic and ecological conditions ultimately resulted in social deterioration and in emigration and circular labor migration by local people (woman, Colonia Yucatán, May 2014).

Many factory workers who emigrated kept their land in the area. Until 1992, they were not allowed to sell it, as the owner of the land remained the Mexican state. After the law of 1992 allowed them to sell the private land, some of them still kept it as a risk-reduction strategy for the market or for their own security (man, Colonia Yucatán, February 2014). Economically speaking, the land is not of high value as private property, as people work mostly in the tourism industry on the Caribbean coast and prices for land in the hinterlands and for agrarian products are relatively low.

Migration, especially circular labor migration, is ongoing in the study area and can be seen as an adaptation strategy, used in particular by the younger generation. In the population that remains in Colonia Yucatán, there is a surplus of young women. This can probably be explained by the fact that young men are emigrating to work in construction or in tourism-related jobs on the coast, while women and children are left behind in the countryside. For the older generation, men are in the majority, and most of are still farming, an activity that is no longer attractive for the younger generation. Approximately three-quarters of the residents of Colonia Yucatán belong to the working population; however, just over half are economically active (Centro de Salud de Colonia Yucatán 2013: 20). Agriculture including livestock production is the main source of income in this area. Closer to the north coast, fishing is the main income source.

Large-scale agriculture and high crop productivity are generated almost exclusively by people from other Mexican states, for example Michoacán. They are able to buy land in Yucatán cheaply and grow products like fruits and vegetables for export. Other residents in the region are shop employees and merchants or farm laborers. Many people have kept their private land for security, in case they, for instance, lose their jobs and have to live off the sale of the land or from cultivating it. Some of them have even considered partially or completely reforesting their property (various interlocutors, Colonia Yucatán, 2013 and 2014), although the small size of the parcels usually makes this impractical. As climate change has become more evident, for several years the state has provided incentives for forestry. In the study area, this is not happening for institutional reasons: The plots are too small, and collective-action options are not supported. A proposal by the community to combine parcels in order to obtain the support of state forest programs was rejected by the Forestry Commission (man, Colonia Yucatán, May 2014).

Another hurdle for successful afforestation is the lack of water. Because the region has long dry periods and climate fluctuations, tree planting without irrigation (which many farmers cannot afford) is not possible. State support for provision of irrigation systems would be beneficial. This is particularly important in the north of Yucatan with its limited surface water.

The lack of incentives for setting up a new institutional framework is affected by various impacts of climate change. Failing rains, droughts, and fires add substantially to the reluctance to reforest the area, as the danger that one's own small forest will dry up and be burned is too high. Many plots are badly maintained and burn easily, or have been damaged by storms. The main damage after a fire is crop loss. For many peasants this means a loss of livelihood, as agriculture is the basis of their survival. In an area where reforestation would be ecologically valuable, existing forested and reforested areas continue to be damaged or destroyed. Allowing the natural regeneration of vegetation, in particular the forest, is hardly possible, and the environmental degradation progresses. This in turn makes an ecosystem vulnerable to storms and fires, in a kind of positive feedback loop from the point of view of ecosystem theory: Institutional disincentives for afforestation (small private land parcel sizes and exceedingly high costs for collective investment in maintenance and protection of forests and fields) lead to neglect, which further increases the likelihood of drought and fires, which further decreases the interest in afforestation and revision of the institutional structure.

While the National Forest Commission (the official state organization that manages the forest) identified traditional slash-and-burn agricultural techniques as the cause of the fires, more research is needed to understand where the fires come from, who starts them, and under what conditions. In any case, it is clear that there would be fewer fires if incentives were provided for forest maintenance and reforestation. The Yucatán's frequent hurricanes aggravate the fire outbreaks because they leave behind a large amount of dead wood, which serves as an explosive fuel for fires (National Forest Commission staff member, Colonia Yucatán, April 2014). The area's increasing deforestation and resulting vulnerability to climate change outcomes (e.g. drought, fire, and irregular rainfall) make it even less attractive for local people, and thus more and more smallholders are selling their land. One of the results is increasing concentration of land ownership in the hands of people with more bargaining power and agricultural technology.

Another local adaptation strategy is that farmers increase, or even exclusively practice, intensive livestock raising (man, Colonia Yucatán, March 2014). Cattle fattening is economically more secure, less vulnerable to climate fluctuations, and more lucrative—another example of the institutional and environmental barriers that counteract government incentives for reforestation. Intensive livestock raising leads to further environmental problems like deforestation, soil degradation, and loss of biodiversity. The way to better livelihood conditions and stabilization of the ecosystem would probably be collective land management and a partial reintroduction of the time-tested Maya agroforestry system.

One interesting exception to this development, and a new positive institutional change, is a smaller plot of *ejido* land held in common for the moment. The 90-hectare property was divided in the 1978 agrarian reform between 35 *ejidatarios* (ejido members). As no agreement has been reached as to who will use which part of that land and for what purpose (e.g. agricultural cultivation), it remains abandoned. Poaching occurs on that land, as well as the collection of firewood (man, Colonia Yucatán, June 2014). However, a positive effect is that the forest on this land has remained intact and serves as protection for the ecosystem and for biodiversity and as a fire barrier.

Conclusion

Environmental changes and political decisions, which can both be analyzed from an institutional change perspective, allowed the hinterland of the Yucatán coast to become marginalized. Climate change and natural risks, such as wildfires and exceptionally long droughts, aggravate this situation. Meanwhile, the Caribbean coastal region has become an economic center. Controversial development policy, which benefits the coastal areas but not the hinterland, sustains the marginalization of the hinterland and aggravates the regional situation. The economic role of tourism on the Caribbean coast has virtually no effect on the development of the hinterland (Brenner and Aguilar 2002: 500).

The current situation is the reverse of that of 50–60 years ago. At that time, the hinterland, which produced wood for national and international interests, could be seen as the economic center, with plentiful work and livelihood opportunities. Cancún and the Riviera Maya did not exist as the booming tourism centers they are today.

For the peripheral regions, a stable population and sustainable long-term economic management of forests would be of great benefit. Measures to reduce the risk from disasters such as fires and droughts are another beneficial way that governmental and nongovernmental organizations could invest more in the rural hinterland. Forests can serve as protection against hurricanes, fire, and loss of biodiversity.

A lesson from this study is that institutional changes linked to incentives must be studied. If institutions work well, they promote sustainable use and positive strategy to cope with the dangers of climate change. In the Yucatán, changes in land-ownership regimes led, after intensive use by a large private owner, to the creation of small parcels with low economic potential and low incentives for maintenance. Climate change conditions created a further disincentive by increasing the risk of losing any benefits gained by investing in reforestation. This has led to a feedback loop in which low maintenance and high risk reinforce each other.

One important option for improving this situation would be collective action and institutional change in areas held as private property. This would trigger the creation of institutions which would balance costs and benefits. Interestingly, the only remaining forested land in the area is communally held. This suggests that one way to help reforest the area and create a more resilient institutional setting would be to give incentives to transfer small private plots to communal tenure and, in a participatory fashion, establish new rules for forest maintenance and management.

Studies that take into account historical and institutional change are important in order to understand what decisions were made, by whom, and with what consequences for regional development. They must be considered not only as comparative studies, but also as sources of information for future political and economic decisions—for example, in reducing communities' vulnerability to extreme climatic events.

References

Alscher S (2008) Von Hurrikanen, Hotels und Hoffnungslosigkeit: Naturkatastrophen und Umweltmigration in Mexikos Südosten. Peripherie 112(28), Münster

Ballesteros-Barrera C et al. (2011) El impacto potencial del cambio climático sobre los Agroecosistemas. El caso del cultivo del maíz, proyecciones al futuro. In: Aragón G.A.D. Jiménez GYM. Huerta L (eds) Manejo Agroecológico de Sistemas vol II. Publicación especial de la Benemérita Universidad Autónoma de Puebla, Puebla, México. pp 1–14

Benjamin RS (1951) Lol lamaban 'el loco': Como un solo hombre reconquistó a Yucatán. Selecciones de Reader's Digest 21(122):21–25

Bracamonte y Sosa P (2003) Los Mayas y la tierra. La propiedad indígena en el Yucatán colonial. Coeditor(es): Centro de Investigación y Estudios Superiores en Antropología Social (CIESAS) y el Instituto de Cultura de Yucatán. Mérida, Mexico. ISBN: 970-701-356-7

Brenner L, Aguilar, AG (2002) Luxury tourism and regional economic development in Mexico. Prof Geogr 54(4):500–520, Association of American Geographers

Centro de Salud de Colonia Yucatán (2013) Diagnóstico de salud y estudio de comunidad 2013, Colonia Yucatán, unpublishesd

Cox M et al. (2010) A review of design principles for community-based natural resource management. Ecol Soc 15(4): 38. [online] URL: http://www.ecologyandsociety.org/vol15/iss4/art38/

Diemont SAW et al (2011) Comparisons of Mayan forest management, restoration, and conservation. Forest Ecol Manage 261:1696–1705. doi:10.1016/j.foreco.2010.11.006

Ensminger J (1992) Making a market. Cambridge University Press, Cambridge

Flores-Delgadillo L et al (2011). A sustainable system of a traditional precision agriculture in a Maya homegarden: soil quality aspects. In: Soil & Tillage Research 113:112–120, Elsevier B.V, doi:10.1016/j.still.2011.03.001

Ford A, Nigh R (2009) Origins of the Maya forest garden: Maya resource management. J Ethnobiol 29(2):213–236

Giorgi F (2006) Climate change hot-spots. Geophys Res Lett 33(8):0001. doi:10.1029/2006GL025,734

Gómez-Pompa A (1987) On maya silviculture. Mexican Stu/estudios Mexicanos 3(1):1–17. doi:10.2307/4617029

Góngora Salas Á (2001) Northeastern Yucatán project: archaeological survey in the Northeastern corner of Yucatán, México. Report on field season 2000. First stage of archaeological surveys in the northeast corner of the Yucatán Peninsula. Foundation for the advancement of mesoamerican studies, Inc. (FAMSI)

Haller T (2007) Understanding institutions and their links to resource management from the perspective of new institutionalism. NCCR North-South Dialogue, No. 2. Bern: NCCR North-South. http://www.north-south.unibe.ch/content.php/publications/filter/0?search=Haller&advancedsearch

Haller T (ed) (2010) Disputing the floodplains: institutional change and the politics of resource management in African floodplains. In: Brill, African Social Studies Series

Haller T (2013) The contested floodplain. institutional change of the commons in the Kafue Flats, Zambia. Rowman & Littlefield, Lanham, Lexington

Haller T (2015) Climate change impacts, institutional resilience and livelihoods vulnerability: a comparative analysis of climate change and adaptation strategies in Asia, Africa and Latin America. In: Rosendhal J, Zanella M, Weigelt J (eds) Pro-poor resource governance under changing climates. Potsdam: IASS. pp 160–181. http://publications.iass-potsdam.de/pubman/item/escidoc:936891:4/component/escidoc:936892/IASS_Study_936891.pdf

Karmalkar A et al (2011) Climate change in Central America and Mexico: regional climate model validation and climate change projections. Clim Dyn 37(3–4):605–629. doi:10.1007/s00382-011-1099-9

Mexico News Daily (2015) Yucatán vulnerable to climate change effects. Importance of mangroves highlighted at Cancún conference. Tuesday, October 20, 2015, Source: El Universal (sp). http://mexiconewsdaily.com/news/yucatan-vulnerable-to-climate-change-effects/ [29.10.15 09:25:17]

Meyers RK, van Lear DH (1998) Hurricane-fire interactions in coastal forests in the South: a review and hypothesis. Forest Ecol Manage 103:265–276

Moßbrucker H (1994) Agrarkrise, Urbanisierung und Tourismus-Boom in Yucatán, Mexiko, Kontoversen 7. Münster, Hamburg

North D (1990) Institutions, institutional change and economic performance. Cambridge University. Press, Cambridge

Ostrom E (1990) Governing the commons. the evolution of institutions for collective action. Cambridge University Press, Cambridge

Pagdee A, Kim Y, Daugherty PJ (2006) What makes community forest management successful: a meta-study from community forests throughout the world. Soc Natural Res 19(1):33–52

Pérez Castañeda JC (2002) El nuevo sistema de propiedad agraria en México. México D.F

Quezada S (1998) Political organization of the Yucatecan Mayas during the eleventh to sixteenth centuries. In: Schmidt et al (ed) Maya civilization, Thames and Hudson, London

Quezada S (2010) Breve historia de Yucatán. El Cologio de México, Fideicomiso Historia de las Américas, México D.F

Ribot J (2010) Vulnerability does not fall from the sky: toward multiscale, pro-poor climate policy. In: Mearns R, Norton A (eds) Social dimensions of climate change: equity and vulnerability in a warming world. World Bank, New York, pp 47–74

Roys R (1957) The political geography of the Yucatán Maya. C.I.W., Publication, Washington D.C

Ruiz Silva JA (Compilador) (2011) Colonia Yucatán. La historia reciente de un pueblo maderero. Fundación Amigos de Colonia Yucatán, A.C., México

Ruiz Silva JA (2013) Colonia Yucatán: Decadencia y migración. La historia de sus hombres y mujeres exitosos. México

Secretaria de la Reforma Agraria (1978) Diario Oficial 10(9):46–48

Villarino T, Araceli Rubi (1984) Estudio de comunidad de la Colonia Yucatán, Yucatán, México. Universidad de Yucatán, Mérida, México

Chapter 11
Designing for Future Uncertainties: Comparative Studies of Two Adaptive Strategies in Urban Design in New York and Sweden

Sander Schuur

Abstract The world has always been exposed to stimuli and change. Due to urbanization and globalization these stimuli have an increasing impact on a larger scale. The magnitude and characteristics of these stimuli in the future are, to a large extent, unknown. As a direct result of globalization and urbanization, the contemporary cities are increasingly vulnerable to stimuli such as climate change and socio-economic risks, creating the immediate and constant necessity to design and re-design cities in order to be adaptive to future change. The paper identifies design parameters that can inform contemporary design for an increased adaptive capacity. Through a focused comparison of two adaptive strategies in urban design, located in New York and Sweden, design parameters can be analyzed. The paper will strengthen the argument for developing an alternate and better informed design process to increase adaptiveness; resulting in minimization of the devastating impact of future stimuli, and by doing so increasing the sustainability of cities through architecture and urban design.

Keywords Adaptive capacity · Architecture · Climate change adaptation · Design parameter · Future uncertainties · Public space · Social sustainability · Transformation · Urban design

Introduction

The world has always been and will continue to be in a state of constant change. Its inhabitants have continuously been exposed to stimuli and change to which they have to adapt for survival. This adaptation consists of protecting oneself against these stimuli or abandoning habitats. In his work On the Origin of Species, Darwin argued that better or more useful adaptations for the immediate, local environment

S. Schuur (✉)
White Arkitekter AB, Östgötagatan 100, 11692 Stockholm, Sweden
e-mail: sander.schuur@white.se

W. Leal Filho and J.M. Keenan (eds.), *Climate Change Adaptation in North America*, Climate Change Management,
DOI 10.1007/978-3-319-53742-9_11

increase the likeliness of survival (Darwin 1868). Darwin indicates more useful or better adaptation, a characteristic that enhances the survival, as being the "fittest" in his theory "natural selection, survival of the fittest". Within biology the understanding that an adaptive trait enhances fitness is generally accepted (Futuyama 2009).

Contemporary global climate change has proven to be one of the most challenging stimuli the world population has to adapt to for survival. Future climate change is generally believed to increase the impact of its stimuli on the environment. Although the survival of the entire world's population is not directly threatened, climate change has and will have a dramatic impact on both physical and mental health. Climate related disasters such as Hurricane Sandy can cause vector borne diseases, injuries and deaths, and are also associated with stress-related psychiatric disorders (Kumar Padhy et al. 2015). The world population has to adapt to climate change to ensure health and survival.

This paper, based on architectural standpoints, intends to be propositional by suggesting urban design proposals to be a relevant contribution to the discussion on adaptive strategies. It suggests the ability of design and the built environment to increase the adaptive capacity of the city. Based on executed studies of two projects and written from the perspective of the designer, the paper indicates the role of design for ensuring adaptation to climate change, as well as other future stimuli. Case 1 is an urban design that is adaptive to climate change and case 2 functions as a comparison being adaptive to man-made change.

In line with Darwin's reasoning, the likeliness of survival is proportional to adaptiveness to stimuli and change of the environment. Due to contemporary urbanization and globalization these stimuli now have an increasing impact on a larger scale. The paper considers adaptation, as described in the first paragraph, as protecting oneself against stimuli and includes the willingness of abandoning habitats in order to survive. In contrast to Darwin's theory arguably excluding the possibility of design, as it explains the apparent design of the living world without resource to a supernatural, omnipotent designer (Futuyama 2009), this paper states the ability of design to ensure an increased adaptive capacity.

Urbanization

Since 2014, for the first time in history, the majority of the world's population lives in urban environments. The United Nations predict the percentage of the world's population living in cities will exceed 66% by 2050 (United Nations 2014). A direct result of the accelerating urbanization has been that the contemporary urban environments house the majority of the world's population, contain most of our infrastructure and are the centers of our economy. This concentration of critical assets makes cities specifically vulnerable to change and stimuli.

Historically, cities were built along rivers, river mouths, on coastlines and near present natural resources for subsistence resources, ecological benefits and logistical reasons. Various estimations contradict in exact numbers, but agree that about half of the world's population continues to live in coastal areas, defined as within 100 km from the coast. A rough estimation indicates that more than 200 million people are living in coastal areas less than five meters above sea level (Bollmann et al. 2010). A vast majority of cities are located in coastal areas including eight of the top ten largest cities in the world (UN Atlas of the Oceans). Urbanization will continue to increase the percentage of the world's population living in coastal areas. The location of cities in coastal zones places the concentration of critical assets in areas specifically vulnerable to (climate related) change and stimuli.

A well-functioning city is defined by the city's ability to provide basic services to its citizens. Urbanization leads to an often rapid increase of population in cities, stressing existing supporting functions and structures, as well as existing balances. Fast growing cities; at a pace of 4% or more (World Economic Forum 2015), can render supporting functions and structures insufficient, risking social instability. The inability of a city to provide basic services is referred to as fragility, and the intensity of fragility is being conditioned by the accumulation of risks (World Economic Forum 2015). In other words: fragile cities are more vulnerable to change and stimuli.

We could argue that urbanization leads to an increased effect of change and stimuli. When Superstorm Sandy made landfall in New York City on October 22nd, 2012, the increased impact of stimuli on the urban environment became very apparent. The damage caused by Superstorm Sandy in New York City was proportional to the large concentration of critical assets.

During Superstorm Sandy the naturally sheltered harbor, that helped New York City grow in significance as a trading port, failed to protect. The "New York Bight", a natural curve in the open shore line, trapped the storm surge causing the water to spill on land (Silverman 2012). Being located directly on the coast proved to be hazardous, especially with a large amount of vital facilities, not the least the financial district with the New York Stock Exchange and Wall Street, being located directly on the city's edge and coastline.

Superstorm Sandy also brought to light existing social inequities and imbalance in New York City. Lower income areas containing poorly constructed and maintained buildings were damaged more severely and lower income households with fewer resources available for recovery have a longer recovery time (Lindell and Prater 2003). The fragility of the city increased the devastating impact of the storm.

Globalization

Globalization is a worldwide movement towards an increasing integration through establishing and enabling global networks. This movement results in an increased interconnectedness and interdependence representing a fundamental change from a world of individual and independent states to a world of state interdependence as described by the World Health Organization (WHO 2016). Globalization has increased rapidly in recent years and with the global interdependence the impact of stimuli such as climate change risks increases dramatically. Although a stimuli as a natural disaster can be local, the impact area can extend globally. Major disasters have the potential to be disruptive beyond the local damage, as supply chains, networks and markets around the world will be affected. The March 2011 earthquake in Japan even shifted the earth's axis, shortening the length of the day (The One Brief 2015).

Superstorm Sandy can be of reference once again to indicate the increasingly large impact of local stimuli. Due to the extensive flooding of southern Manhattan, the storm closed Wall Street for two trading days, directly disturbing the global financial system. Data quantifying the effect of this impact is seemingly not available, but Wall Street being the trading hub of the second-largest trading center in the foreign exchange market indicates its significance (Picardo 2014).

Superstorm Sandy caused a black out in local communication networks in a time when needed, but also effecting the world's Internet traffic; traffic that neither originated from nor was destined to areas affected by the storm. As a result of outages in New York City Internet traffic was (successfully) shifted away from the city to alternative paths located in Chili, Sweden and India avoiding failures at critical transit points in New York (Madory 2012).

The forced closing of New York's three main airports after flooding inundated the runways, disrupted the busiest U.S. aviation market handling 300,000 passengers a day and extended the impact of Sandy far beyond the New York region. The vulnerability of the urban environment is not just a problem for the city itself but stretches far beyond the city's physical boundaries.

Designing for Future Uncertainties

The nature, magnitude and characteristics of the stimuli and change that will challenge our cities is uncertain and to a large extent unknown. Superstorms as Sandy will become more frequent and more powerful when sea levels rise and sea water temperatures increase. A study by the Pennsylvania State University indicates that storms of a ~2.25-m flood height had a 500-year return period during the pre-anthropogenic era (A.D. 850–1800) and that this frequency has increased to ~24.4-year in the anthropogenic era (A.D. 1970–2005) (Reed et al. 2015), an increase that is generally believed to continue. A study by the University of Maryland's Earth System Science Interdisciplinary Center (ESSIC) indicates that a warmer Atlantic Ocean could substantially increase the impact of a future Superstorm (Kim et al. 2016). Some scientists agree however that the frequency, magnitude and characteristics of future storms are impossible to predict because "the myriad factors affecting storm behaviors are too complex and impossible to simulate"—William Lau, research scientist at ESSIC and senior scientist emeritus at NASA's Goddard Space Flight Center noted in an interview on the website Science Daily in 2016.

The stimuli and change that will affect cities are certainly not limited to the risks associated with climate change. Socio-economic stimuli such as the current peak in immigration in Europe, triggered by the civil war in Syria, took cities by surprise. Sweden received 162,877 applications for asylum in 2015, which is 1.7% of the total population and more than double of the amount it received in 2014. Swedish cities and population struggle to house and integrate the immigrants as the cities were not sufficiently prepared for these stimuli.

Considering the dynamics of the contemporary urban environment as a result of urbanization and globalization, it is likely that the change in characteristics and magnitude of future stimuli will be dramatic. A dramatic change of stimuli can lead to an entirely different state of operations that pushes the urban environment beyond its threshold of maintaining fundamental structures and functions and requires radical transformation (Keenan 2014). To be able to meet these future stimuli, the contemporary urban environment is in urgent need for an increased adaptive capacity: to be able to adapt in order to protect itself and its inhabitants against these stimuli, and to do so continuously.

Adaptive Capacity
This paper places adaptation in the context of mitigation in order to clarify adaptation and the requirements for adaptive capacity. Mitigation is described as reducing or preventing the (negative) impact of stimuli by reducing or preventing the factors causing these stimuli. Mitigation to reduce climate change for example includes the reduction of greenhouse gases, either by reducing the sources of these gases or enhancing the "sinks" that accumulate and store these gases according to NASA (2016). Adaptation in the context of this article is the ability to adapt to change and stimuli that are seen as evident, although their nature and characteristics being unknown. Starting point is the awareness that stimuli will occur, followed by the requirement for an understanding of the necessity to adapt to these stimuli, as well as a willingness to do so, and to do so constantly. We have to note that the actual acts of adaptation are often also acts of mitigation, making the two terms not easily separated (Keenan 2014).

Methodology

Through a focused comparison of two adaptive strategies in urban design, the article identifies design parameters that can inform contemporary and future designs to establish an increased adaptive capacity. The comparative study is a descriptive comparison with the purpose of identifying, analyzing and explaining differences and similarities of the two cases and to identify and describe design elements leading to adaptation. This comparative method allows for a first step to ascend from an explanatory level to a more advanced level of theoretical models.

The two reference projects are winning submissions of international design competitions by White, excluding work by others to ensure full access to study material. The cases have a high level of experimentation that facilitates testing and bringing forward new concepts and ideas. The concepts and strengths of the cases are formulated clearly and concisely in brief and comprehensive documentation—characteristics well suited for comparative studies.

The paper compares two projects with relevant approaches of adaptive urban design; one to climate change and the other to man-made change. The man-made change, in the reference project caused by mining adjacent to the city, is a known factor and characterized by urgency, while climate change is more uncertain. The paper studies how the design proposals vary in approach to be adaptive to known and unknown stimuli and how they can inform each other in possible future projects. The first project is located in the New York metropolitan area and the second in the arctic town of Kiruna, Sweden.

The first case is the FAR ROC [For a Resilient Rockaway] two-phase design competition that aimed to explore innovative strategies for the planning, design and construction of a resilient and sustainable development in the neighborhood Arverne East, on the Rockaway Peninsula, NY. The competition was hosted by the NYC Department of Housing Preservation and Construction [HPD], the American Institute of Architects New York Chapter [AIANY], Enterprise Communities, and the client team consisting of three private developers. Controversially, the competition questioned whether areas of geographic vulnerability should be rebuilt, maintained and defended, or simply abandoned. It aimed to solicit creative ideas for resilient development strategies that could be implemented throughout New York City and vulnerable communities everywhere (FAR ROC Design Brief 2013).

In 2012 the City of Kiruna invited ten architect offices after a Request for Qualification [RFQ] process to provide a vision, strategy and design for a new city center for Kiruna. The ground deformation caused by the iron ore mine owned by LKAB has long affected the city, necessitating the re-localization of many of the city's functions, and in a long-term perspective will affect the city now being planned for. The competition "New Kiruna" aimed for a design for a diversity of people, being capable of meeting new challenges and of developing into a place offering the best possible prospects of urban development (New Kiruna Design Brief 2012). White's proposal "Kiruna 4-ever"[1] was announced as the winning submission in March 2013.

Key focus of the comparative study is to identify, analyze and explain similarities of strategies and design solutions that generate an increased adaptive outcome. This may or may not have been intentional during the design process and thus read as so in reflection of this study. The study aims to formulate design parameters to inform future design process for an increased adaptive capacity.

Case 1

White's competition entry titled "Small Means and Great Ends"[2] was announced the winning design solution of the FAR ROC design competition among 117 submissions from over twenty countries. In a public event on October 23rd 2013, roughly one year after Superstorm Sandy made landfall in the Rockaways,

[1]Kiruna 4-ever, Competition Submission by White in collaboration with Ghilardi + Hellsten, Spacescape and Vectura Consulting Excerpt: http://en.white.se/projects/kiruna/

[2]Small Means and Great Ends, Competition Submission, White and Arup in collaboration with Gensler Excerpt: http://en.white.se/projects/small-means-and-great-ends/

the design proposal was presented on site to the community of Far Rockaway and the media.

Site plan, illustration by White

Social Sustainability

The starting point and main goal of the design proposal "Small Means and Great Ends" was to create an authentic urban development that is a new home for the community of Arverne East. The design proposal argued the necessity for any sustainable urban development to be authentic, robust and socially sustainable as base requirements. The proposed urban strategy creates an inclusive community that will enable all members to flourish. After Hurricane Sandy, the strong sense of community in the Far Rockaways was witnessed growing even stronger and more tight-knit in the face of disaster. The proposal takes that sense of community to another level by providing space for the community to hold a central role in everyday life.

The proposal establishes an inclusive community with a strong identity through an urban design that affords members of the community opportunities to engage, grow and prosper. It creates an integrated solution that positions the community to provide leadership throughout the project's development, during everyday life, and in the event of a storm and its recovery afterward.

Public place forms the most essential element in urban design. The design proposal is therefore organized around a series of public spaces, being the social nodes of the development embedded in a recognizable continuation of the surrounding grid layout. As the colorful community of Arverne East comprises a wide diversity of individuals, the design establishes a network of varying public spaces of differing scales. The proposed public places vary from a protected nature preserve, landscaped parks, community gardens, a town square, to a wetland park filled with program for children. At the center of the development a new town square, in addition to being a main attraction for visitors, offers the community a new identity. Shared courtyards are proposed within the building blocks providing a safe, controlled outdoor space for communal use and social interaction.

In addition to a diversity in public space, the urban plan includes a mixture of housing types and public and commercial functions and services that provide equal opportunities for business entrepreneurship and engage residents in decision-making processes that directly affect their lives. The variety of housing types caters to the community; property types were designed to give ownership to young individuals, couples, families, and older generations alike.

In line with the vision to place the community in a central role, the proposal includes possible means to ensure and maintain affordability and continuous funding of community initiatives, such as a community land trust [CLT], on site revenue and business opportunities for community members. The proposal aims to extend the definition of sustainability into land development and ownership that will strengthen local resilience through community participation and a true sense of ownership, accountability and belonging.

Interact and Adapt

The design proposal acknowledges that extreme environmental conditions are inevitable in Far Rockaway and proposes that the community of Arverne East must be aware of and accept future uncertainties, rather than to ignore and fight them. In essence, the community must engage with the environment developing a symbiotic relationship. The design aims to provide a lively environment that interacts, rather than counteracts, with the sea and responds to natural forces.

In order to establish this interaction, the design, instead of delivering a one package solution, includes a series of small, affordable, and smart interventions. This allows for greater flexibility and adaptive capacity as these interventions or design solutions can be tested, adjusted, and redesigned as required, not only in case climate change stimuli, but even during the development process. In the event of a natural disaster, the overall development strategy is designed to reduce and control damage by dispersing and redirecting the stormwater as it surges; to maintain access to and operation of essential buildings and infrastructure; to ensure a quick recovery; and to empower the community to take a leadership role before, during, and after the event.

First in line to mitigate a storm surge are sandbanks that form a buffered ecological zone along the shoreline. Besides dissipating the ocean's energy during a storm, a new landscape will take shape with more shallow and calmer water for leisure, forming at the same time a dynamic defense against coastal flooding and erosion. A "kinked" boardwalk further helps to dissipate forces and directing a possible storm surge while providing a diverse and unique experience of the shoreline. Two parks cut through the urban layout, acting as water retention and detention basins, directing stormwater away from housing and other essential functions to control and reduce damage and to ensure their accessibility. Design solutions such as positioning installations on higher floors and foldable facades further reduces damage.

To design proposal ensures a quick recovery after the event of a storm and positions the community to react quickly and respond in a way that best suits the situation. The design includes training of community members and providing them the resources to lead the recovery as they are already on site and the most engaged to do so.

Illustration, copyright and credits to MIR

Adaptive Capacity

The design proposal 'Small Means and Great Ends' aims to establish an urban development for the community of Arverne East that is first of all robust and inclusive. These proven qualities form a solid functioning base, rather than a total new experiment. Adaptive capacity is ensured through a series of small, affordable and smart interventions allowing for greater flexibility. The interconnected design solutions build a network of adaptive measures that together, supplemental and providing backup, mitigate stormwater allowing for continuous adjustment and redesign when required.

Case 2

In March 2013 the jury of the 'New Kiruna' competition unanimously decided to appoint White's competition entry titled 'Kiruna 4-ever' as the winning design among 10 shortlisted submissions. The understanding of the proposal that urban development is not only a physical process, but to the same extent a process of involving the inhabitants in the transformation of Kiruna, was stated essential by the jury (New City Center Kiruna, Jury Statement March 2013).

Site plan, illustration by White

Social Sustainability

White's design proposal 'Kiruna 4-ever' indicates three key aspects in its strategy and vision: a step-by-step process, the inclusion of all stakeholders and the aim to be a model city. The goal for Kiruna is to be a welcoming global city with a unique mix of people and culture in close relation to the surrounding artic landscape and mine, and a global meeting place for urban development. It states the ambition to be the most democratic urban transformation worldwide.

The proposal acknowledges that Kiruna's future depends on its residents' knowledge, willpower and engagement. To be the most democratic urban transformation, the proposal formulates various strategies to include the population of Kiruna in the envisioned process of urban transformation from design phase to execution and beyond.

The attractiveness of a city is defined by its meeting places and public space; these places are the crossing points in a city's networks and infrastructures establishing contact and interaction between people. The design envisions a network of public spaces and events that are fully connected, providing a strong structure throughout the future urban transformation to ensure an integrated and connected city. The structure, or backbone of the future development, consists of a central street surrounded by three blocks of buildings providing good conditions for easy orientation. The New Kiruna is more densely built and includes a wider variety of functions which will benefit people, industry and ecology making an attractive and sustainable city. It aims to be a well-planned urban environment with a healthy mix of socio-economic and cultural functions and expanded outdoor living, creating public spaces that support well-being, are accessible and safe, to strengthen the social cohesion.

Interact and Adapt

The first sentence of White's proposal 'Kiruna 4-ever' reads: "There is nothing permanent except change". It acknowledges that a city is never finished as it is in a process of constant change. The proposal sees the urgent necessity to change as a great opportunity and argues the New Kiruna to plan far beyond the 2033 time horizon stated in the competition brief.

The ground deformation caused by the expansion of LKAB's iron ore mining activity forces the city to move. The entire town center and its surrounding neighborhoods are to be demolished and rebuilt two miles further east at the cost of LKAB. In Kiruna the mine defines, dominates and sustains the city; the continuance of its operations is vital to both the local economy and the community.

White's vision provides the city of Kiruna with clear tools and strategies for the long term to grow in a flexible manner. It includes a step-by-step development through the consolidation of new public spaces or social nodes that together form a network along a central street, the so-called 'east-west corridor'. These new public spaces are the focal points of new development towards the east. The design proposal acknowledges the dramatic impact and scale of relocating an entire city. Thus the argument that a step-by-step approach of little and less dramatic steps, allowing for a continuous connection between the new development and the existing city, should be seen as the development of Kiruna rather than moving the entire city.

By ensuring an attractive and safe public space with a strong and clear focus on pedestrians of the new 'east-west corridor', the municipality is to set the direction for future transformation and establish a communal willingness to follow. The provision of social nodes, such as parks, schools and pre-schools, guide the eastern movement of the development. The limitation of three blocks of buildings along the east-west corridor allows for excellent accessibility to the city center and the surrounding artic landscape, providing an attractive and high quality housing environment.

Through the design of public space and of the overall urban layout, the proposal is inclusive and ensures a willingness among its residents to take part in the urban transformation. Or in the words of Peter Niemi, Kiruna's municipal chief executive: "when we move, we move together" (Perry 2015).

Illustrations by White

Adaptive Capacity

The competition entry 'Kiruna 4-ever' is both a vision and a strategy how to move the city of Kiruna. It is inclusive, the most democratic form of urban transformations, and ensures a willingness of its residents to participate. It proposes a clear and strong urban structure along a central east-west corridor with a string of public spaces that are the social nodes of the community. The proposed step-by-step approach allows for varying speed of development and adjustments during the development. It provides a solid backbone for development that can continue far beyond the 2033 time horizon as suggested in the competition brief.

Comparisons

Although the changes affecting the case studies vary, being unknown and known, both design proposals originate from the awareness that future change and stimuli are inevitable and their characteristics uncertain. Both submissions acknowledge the necessity of an adaptive capacity in the design and development phases, and for the realized urban environments.

The cases strongly focus on social sustainability, aiming to be inclusive in the design and development as well as the realized projects. In order to be adaptive to change and stimuli a robust urban development is beneficial. 'Fragility' as described in the introduction makes cities more vulnerable to change and stimuli, while a robust inclusive urban plan reduces fragility. The submission 'Kiruna 4-ever' includes strategies and events to include local residents in the process of urban transformation and both cases include design elements and strategies in order to be inclusive.

Both cases position their community as an essential and central part of the development: case 1 includes training of community members to lead recovery after the event of a storm and case 2 formulates various strategies to include the population and to ensure a willingness to participate in the urban transformation.

The starting point common to both cases is the design of a robust, tested and proven urban structure: case 1 with a recognizable continuation of the surrounding grid layout and case 2 with an east-west corridor surrounded by three blocks of buildings respectively. Both proposals include a variety of public spaces as social nodes for a diversity of people constituting the communities of Arverne East and Kiruna. Both aim to establish a new and strong identity to which their communities relate.

The proposed 'small means' and 'step-by-step' approach allow for greater flexibility and adaptive capacity during development and for the realized project. The smaller design solutions and components can be tested, modified and redesigned in all phases of development to increase the adaptive capacity of the design. In the case of the 'Small Means and Great Ends' proposal, these components can be altered and adjusted after a storm event to better protect the urban development during possible future storms. In 'Kiruna 4-ever' the components allow for adjustment during construction phase and for continuation of the urban transformation beyond the envisioned two miles during the competition phase when ground deformation demands so.

The new location of the city center of Kiruna, however, is not entirely safe from mining activity either. Eva Ekelund, head of the department of land and development, notes that "iron is under the new town center, too." It is however generally believed to be too expensive for LKAB to move the city once again (Perry 2015). Also, global demand of iron ore has an immediate effect on the intensity of the mining activity and therefore the speed of ground deformation. A scenario of a lesser demand of iron ore leading to lower iron ore prices raises doubts whether LKAB can continue to finance the ongoing urban transformation. In Arverne East, large upfront investment requirements raise questions on the feasibility of new development based primarily on affordable housing. The predicted sea level rise and expected increase in occurrence of storm events will likely raise insurance fees far beyond the resident's abilities and resulting in a non-sustainable development.

Conclusion

The urban proposals compared in the paper are 'robust' urban designs that include a series of design solutions or 'components' that can be seen as 'adaptive traits', or useful adaptations for stimuli and change increasing the likeliness of survival. In the context of the paper, a 'robust' urban design is defined as tested and proven urban layout that is socially sustainable and reduces 'fragility'.

An adaptive trait increases the likeliness of survival allowing for the argumentation that the introduction of multiple integrated traits or components lends to a plausible increase in the likeliness of survival. Multiple integrated design components enable urban environments to avoid dependency on one component or solution only; smaller integrated components used in urban design are more easily adjusted when the requirement arises, which increases the flexibility or adaptive capacity to climate change and other stimuli.

The described 'designing for future uncertainties' directly relates to the concept 'likeliness of survival' as described in the first paragraph. Firstly, designing to meet uncertain stimuli and change, in the context of the reference cases, is based on a (proven) likeliness of occurrence and characteristics of future stimuli and change. The ability to adjust components within the design accordingly to these anticipated stimuli allows for an increased ability to adapt. Secondly, starting with a 'robust' and socially sustainable urban design that is better able to cope with unforeseen stimuli and change, components that are spatial solutions allow for radical transformation when the urban environment is pushed beyond its threshold of maintaining fundamental structures.

The paper states the ability of design to ensure an increased adaptive capacity of the urban environment, making design (often lesser considered but nonetheless significant) a contribution to the discourse of adaptation. The compared urban proposals both include multiple integrated components that each are designed optimally to their envisioned function within the urban environment. Design in the context of this paper relates to the architectural process and outcome and is not limited to visual appearance.

The identified design parameters in this paper are limited, but provide a basis for further studies as well as an indicator of elements that can inform contemporary and future design to establish an increased adaptive capacity. They include ensuring a 'robust' and socially sustainable urban design and the incorporation of a multiple of integrated design components, that together ensure inclusive design, adaptive capacity and spatial solutions allowing for radical transformation. The design parameters are in effect during both the design and development phases, and for the realized urban environments. The inclusion of the identified design parameters in the contemporary and future design process will increase the adaptive capacity and sustainability of our cities through architecture and urban design.

References

Bollmann M et al (2010) World ocean review, living with the oceans, Maribus, pp 68–73

Darwin C (1868) On the origin of species, by means of natural selection, or the preservation of favoured races in the struggle for life, 5th edn. John Murray London

Futuyama D (2009) Evolution, 2nd edn, Chapter 11 Natural selection and adaptation, Sinauer Associates, Inc., pp 279–301

Kim K et al (2016) What would happen to Superstorm Sandy under the influence of a substantially warmer Atlantic Ocean?, AGU Publications Geophysical Research Letters, January 2016, pp 802–811. doi:10.1002/2015GL067050

Keenan JM (2014) Material and social construction: a framework for the adaptation of buildings. Enquiry ARCC J, pp 176–185

Kumar Padhy S et al (2015) Mental health effects of climate change. Indian J Occup Environ Med 19(1):3–7

Lindell M, Prater C (2003) Assessing community impacts of natural disasters. Nat Hazards Rev:178–180

Madory D (2012) Hurricane Sandy: global impacts, http://research.dyn.com/2012/11/sandys-global-impacts/. Accessed 1 Mar 2016

NASA (2016) Responding to climate change. Solutions, mitigation and adaptation, http://climate.nasa.gov/solutions/adaptation-mitigation/. Accessed 24 Feb 2016

New Kiruna Design Brief (2012), http://www.kiruna.se/PageFiles/5559/Microsoft%20Word%20-%20Eng_Prekvalificeringsprogram_2012-04-12.doc.170.pdf?epslanguage=sv. Accessed 2 Mar 2016

Picardo E (2014) Why wall street is a key player in the world's economy http://www.investopedia.com/articles/investing/100814/wall-streets-en/. Accessed 7 Mar 2016

Reed A et al (2015) Increased threat of tropical cyclones and coastal flooding to New York City during the anthropogenic era. In: Proceedings of the National Academy of Sciences of the United States of America (PNAS), pp 12610–12616

Silverman R (2012) Why New York City is the worst place for a hurricane, National Geographic News, http://news.nationalgeographic.com/news/2012/11/121101-new-york-ci/. Accessed 1 Mar 2016

The FAR ROC Design Brief (2013), http://www.farroc.com/wpcontent/uploads/2013/04/downloads_farROC_designbrief.pdf. Accessed 2 Mar 2016

The Guardian (2015) Kiruna: the arctic city being knocked down and relocated two miles away, https://www.theguardian.com/cities/2015/jul/30/kiruna-the-arctic-city-being-knocked-down-and-relocated-twomiles-away. Accessed 2 Mar 2016
The One Brief (2015) The impact of natural disasters on the global economy, http://www.theonebrief.com/the-impact-of-natural-disasters-on-the-global-economy/. Accessed 2 Mar 2016
UN (2014) World urbanization prospects, the 2014 revision, highlights, The Department of Economic and Social Affairs of the United Nations Secretariat, United Nations, pp 1–2. http://esa.un.org/unpd/wup/Publications/Files/WUP2014-Highlights.pdf. Accessed 25 Feb 2016
WHO (2016) Globalization, http://www.who.int/trade/glossary/story043/en/. Accessed 1 Mar 2016

Chapter 12
Designing with Risk: Balancing Global Risk and Project Risks

Matthijs Bouw

Abstract The World Economic Forum, in their yearly global risk report which highlights the most significant long-term risks worldwide, consistently indicates that the 'failure of climate change mitigation and adaptation', as well the interconnected 'water crises', 'greater incidence of extreme weather events' and 'food crises', are among the top 10 global risks of highest concern. It is evident that the built environment will play a central role in fostering resilience towards such risks. Within the production of the built environment, at the same time, there is a concern with an altogether different set of risks, those related to the feasibility of a project. 'Project resilience', the capacity of a project to cope with shocks and stresses that are related to its feasibility, is often in conflict with 'global' resilience goals. Successful implementation of resilience projects in the built environment, such as 'The Big U' (of which the author is a co-design- lead), depends on designing the right balance between the two. In this paper, based on the author's work on 'The Big U' and its successor projects, as well as on two interdisciplinary seminars at PennDesign, 'Designing with Risk', the author presents the research into this question, and propose that designers can have agency in balancing the two risk types in resilience projects.

Keywords Climate-adaptive design · Urban resilience · Community engagement · Risk management · Adaptation pathways · Big U · Rebuild by design · New York City

Introduction

After Hurricane Sandy, New York City (NYC) has started implementing numerous projects that address the increased climate change risks the city is facing. These projects are part of a larger strategic approach based on the notion of 'flexible

---

M. Bouw (✉)
School of Design, University of Pennsylvania, Philadelphia, USA
e-mail: mbouw@upenn.edu

W. Leal Filho and J.M. Keenan (eds.), *Climate Change Adaptation in North America*, Climate Change Management,
DOI 10.1007/978-3-319-53742-9_12

adaptation pathways' (Rosenzweig and Solecki 2014). This approach acknowledges the uncertainties relating to climate change and the need for a multi-layered and iterative response in a complex adaptive system such as a city. Each project, from that perspective, is itself a mechanism of action in a long term process.

Based on the author's experiences as co-lead of the design team for the Big U in Rebuild by Design (and subsequent follow-up projects), as well as through his research at PennDesign, this paper describes how climate adaptation projects need to navigate between addressing 'global risks', such as those resulting from long-term climate change, and 'project risks', that are the result of the complex political, financial, cultural and technical environment of the project, resulting in risks to its feasibility. Project Risks are considered in the shorter term. There is a tension between these two goals that can be both productive and unproductive.

The success of the Big U concept in triggering responses to climate change that fit in with NYC's 'flexible adaptation pathways' (FAP) approach is path dependent (Boschma 2015; Sjöstedt 2015) on a number of crucial initial design decisions. These decisions continue to influence the planning and implementation process of the project. The paper explores the agency that design can have in balancing the management of Global Risks and Project Risks as proxies for divergent time horizons, geographies and political interests. The paper suggests that one should not design for a static set of risks but should be designing the adaptive capacity itself, such that future iterations or designs can be accommodated at a point of time when different sets of risks may manifest.

Global Risks and Project Risks

The World Economic Global Risk Report 2016, based on a survey of more than 750 experts and decision makers from the WEF community, lists the failure of climate change mitigation and adaptation as the most impactful risk for the years to come (WEF 2016). In the report, the writers explore what they call Global Risks, "uncertain event[s] or condition[s] that, if it occur[s], can cause significant negative impact for several countries or industries within the next 10 years" (WEF 2016, p. 8). The failure of climate change mitigation and adaptation is linked to societal risks such as water crises, food crises, profound social instability and large scale involuntary migration as well as to environmental risks such as extreme weather events, natural catastrophes and biodiversity loss and ecosystem collapse, each in themselves high on the list of Global Risks.

Many of these risks will increase. Climate change is resulting in sea-level rise and increased storm intensity, the extent of which is still uncertain. Projections of sea level rise on the East Coast of the United States vary from 0.3 to 2 m (the latter under a scenario in which the polar icecaps collapse) by 2100 (Hauer et al. 2016; Vermeer and Rahmstorf 2009). The extent of the sea level rise depends on many factors, such as the success of the Paris climate mitigation agreements and planetary feedback mechanisms.

While the magnitude of the effects of climate change is uncertain, it is clear that urban environments have a great impact on climate change itself, and will play a great role in climate change mitigation and adaptation. Urban areas, home to more than 50% of the world's population, account for more than 70% of CO_2 emissions (Seto et al. 2014, p. 927). Improvements in building design, land-use and transportation planning, as well as the transition of urban energy systems, should all contribute greatly to the reduction of CO_2 emissions (Seto et al. 2014).

At the same time, it is in the urban environments that most measures need to be taken to reduce the impact of the risks, through mitigation of (potential) damages or through building resilience. Resilience can be defined in different ways. In a more limited definition, from the domain of engineering, resilience refers to the ability to return to its original form after being impacted by a stressor, such as those induced by climate change. In the field of complexity science, the concept of resilience is defined more broadly including the adaptability or transformational capacity of a complex system under stress (Martin-Breen and Anderies 2011). Resilience, in this latter definition, has the potential to not only play a role in risk mitigation, but also, through the ability of a complex adaptive system to transform, to mitigate and adapt its stressor.

The different definitions of resilience have made the term somewhat of a catch-all name for many strategies and projects that result from the need to mitigate and adapt from climate change and other risks, especially in the professional environment. There is an inherent messiness to the use of the term as it is used for such resilience strategies and—projects, a messiness that is holding back the development of climate change response initiatives (Keenan et al. 2015). To a large extent this is because these strategies and projects themselves become agents in a complex adaptive system, and reduce, or even increase, the uncertainty. The strategies and projects become part of the feedback mechanisms. This makes planning and building resilience projects highly complex, and is, in part, the reason that resilience projects have been reduced in dimension.

The Dutch Approach

The Netherlands' Delta Works is a case in point. After a major flood in 1953 that inundated much of the Southwestern Rhine/Meuse/Scheldt delta and killed approximately 2000 people, the Netherlands set forth a massive flood measure to reduce the risk of inundation to 1:10,000, meaning one flood event per 10,000 years. Devised by the engineers of Rijkswaterstaat, the Dutch Department of Waterways and Public Works, the Delta Works were originally envisioned as a series of dams which would close off all the estuaries with the exception of the Nieuwe Maas, and the Oosterscheldt, such that the ports of Rotterdam and Antwerp would remain accessible. Complemented by heightened dikes, this massive plan, at a cost of more than 15 billion US$ (in current terms) would protect the Rhine/Meuse/Scheldt delta against future floods and reduce the risk of inundation.

After most of the dams were constructed, it became clear that the original plans needed to be ameliorated. The closed-off estuaries suffered from ecological degradation (Ysebaerta et al. 2016). At the same time, the ecology of the delta was increasingly valued by scientists and the public. As a response, the final dam, at the Oosterscheldt, was built as only partly closed, with operable panels that could be opened to let the tidal flows do their work. The Delta Works is an example how large scale projects in which one set of goals (flood protection) executed by one type of experts (civil engineers) can have adverse large scale effects in other realms.

It is in part the awareness that flood protection measures are part of a complex adaptive systems that have inspired later generations of Dutch flood protection protects, in which the focus has shifted from a singular one, protection, or damage mitigation, to multiple goals, which include ecological value and spatial quality. Additionally, programs such as 'Ruimte voor de Rivier' (Room for the River) also allow for a more flexible approach to flood protection standards, ultimately resulting in a series of smaller, more adaptive solutions (Wesselink et al. 2007).

Another reason resilience projects have been reduced in dimension is the management of the associated project related risks. Adverse environmental impacts, such as seen in the Delta Works but also in the Oresund and Great Belt links in Scandinavia (in themselves not resilience projects but built within a similar environment), have led to significant adjustments in projects, with large cost overruns as a result of poor project risk management. In general, many large projects in the past decades have had sizeable cost overruns. This, in return, has reduced the ability of the body politic to propose such large scale projects (Flyvbjerg et al. 2003). Even after a catastrophic event, which might create the necessary political momentum, it is difficult to sustain (politically and financially) the big interventions a large scale resilience project might warrant.

New York City's Approach

New York City is a good example where resilience projects have been re-thought from the large scale to the smaller scale. In the late 1990s, it became clear that New York was extremely vulnerable to coastal flooding (Hill 1996; Rosenzweig and Solecki 2001). Since then, various proposals have been produced for storm surge barrier systems, similar to the Oosterscheldt barrier. One such example was Arcadis's 2006 proposal for a storm surge barrier in the Verrazano Narrows, which would connect Staten Island with Brooklyn. Combined with barriers in Arthur Kill and on the East River, such a system would protect much of New York City against coastal flooding. The cost of the Verrazano barrier was estimated to be approximately \$6.5 billion (Arcadis 2009). After Hurricane Sandy, in October 2012, such proposals, sometimes with different locations of the barriers (such as one between Sandy Hook and Breezy Point in the Rockaways), were re-iterated by various engineers, and were given prominent attention in the media (Lynch 2012).

In 2008, New York City Mayor Bloomberg convened a panel the First New York City Panel on Climate (NPCC). Rather than focusing on a few large projects, such as the storm surge barriers proposed earlier, the NPCC proposed a risk-management approach called 'Flexible Adaptation Pathways' (FAP) (NPCC 2010). In this approach, climate change responses evolve over time as understanding about impacts and effectiveness of mitigation, resilience and adaptation measures develops ("learn, act, then learn some more"). Given this thinking, the NPCC noted that proposed storm surge barriers might be a relevant solution in the long term, but given the high economic, environmental, and social cost, they would require extensive study before being regarded as appropriate for implementation (Rosenzweig et al. 2011).

The subsequent Special Initiative for Rebuilding and Resiliency (SIRR) has proposed a similar iterative and multi-layered approach to climate change (City of New York 2013). In response to Hurricane Sandy, which made landfall on October 29 2012 and which caused extensive losses to the city including 43 deaths and more than 19 billion in damages, the SIRR-report outlined a large number of small-to-medium scale flood protection initiatives. Many of these were green infrastructure projects, such as expanded dune systems, wetlands and off-shore breakwaters. Locally, the hardening of critical infrastructure, floodwalls and the raising of bulkheads were proposed. Building codes and standards were to be adapted to the new condition. Further, the report explicitly debunked the idea of the flood barriers: "For some observers, the idea of constructing a single piece of engineering offers the appeal of seeming simplicity, as compared to a suite of more targeted, localized protections. However, the construction of such harborwide storm surge barriers actually presents many complications." (City of New York 2013, p. 49). It then lists the high costs, the long time it takes to plan, design and permit and the possible environmental and urban impacts. Of the many smaller initiatives, one physical coastal protection project stands out in its scale; an Integrated Flood Protection System for Lower Manhattan, encompassing some 10 miles of low-lying coastline, and including a multi-purpose levee in the Financial District. The first phase of this project alone, on the Lower East Side, was estimated to cost more than $300 million.

The damages from Hurricane Sandy on Lower Manhattan were extensive. The coast below West 57th street down to the Battery, the southern tip, and up to East 42nd historically consists of landfill on former marshland that has been used for port activity and, later, housing. The FEMA Flood zone, which wraps around the island, extends up to 5 blocks inland on the Lower East Side. With a storm tide at the Battery, on Manhattan's southern tip, of 4.3 m above mean low water, Sandy resulted in extensive flooding in this zone. The water levels were such that the subway and tunnels also flooded, and that the ConEd electricity plant at East 14th street was knocked out, causing a blackout in much of Lower Manhattan that lasted up to two weeks (SIRR 2013). On the Lower East Side, much of the New York City Housing Authority (NYCHA) public housing was flooded, causing extensive damages to the electrical systems and boilers, and greatly increasing the amount of apartments suffering from mold, with the associated health risks. Hurricane Sandy substantially increased the challenges for NYCHA to preserve the housing on the Lower East Side, home to 32,000 residents, which was already suffering from disrepair because of a financially strained housing authority (ALIGN 2014).

The Big U

The recommendation of the SIRR report to develop an integrated flood protection system for Lower Manhattan became the starting point for the BIG Team in Rebuild by Design (2014), a planning and design competition launched by the Hurricane Sandy Task Force "to promote resilience in the Sandy-affected region" and to "promote innovation by developing regionally-scalable but locally-contextual solutions that increase resilience in the region." The competition set aside HUD (U.S. Department of Housing and Urban development) Community Development Block Grant Disaster Recovery (CDBG-DR) funding specifically to incentivize the implementation of winning projects and proposals (HUD 2016).

During the BIG Team's design process, it became clear to the team that in order to design an implementable and effective project for an integrated flood protection system for Lower Manhattan (the Big U), the team had to understand the project itself as part of the complex adaptive system of New York, adjust and reduce the dimension of the project such that the City can 'digest' it, and, as such, to frame the project within FAP approach that NYC has embraced based on the NPCC recommendations (Fig. 12.1).

Multi-dimensionality

Climate change adaptation measures, according to the NPCC, need to be considered within the cultural, political, economic, environmental and developmental contexts in which climate change occurs. In 2006, Mayor Bloomberg established the Office of Long Term Planning and Sustainability (OLTPS), part of the Mayor's Office, with the goal of developing and implementing a comprehensive plan to create a 'greener', more sustainable city by 2030. The resulting plan, PlaNYC, recognized the importance of promoting both climate change mitigation and adaptation. In the 2011 update to the plan, the City described adaptation strategies in a wide variety of dimensions, including public health, green buildings, natural systems, the waterfront and public engagement, with a practical focus on the resilience of critical infrastructure (Rosenzweig and Solecki 2014, p. 401). In Bloomberg's successor De Blasio's follow-up plan, One New York; the Plan for a Strong and Just City

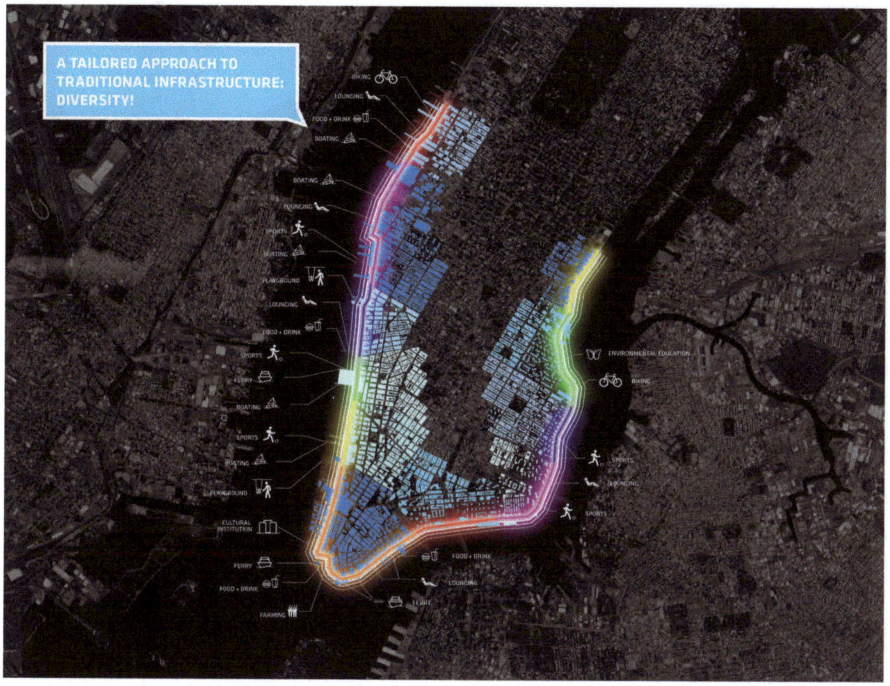

Fig. 12.1 The Big U, concept diagram (own figure)

(OneNYC), this multidimensional aspect was broadened to include the four interrelated dimensions of resilience, growth, sustainability and equity (City of New York 2015).

In the Big U proposal, this multi-dimensionality is best expressed through the BIG Team's introduction of the idea that the project should also be 'social infrastructure'. Taking a cue from the High Line, where a decommissioned piece of infrastructure is given a social function, the team asked itself if it is possible to "conceive of our public infrastructures to come with intended social side-effects from day one" (Ingels 2012). In the Big U proposal, this idea is translated in many different ways, linked to the different geographies of the project, such as an improved waterfront park with better access from the public housing, sports facilities, a new public school with flood protection built in, stormwater flooding protection measures that improved public space in the nearby housing, and community centers. Many of these ideas for social infrastructures were developed in an intensive public engagement process with the local communities. A second idea within the Big U's proposal for this multidimensional approach to adaptation was the development of 'resilient community districts', in which resilience measures (and the associated social infrastructures) were not only designed from the local perspective, but also, when possible, linked to, and in part financed by, future urban development (Rebuild by Design 2014) (Figs. 12.2 and 12.3).

Fig. 12.2 Examples of 'social infrastructure' in the Big U proposal (own figure)

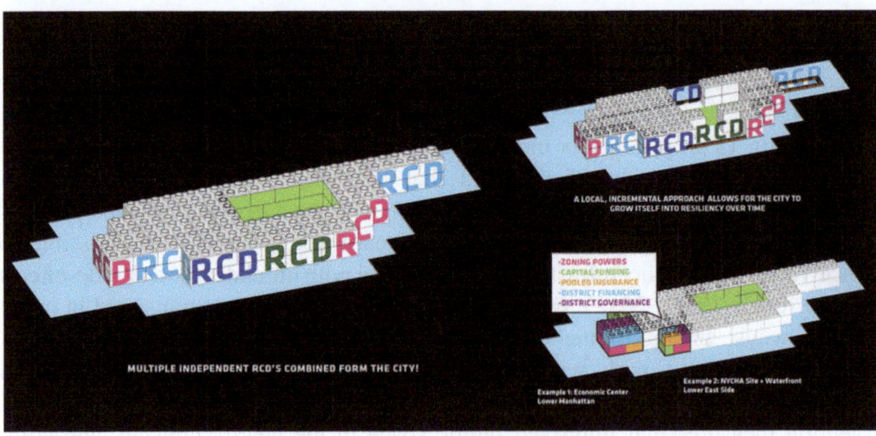

Fig. 12.3 Illustration of the idea of resilient community districts in the Big U proposal (own figure)

Intertemporality

Another important component of NYC's approach to climate change adaptation is the notion that adaptation will evolve through time. A locked-in and too rigid set of measures brings risks similar to those of large projects as described above. New climate science, actual sea-level rise, political and societal changes, adverse side-effects and technological developments all influence decisions about future climate adaptation projects. One way in which New York City has instituted means to monitor path dependency and to avoid lock-ins is the establishment of the NPCC and the Climate Change Adaptation Task force as ongoing bodies, required to regularly update the comprehensive resiliency plan (Rosenzweig and Solecki 2014, p. 402).

In order to accommodate this need for future adaptation, the Big U proposal, rather than conceiving of the flood protection system as one 10-mile long infrastructure, approaches the system as a series of self-contained compartments, 'little U's', with upland connections to higher ground at regular pinch-points. As such, the 'little U's' can each be planned and built independently from each other. When only part of the system is built, or when one of the compartments is breached, the system still provides safety for the (other) finished compartments. This compartmentation also allows the system to be easier adapted incrementally, and to respond better to local issues in building a social infrastructure (Fig. 12.4).

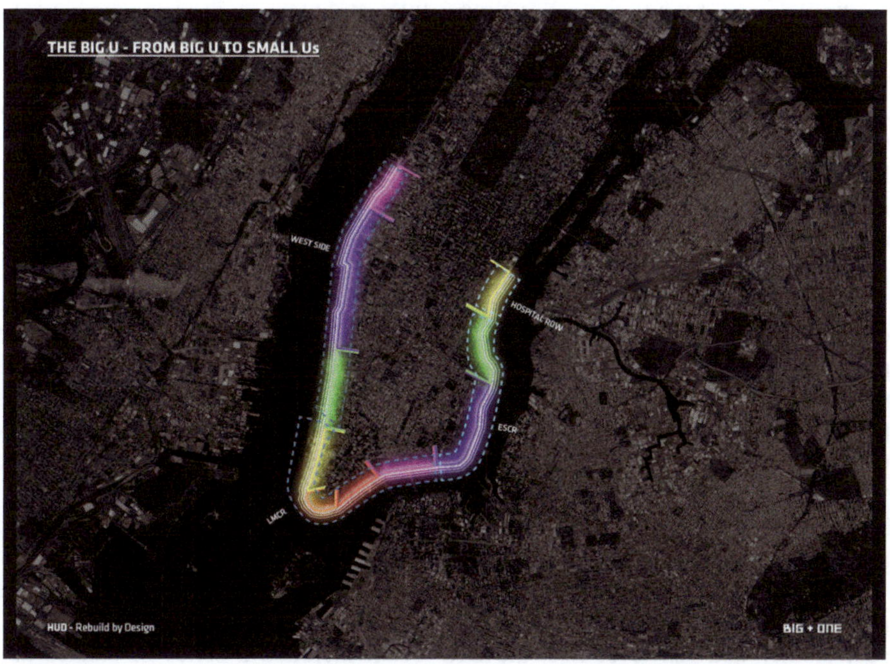

Fig. 12.4 Diagram illustrating 'Little U's', the distinct compartments in the Big U proposal (own figure)

From Design Concept to Implementation

In the case of the Big U, the Rebuild by Design competition's focus on implementable projects, backed by $1 billion of CDBG-DR funding, resulted in a proposal that was as much driven by considerations about the Project Risks that might affect implementation, such as funding requirements, community buy-in, the City's climate change approach and its capacity to handle big projects, as it was about global climate risk mitigation. And in the process of balancing the two types of risks in the proposal and its further elaboration, it is discovered that Project Risk reduction makes it easier to add more dimensions and make the project more responsive to new insights.

The funding requirements had a strong impact on the location, size and nature of the initial Big U proposal. The HUD administered CDBR-DR funding that the U.S. Congress authorized (U.S. Congress 2013) for what became the Rebuild by Design competition consisted of flexible grants to help cities, counties, and States recover from Presidentially declared disasters, especially in low-income areas. Since CDBG-DR funding had to focus on vulnerable populations, the team decided to focus first on New York's Lower East Side, where much of Manhattan's public- and low-income housing was located in the floodplain. Funding also influenced the team in determining the project's individual compartments. With the assumption that the 1 billion US dollars shall be roughly divided equally between New Jersey, New York State, and the City of New York, the team reasoned that any specific funding request should not exceed $400 million. In this context, the strategy of the 'little Us' was not only a condition for eventual success, it also neatly coincided with the size of a first compartment to be implemented at the Lower East Side. A third result of the funding requirements that the team considered was the timetable of the funding: with all possible waivers and extensions, the funding needed to be spent within approximately 5 years (Office of the Federal Register 2014). This requirement was translated by the BIG Team in a proposal that carefully avoided any lengthy regulatory processes, such as those induced by building in the water, or complex planning processes, such as those resulting from building on private land.

In the BIG Team's reasoning, the combination of a conceptual vision for the entire waterfront of Lower Manhattan (the Big U) with a readily implementable plan (one 'little U' between 23rd street and Montgomery street) based on the momentum and funding availability after Hurricane Sandy, would make it possible to make a start with building resilience such that it would trigger the implementation of additional 'little U's' and would allow for further adaptation of the initial 'little U' along the lines of the Flexible Adaptation Pathways strategy in the future. The first implemented project would, as such, become a powerful agent for change and resilience building in the complex adaptive system that is New York City.

In order to demonstrate and explore possibilities for future adaptation of the project, specifically those related to the dimensions of (housing) equity, health, accessibility and sea level rise, the BIG Team included a series of studies and 'toolboxes' that showed future iterations and expansions of the project. These were

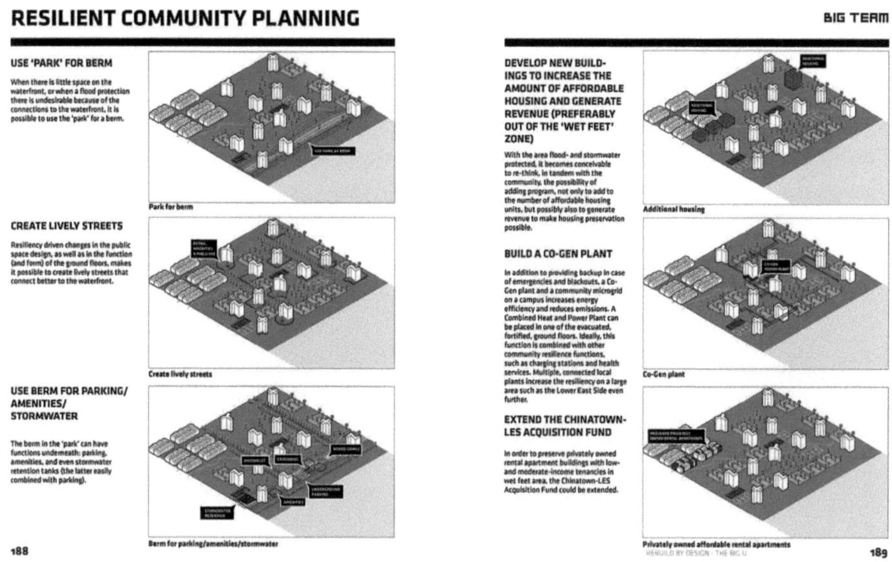

Fig. 12.5 Examples of 'toolboxes' in the Big U proposal (own figure)

explicitly not conditional at this stage, and were drawn without any geographic specifics, such as not to become an obstacle to implementation of the core project (Figs. 12.5 and 12.6).

East Side Coastal Resiliency

Quicker than expected, the Big Team's first strategic goal, to design the project such that it would trigger additional investment in the Big U concept, seemed achieved. After being awarded $335 Million in CDBG-DR funding as a result of the Rebuild by Design competition, New York City launched the East Side Coastal Resiliency project (ESCR) to implement the Big U compartment between East 23rd street and Montgomery street, with a scheduled construction start in late 2017. Shortly after the announcement, residents from the neighboring Two Bridges area, south of Montgomery street, expressed their displeasure about not being included in the first compartment. Also Community Board 1, which covers Manhattan's Financial District, said they hoped to be included in the Big U (Malesevic 2014). These reactions from the community have helped the City expedite their search for funding for future compartments, and have now resulted in the Lower Manhattan Coastal Resilience Project, which is run by the New York City Economic Development Corporations, with an additional $300 million secured for the implementation of future compartments.

The success of the second strategic goal, to make the project itself so digestible by city officials and the community that it can easier integrate other dimensions than

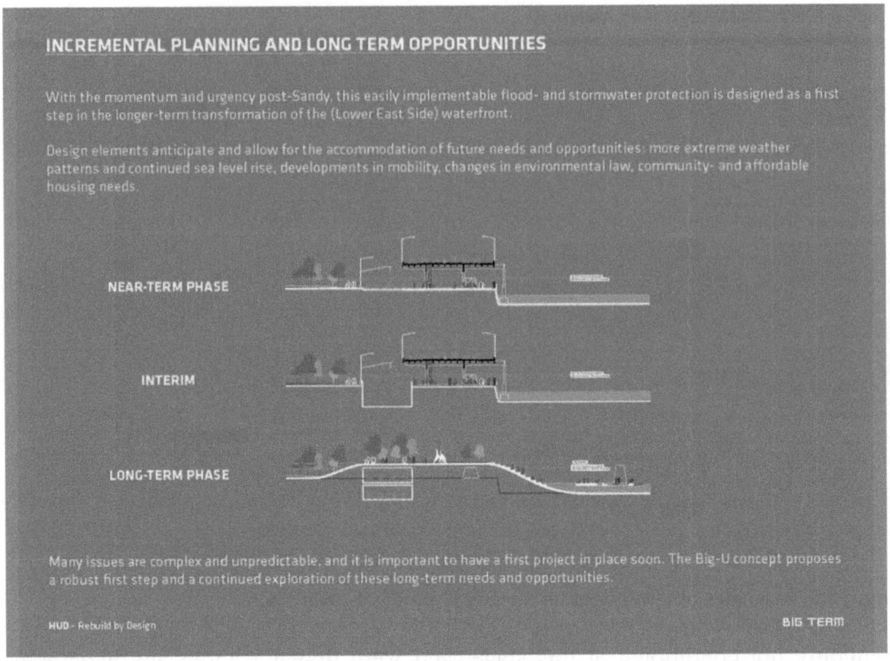

Fig. 12.6 Examples of studies for long term adaptation in the Big U proposal (own figure)

flood protection, is more difficult to assess. At the City's presentation of the initial concept for ESCR in late 2015, it became clear that many of the original features of the design, such as an improved park experience for the local community and improved access from the public housing, both designed in close collaboration with the local community, were still part of the design, although there were concerns about the eventual feasibility. Similarly, the integration of Solar 2, a self-sustaining, solar-powered center for green energy, arts and education, in the project strengthens its climate mitigation credentials (The Lo-Down 2015). At the same time, a high-level integration with resilience efforts by other actors in the area, such as the resilience measures in the NYCHA housing and the protection of the Con-Ed substation, seems to be pursued only in a very limited way at this time. A reason behind that can be that the strict HUD timetable does not allow for time to be expended on coordination efforts.

Conclusions

In tackling the Global Risks related to climate change, New York City, acutely aware of the Project Risks associated especially with large projects, has developed an approach to climate change adaptation that favors multiple smaller initiatives.

These initiatives, so is the City's thinking, can easier address the multiple dimensions of climate change in the city, be adaptive over time, and be more responsive to local issues and local actors. NYC's approach of FAP is based on an understanding of the city as a complex adaptive system in which each initiative generates feedback loops that can be assessed and result in new initiatives, or in the adaptation of existing ones.

The Big U proposal for the flood protection of Lower Manhattan demonstrates the role that design can play in this process. First it does so by integrating multiple dimensions through the notion of 'social infrastructure' in a public design process. And secondly it does so by dimensioning and placing the proposal such that the Project Risks are reduced and as such more manageable by the city. In the Big U, this project risk reduction has made the implementation of a signature project in the near term foreseeable, which has subsequently triggered other climate adaptation projects in the area.

Within the FAP approach, however, the big question is how the initial momentum can be preserved and how future initiatives can slowly transition from hazard mitigation to resilience to adaptation. It can be argued that working at a local scale simultaneously prevents issues from being addressed at a larger scale. At the same time, the strong (visual) image of the Big U as a concept for the flood protection of Lower Manhattan is already generating momentum for similar solutions elsewhere. However, NYC is also very much driven by Project Risks. This has led to a project that might not reach its full potential for the integration of the different dimensions, which can be considered a warning sign. But as the FAP approach is also very much the approach of learning systems, this worry might be premature. In any case, designing the adaptive capacity itself, such that future iterations or designs can be accommodated at a point of time when different sets of risks may manifest, should remain central to New York City's climate adaptation approach and to the projects therein.

Methodology

The BIG-Team (2014), which entered the Rebuild by Design (Rebuild by Design 2014) competition in July 2013, was initiated by the author and co-led by the office of which the author is its founding principal, One Architecture, and BIG (Bjarke Ingels Group). It further consisted of Starr-Whitehouse, James Lima P+D, Arcadis, Buro Happold, Level Engineering, Green Shield Ecology, AEU Consultancy, Project Projects and Parsons School of the Constructed Environments. The author fulfills the role of principal in the team.

Rebuild by Design was a four phased design competition. In the second phase of the competition, the analytical phase, the BIG-team focused its attention on New York City, and Lower Manhattan in particular. In this phase, the team explored

New York City's climate change policy mostly through site visits, desk top and design research, as well as through site visits. At the end of this phase, the concept of the Big U was developed and presented as one of three 'design opportunities'. The Big U was subsequently selected by the Rebuild by Design jury for further development. In the third phase of the competition, the design phase, the team collaborated with the Mayor's Office of Recovery and Resilience (ORR) on the design project, and had a total of 49 meetings with stakeholders, such as City agencies and local community groups. The design project had its primary focus on three sections, or compartments, of the Big U concept, running from East 23rd street to The Battery. The resulting design was presented to the Jury in April 2014 and announced a winner of the competition in June 2014. It was also announced that New York City was awarded $ 335 million to implement the proposed first phase of the project, the compartment between East 23rd street and Montgomery Street. The funding agency, HUD, specified the requirements for implementation in a federal notice in October 2014 (Office of the Federal Register 2014).

New York City designated the Department of Design and Construction (DDC) as the implementing agency, with ORR and the Department of Parks and Recreation (DPR) as initial partners, for the first phase of the project in the Fall of 2014, in and called this first phase the East Side Coastal Resiliency project (ESCR). Five members of the original BIG-Team (BIG, One Architecture, Starr-Whitehouse, James Lima Planning + Development and Arcadis U.S.) are part of the consultant team DDC procured to develop the design for ESCR. In this consultant team, the author continues to fulfill the role of principal. The team has weekly meetings with City Agencies, and frequent meetings with stakeholders and the community. In addition to ESCR, the City is also working on the Lower Manhattan Coastal Resiliency project (LMCR). This planning study, led by New York City's Economic Development Corporation (EDC) in partnership with ORR, started in June 2016. In the consultant team for LMCR, three of the original BIG-Team members (BIG, One Architecture and James Lima Planning + Development) are represented. In both the ESCR and LMCR project, the author is the principal designer of One Architecture. While the author has attempted to base this paper strictly on publicly available material about the three projects, in part because of the non-disclosure agreements that are part of his ongoing consultancy contracts with NYC, his intimate knowledge of, and role in, the decision making processes within the BIG-Team and the City is reflected in this paper.

Concurrent to the design work on ESCR, the author has developed and twice taught a graduate seminar 'Designing with Risk' at PennDesign (the first year together with Ellen Neises). A central element of this seminar was the possible tension between Global Risks and Project Risks. Much of the general thinking about this tension has been developed in conversations with guest lecturers in the seminar, in particular Michael Berkowitz, Andrew Salkin, Marilyn Jordan Taylor, Richard Weller and Peter Hendee Brown, and with students.

References

Alliance for a Greater New York (ALIGN), Alliance for a Just Rebuilding (AJR), Community Voices Heard, Faith in New York, Families United for Racial and Economic Equality (FUREE), Good Old Lower East Side (GOLES), New York Communities for Change, and Red Hook Initiative (2014) Weathering the storm, rebuilding a more resilient New York City housing authority post-sandy, report, March 2014

Arcadis NV (2009), ARCADIS presents conceptual design for storm surge barrier to protect the New York Metropolitan area, from http://www.prnewswire.com/news-releases/arcadis-presents-conceptual-design-for-storm-surge-barrier-to-protect-the-new-york-metropolitan-area-61649597.html. Accessed 4 July 2016

Boschma R (2015) Towards an evolutionary perspective on regional resilience, regional studies. 49, Issue 5

City of New York (2013) A stronger, more resilient New York, SIRR report, via: http://www.nyc.gov/html/sirr/html/report/report.shtml

City of New York (2015) One New York; the plan for a strong and just city

Flyvbjerg B, Bruzelius N, Rothengatter W (2003) Megaprojects and risk: an anatomy of ambition. Cambridge University Press, Cambridge

Hauer ME, Evans JM, Mishra DR (2016) Millions projected to be at risk from sea-level rise in the continental United States. Nat Clim Change

Hill D (ed) (1996) The baked apple? Metropolitan New York in the greenhouse. New York Academy of Sciences 1996, New York

HUD (Department of Housing and Development) (2016) website, http://portal.hud.gov/hudportal/HUD?src=/sandyrebuilding/rebuildbydesign. Accessed 13 Aug 2016

Ingels B (2012) Rethinking social infrastructure, CNN, sun April 22, 2012, from http://www.cnn.com/2012/04/22/tech/rethinking-social-infrastructure/index.html. Accessed 4 July 2016

Keenan JM, King DA, Willis D (2015) Understanding conceptual climate change meanings and preferences of multi-actor professional leadership in New York. J Environ Policy Plann

Lynch D (2012) Engineers lead calls for New York flood barrier, New Civil Engineer, from http://www.newcivilengineer.com/engineers-lead-calls-for-new-york-flood-barrier/8638544.article. Accessed 14 July 2016

Malesevic DS (2014) 2 years after Sandy, who will get shelter from the storm? Only some downtown, http://www.downtownexpress.com/2014/10/29/2-years-after-sandy-who-will-get-shelter-from-the-storm-only-some-downtown/. Accessed 15 July 2016

Martin-Breen P, Anderies JM (2011) 'Resilience: a literature review' Bellagio initiative. IDS, Brighton

New York City Panel on Climate Change (NPCC) (2010) Climate change adaptation in New York City: building a risk management response. In: Rosenzweig C, Solecki W (eds) Prepared for use by the New York City climate change adaptation task force. Annals of the New York Academy of Sciences, New York

Office of the Federal Register (2014) Third allocation, waivers, and alternative requirements for grantees receiving community development block grant (CDBG) Disaster recovery funds in response to Hurricane Sandy. Federal Register/Vol 79, No. 200/Thursday, 16 Oct 2014

Rebuild by Design (2014) http://www.rebuildbydesign.org, and www.rebuildbydesign.org/project/big-team-final-proposal/. Accessed 4 June 2016

Rosenzweig C, Solecki W (ed) (2001) Climate change and a global city: the potential consequences of climate variability and change. Metro East Coast. Report for the U.S. Global Change Research Program, Columbia Earth Institute, National Assessment of the Potential Consequences of Climate Variability and Change for the United States

Rosenzweig C, Solecki WD, Blake R, Bowman M, Faris C, Gornitz V, Horton R, Jacob K, LeBlanc A, Leichenko R, Linkin M, Major D, O'Grady M, Patrick L, Sussman E, ohe G, Zimmerman R (2011) Developing coastal adaptation to climate change in the New York City infrastructure-shed: process, approach, tools, and strategies. Clim Change 106:93–127

Rosenzweig C, Solecki W (2014) Hurricane Sandy and adaptation pathways in New York: lessons from a first-responder city. Global Environmental Change 28

Seto KC, Dhakal S, Bigio A, Blanco H, Delgado GC, Dewar D, Huang L, Inaba A, Kansal A, Lwasa S, McMahon JE, Müller DB, Murakami J, Nagendra H, Ramaswami A (2014) Human settlements, infrastructure and spatial planning. In: Edenhofer O, Pichs-Madruga R, Sokona Y, Farahani E, Kadner S, Seyboth K, Adler A, Baum I, Brunner S, Eickemeier P, Kriemann B, Savolainen J, Schlömer S, von Stechow C, Zwickel T, Minx JC (eds) Climate change 2014: mitigation of climate change. Contribution of working group III to the fifth assessment report of the intergovernmental panel on climate change. Cambridge University Press, Cambridge, United Kingdom and New York, USA

Sjöstedt M (2015) Resilience revisited: taking institutional theory seriously. Ecology and Society 20(4):23

Special Initiative for Rebuilding and Resiliency (SIRR) (2013) A strong, more resilient New York. The city of New York. Available at: http://www.nyc.gov/html/sirr/html/report/report.shtml. Accessed 4 Mar 2017

The BIG Team (2014) Consisted of co-leads BIG (Bjarke Ingels Group) and One Architecture, with Starr-Whitehouse, James Lima P + D, Arcadis, Buro Happold, Level Engineering, Green Shield Ecology, AEU Consultancy, Project Projects and Parsons School of the Constructed Environments

The Lo-Down (2015) Residents voice hyperlocal concerns about East river Levee Plan, in http://www.thelodownny.com/leslog/2015/10/residents-voice-hyperlocal-concerns-about-east-river-levee-plan.html. Accessed 15 July 2016

U.S. Congress (2013) Disaster relief appropriations Act, 2013 (P.L. 113–2, enacted January 29, 2013)

Vermeer M, Rahmstorf S (2009) Global sea level linked to global temperature. Proc Natl Acad Sci USA 106:21527–21532

Wesselink AJ, Bijker WE, de Vriend HJ, Krol MS (2007) Dutch dealings with the selta. In: Nature and culture (vol 2, No 2), Autumn 2007, pp 188–209

World Economic Forum (2016) The global risk report 2016, Geneva

Ysebaerta T, van der Hoekc D-J, Wortelboerc R, Wijsmana JWM, Tangeldera M, Nolted A (2016) Management options for restoring estuarine dynamics and implications for ecosystems: a quantitative approach for the Southwest Delta in the Netherlands. Ocean Coastal Manag 121, pp 33–48

Chapter 13
Coping with Higher Sea Levels and Increased Coastal Flooding in New York City

Vivien Gornitz, Radley Horton, Daniel A. Bader, Philip Orton and Cynthia Rosenzweig

Abstract The 837 km New York City shoreline is lined by significant economic assets and dense population vulnerable to sea level rise and coastal flooding. After Hurricane Sandy in 2012, New York City developed a comprehensive plan to mitigate future climate risks, drawing upon the scientific expertise of the New York City Panel on Climate Change (NPCC), a special advisory group comprised of university and private-sector experts. This paper highlights current NPCC findings regarding sea level rise and coastal flooding, with some of the City's ongoing and planned responses. Twentieth century sea level rise in New York City (2.8 cm/decade) exceeded the global average (1.7 cm/decade), underscoring the enhanced regional risk to coastal hazards. NPCC (2015) projects future sea level rise at the Battery of 28–53 cm by the 2050s and 46–99 cm by the 2080s, relative to 2000–2004 (mid-range, 25th–75th percentile). High-end SLR estimates (90th percentile) reach 76 cm by the 2050s, and 1.9 m by 2100. Combining these projections with updated FEMA flood return period curves, assuming static flood dynamics and storm behavior, flood heights for the 100-year storm (excluding waves) attain 3.9–4.5 m (mid-range), relative to the NAVD88 tidal datum, and 4.9 m (high end) by the 2080s, up from 3.4 m in the 2000s. Flood heights with a 1% annual chance of occurrence in the 2000s increase to 2.0–5.4% (mid-range) and 12.7% per year (high-end), by the 2080s. Guided by NPCC (2013, 2015) findings, New York City has embarked on a suite of initiatives to strengthen coastal defenses, employing various approaches tailored to specific neighborhood needs. NPCC continues its collaboration with the city to investigate vulnerability to extreme climate events, including heat waves, inland floods and coastal storms. Current research entails higher-resolution neighborhood-level coastal flood mapping, changes in storm characteristics, surge height interactions with sea level rise, and stronger engagement with stakeholders and community-based organizations.

V. Gornitz (✉) · R. Horton · D.A. Bader · C. Rosenzweig
Center for Climate Systems Research, Columbia University/NASA
Goddard Institute for Space Studies, 2880 Broadway, New York, NY 10025, USA
e-mail: vmg1@columbia.edu

P. Orton
Stevens Institute of Technology, Davidson Laboratory, Hoboken, NJ 07030, USA

W. Leal Filho and J.M. Keenan (eds.), *Climate Change Adaptation in North America*, Climate Change Management,
DOI 10.1007/978-3-319-53742-9_13

Keywords Sea level rise · Coastal flooding · Climate change adaptation

Introduction

New York City, the largest city in the United States (population 8.5 million), is a leading center of global finance and commerce. Significant economic assets and dense population exposed to coastal flooding line its 837 km of shoreline. The population within the 100-year flood zone[1] exceeds that of other U.S. coastal cities, including Houston, New Orleans, and Miami (SIRR 2013). The New York-Newark-New Jersey area ranks among the world's ten port cities most vulnerable to coastal flooding (by assets) (Hanson et al. 2011). New York's proximity to the coast, its lengthy, highly-developed shoreline, sea level rise, and consequent increasing frequency of coastal flooding pose growing challenges to the city and its residents.

The city has had a long history of widespread coastal flooding (see Table A.1, Appendix II, in NPCC 2015). In September 1821, a large hurricane generated storm floods up to 13 ft (3.96 m) high that flooded parts of lower Manhattan. In September 1893, a Category 1–2 hurricane completely destroyed a small island resort, Hog Island, off the Rockaway shoreline. The Long Island Express, a Category 3 hurricane, crossed central Long Island on September 21, 1938, severely impacting southern New England. Nevertheless, New York City still suffered substantial coastal flooding, augmented by surge amplification[2] (Coch 2015). Since the early twentieth century, at least six major tropical and extratropical storms have struck the city. The December 1992 nor'easter lasted for three tidal cycles at spring tide, reached a peak flood elevation of 2.1 m NAVD88 at the Battery, flooded parts of lower Manhattan, and nearly completely halted metropolitan New York transportation (Gornitz et al. 2002).

Underscoring New York City's vulnerability to powerful coastal storms, Hurricane Sandy[3] generated the highest surge levels in 300 years on October 29, 2012 (Talke et al. 2014; Orton et al. 2016). The storm caused 43 fatalities, hospital and nursing home evacuations, inundation of subways and tunnels and 17% of the city's land area, power outages affecting nearly two million people, and major transportation disruptions. Over 70,000 buildings were flooded and/or damaged. Storm damages totaled an estimated $19 billion (SIRR 2013). A rare coincidence of meteorological and tidal conditions, together with ongoing sea level rise, exacerbated the extent and severity of the flooding.

---

[1]The area flooded by a storm with a 1% probability of occurrence per year, or average likelihood of occurrence of one per century.

[2]Landward funneling of storm surge by southeast winds due to near right-angle geometry of New Jersey and Long Island shorelines.

[3]Hurricane Sandy had weakened to a tropical storm and merged with a frontal system by the time it reached New York City.

New York City has long pioneered climate change adaptation efforts (Rosenzweig and Solecki 2014; Rosenzweig et al. 2011; NPCC 2010). Following Hurricane Sandy, former Mayor Michael Bloomberg re-convened the New York City Panel on Climate Change (NPCC), a group of climate, social science, and risk-management experts, to advise the city on climate change and adaptation issues, working closely with city and regional agencies. The city subsequently developed a comprehensive adaptation approach to reduce present and future climate-related risks, incorporating NPCC climate change projections. A number of resiliency measures and new projects are now underway (SIRR 2013; NYC Hazard Mitigation Plan 2014; New York City PlaNYC 2014, 2015).

This paper highlights recent NPCC research designed to assist New York City in its present and planned adaptation responses to accelerated sea level rise and associated coastal flooding risks. Other climate change impacts (e.g., health; costs of urban and ecosystems adaptations) are treated elsewhere (e.g., NPCC 2015; Rosenzweig et al. 2011, 2012; Neumann et al. 2015; Estrada et al. 2015). Following the Flexible Adaptation Pathways described in NPCC (2010), the City is preparing for short-term coastal flood risks, while laying the framework for adaptation to higher sea levels and more frequent and severe future flooding. This overview of current NPCC findings outlines essential guidelines in anticipating future climate change hazards. Some examples are given of the City's coastal adaptation activities.

Climate Trends, Sea Level Rise, and Coastal Flooding

New York experiences hot humid, summers and cold winters. Annual temperatures climbed 0.2 °C per decade between 1900 and 2013 (NPCC 2015). Yearly rainfall ranges between 109 and 127 cm and has increased by 2.0 cm per decade between 1900 and 2013. However, annual to multi-decadal variability has grown markedly over the past 40 years (NPCC 2015). Extreme rainfall events are very likely to increase, with greater risk of street and basement flooding.

Before the early 19th century, sea level rose 0.9–1.2 mm/year along the East Coast of the United States (Gehrels and Woodworth 2013), largely due to regional subsidence from glacial isostatic adjustments after the last ice age. Within the last 100–150 years, regional sea level has accelerated relative to the preceding 2000 years (Kopp et al. 2016; Engelhart et al. 2011). Twentieth-century regional sea level rise ranged between 2.4 and 4.1 mm/year in the New York City metropolitan area[4] (NOAA 2015), as compared to the global average rise of 1.7 mm/year (IPCC 2013; Hay et al. 2015).

The NPCC (2015) has developed a multi-component methodology for projecting future sea rise for New York City. Components include oceanographic changes, ice

[4]Around ¼ to ½ stems from glacial isostasy-related land subsidence (NOAA 2015; PSMSL). http://www.psmsl.org/train_and_info/geo_signals/gia/peltier/index.php, posted August 13, 2012.

mass losses corrected for gravitational and glacial isostatic adjustments, and anthropogenic land water storage, for an ensemble of 24 CMIP global climate models and two climate change scenarios (RCP 4.5, RCP 8.5), as well as literature review and expert judgment. Sea level rise, relative to the 2000–2004 base period, is calculated for the 10th, 25th, 75th and 90th percentiles of a model-based distribution and estimated ranges from the literature. Several other studies have taken a similar regionalized approach (e.g., Kopp et al. 2014; Perrette et al. 2013).

The NPCC (2015) projects a mid-range sea level rise at the Battery (southern tip of lower Manhattan) of 28–53 cm by the 2050s and 46–99 cm (mid-range) by the 2080s, relative to 2000–2004. High-end estimates (90th percentile) could reach 76 cm by the 2050s, 147 cm by the 2080s, and 190 cm by 2100 (Fig. 13.1).

Major remaining uncertainties include the extent of ice sheet melting at higher temperatures, changes in storm characteristics, ocean circulation, and their coastal impacts. Data gaps and incomplete representation of physical processes in climate models also introduce uncertainty. In addition, future socioeconomic trends (e.g., population, shoreline development) remain difficult to project.

Flood risk exhibits a strong seasonality. New York's most severe floods occur during hurricane season (June–November), whereas more frequent, smaller flood events happen during cooler months. The three highest storm tides back to 1821 occurred during hurricane season. However, flooding from extratropical cyclones in the cool season account for 15 of the top 20 events (Orton et al. 2016; Talke et al. 2014).[5] Extratropical-cyclone storm surges are typically smaller, last longer, and often cause flooding when they coincide with high tide.

Most recently, Hurricane Sandy produced the highest recorded water level (3.50 m MSL) at the Battery (the southern tip of Manhattan) in nearly 200 years of record-keeping. Several unusual events coalesced to produce this record level: strong easterly winds driving water ashore, low atmospheric pressure, maximum storm surge at high tide near full moon, surge amplification, plus ongoing sea level rise (0.44 m since 1856), which exacerbated the extent of coastal flooding and storm damages (Horton and Liu 2014; Talke et al. 2014).

No consensus has emerged concerning changing frequencies of North Atlantic tropical cyclones (hurricanes) since the 1970s (Peduzzi et al. 2012); however, the number of strongest cyclones (Category 3–5 on the Saffir-Simpson scale) has grown (Elsner et al. 2008; Horton and Liu 2014). Furthermore, maximum intensities of tropical cyclones have shifted northward within the past three decades (Kossin et al. 2014). If these trends persist, flood hazards for mid-Atlantic coastal communities, previously less exposed to such damaging storms, could increase. In addition to intensifying natural hazards, growing coastal populations and increased development will magnify hurricane risks and damages (Mendelsohn et al. 2012; Hallegatte et al. 2013; Estrada et al. 2015).

[5]A tropical cyclone forms at low latitudes over warm ocean water, exhibiting low atmospheric pressures, strong winds, and heavy rains. Severe tropical cyclones are called hurricanes in the U.S. and Caribbean, typhoons in the Pacific Ocean, and cyclones in the Indian Ocean. An extratropical cyclone is a large-scale mid-latitude cyclonic storm, often associated with frontal systems.

Sea Level Rise Baseline (2000 – 2004)	Low-estimate (10th percentile)	Middle range (25th to 75th percentile)	High-estimate (90th percentile)
2020s	+ 2 in 0.05 m	+ 4 in to 8 in 0.10 to 0.20 m	+ 10 in 0.25 m
2050s	+ 8 in 0.20 m	+ 11 in to 21 in 0.28 to 0.53 m	+ 30 in 0.76 m
2080s	+ 13 in 0.33 m	+ 18 in to 39 in 0.46 to 0.99 m	+ 58 in 1.47 m
2100	+ 15 in 0.38 m	+ 22 in to 50 in 0.56 to 1.26 m	+ 75 in 1.91 m

Based on 24 GCMs and two Representative Concentration Pathways. Shown are the low-estimate (10th percentile), middle range (25th percentile to 75th percentile), and high-estimate (90th percentile).

Fig. 13.1 New York City sea level rise projections for the 2020s, 2050s, 2080s, and 2100 (NPCC 2015)

How will tropical storms change in the future? While changes in the total number of hurricanes remain uncertain, the most intense cyclones, with heavier rainfall, are likely to increase (Bender et al. 2010; Knutson et al. 2010). The higher atmospheric humidity in a warmer climate will also likely generate stronger precipitation (Knutson et al. 2010; Horton and Liu 2014). The consequences for the New York metropolitan region remain unclear because of the high variability of individual storm tracks and little knowledge of their future courses.

Extratropical cyclones have not yet demonstrated any consistent large-scale changes and projected changes in activity remain uncertain (Colle et al. 2015). Nevertheless, rising sea level has produced greater coastal flooding, even from minor storms (Strauss et al. 2016; Sweet and Park 2014; Sweet et al. 2014). Heavy precipitation from extratropical cyclones can also magnify coastal flood impacts, because of the storms' large synoptic scale and duration over several tidal cycles that often cause overflowing of stormwater drainage systems.

A static flood model maps future flood zones by adding sea level rise to storm surge and assumes that the water spreads evenly across the area. Using NPCC sea level rise projections (Fig. 13.1), updated Federal Emergency Management Agency (FEMA 2013) flood return plots, and assumed constant storm behavior, flood heights for the 100-year storm (excluding waves) in New York City would rise from 3.4 m in the 2000s to 3.9–4.5 m by the 2080s (mid-range) and 4.9 m in the high estimate (90th percentile). The annual likelihood of such a flood would increase from 1 to 2.0–5.4% (mid-range) and 12.7% in the high estimate, by the 2080s. Future flood zones expand considerably, depending on local topography (Fig. 13.2).

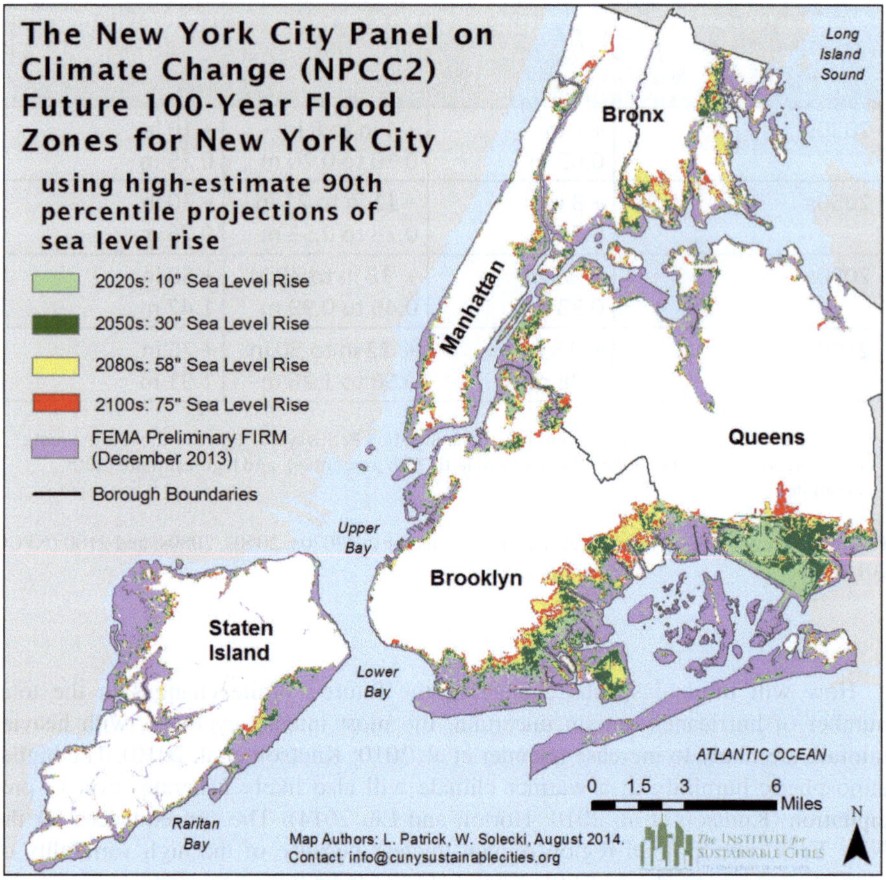

Fig. 13.2 Areas potentially at risk to the 100-year flood by the 2020s, 2050s, 2080s, with projected sea level rise (90th percentile) (NPCC 2015) (Due to limitations in accuracy of the datasets, models, and methodology, this map provides only a general guide to areas exposed to extreme flooding and should not be used to assess actual flood risk, property value, or insurance costs)

Static flood mapping assumes spatially uniform flood elevations over affected areas. However, peak water elevations during a major hurricane or extra-tropical cyclone can vary considerably inland. Therefore, a hydrodynamic model was also applied that explicitly calculates forces acting on the water and its resulting movement inland (NPCC 2015). Hydrodynamic flood modeling uses a physics-based computer simulation of factors that include wind, atmospheric pressure, and friction to calculate flood heights. Results from both static and hydrodynamic modeling were compared for selected test sites around the New York metropolitan area, for the same time periods at the 90th percentile, to emphasize high-end risks. A comparison of both approaches yielded differences in flood heights less than 0.15 m for most localities. Localities displaying larger differences were more geographically widespread in the case of hurricanes (tropical cyclones) than for extra-tropical storms (Orton et al. 2015a).

The 20th century rise in sea level has already noticeably altered the frequency and duration of coastal flood events in New York City, and elsewhere in the U.S., even although basic storm characteristics remain unchanged (e.g., Strauss et al. 2016; Sweet and Park 2014; Sweet et al. 2014). Many neighborhoods already experience "nuisance floods", or frequent street and basement flooding during above-average tides or minor storms. For example, the average number of days of nuisance flooding[6] in New York City increased from 1.8 (1956–1960) to 4.0 (2006–2010). Several other U.S. East Coast cities, such as Atlantic City, Baltimore, and Washington, D.C. have sustained more than 10-fold increases in minor flooding over this period (Sweet and Park 2014).

Vulnerability to Sea Level Rise and Coastal Storms

New York City lies at the center of a much larger metropolitan region, home to 20 million inhabitants spread over three adjacent states—Connecticut, New York, and New Jersey (U.S. Census 2010). Many essential New York City infrastructure components lie along the waterfront. At high risk to coastal flooding are three high-volume international airports, shipping facilities, large segments of a major network of bus and rail transit systems (subways, commuter and intercity trains), many subway, tunnel, and bridge entrances, the majority of city wastewater treatment plants, oil tanks and refineries, and a significant percentage of the power plants and telecommunication networks that lie within current or projected 100-year flood zones (Solecki et al. 2015; NYC Hazard Mitigation Plan 2014; Zimmerman and Faris 2010).

Storm-surge damage to wastewater treatment facilities can produce combined sewer overflows and pollution of waterways (NYC Hazard Mitigation Plan 2014). The location of many key regional oil terminals, pipeline, and refineries within the FEMA 100-year flood zone places them at high risk to storm surge. Disruptions of fuel supply lines could also interfere with power and steam generation, which in turn would impact telecommunication services that rely on electrical power. Buildings damaged by severe coastal erosion and/or permanent inundation in low-lying areas would require costly retro-fitting or even eventual relocation. This example clearly illustrates the cascading impacts of coastal flooding on interdependencies between urban sectors.

[6]The National Weather Service, NOAA, defines a nuisance flood in NYC as one in which water levels exceed 0.65 m above mean higher high water, MHHW. The specific threshold varies by city.

The increasing incidence of minor nuisance flooding creates a growing public inconvenience because of potential damages to low-lying infrastructure (including freshwater and wastewater treatment plants, power stations, communications installations, and transportation infrastructure). Structures not designed to withstand repeated and lengthening saltwater exposure would require more frequent and higher repair or replacement costs (Solecki et al. 2015; Horton and Rosenzweig 2010).

Approximately 400,000 people living in the current 100-year floodplain [delineated by the Federal Emergency Management Agency (FEMA 2013) Preliminary Flood Insurance Rate Maps] are vulnerable to storm surge flooding. Although diverse income groups share a similar exposure to coastal flood hazards in New York City, poor, elderly, and disabled people are less able to cope with natural disasters such as Hurricane Sandy, owing to limited mobility, means of evacuation, and frequently, language barriers, in spite of multi-lingual emergency alerts across diverse media (Buonaiuto et al. 2011; NPCC 2015). Prolonged power outages following Sandy burdened elderly people living on the upper floors of high-rise buildings, dependent on neighbors for water, food, and medications. Several middle-class neighborhoods in Staten Island and Breezy Point, on the Rockaway Peninsula, directly exposed to the combined impacts of surge and waves on older structures, were among the hardest hit. Fires triggered by downed power lines and spread by strong winds across three neighborhoods on Rockaway Peninsula added to the storm damages.

Adaptation to Rising Seas and Storm Surges

The destruction associated with major coastal storms provides coastal-zone managers an experience that foreshadows increasing future risks. Hurricane Sandy in October 2012 produced the greatest flooding and damages in 300 years and acted as a "wake-up call" to further action (Rosenzweig and Solecki 2014).

Following Sandy, New York City initiated a major effort to enhance resiliency to future coastal flooding risks (SIRR 2013; NYC 2014a). Building codes have been strengthened: FEMA's National Flood Insurance Program (NFIP), which covers flood insurance for property owners residing within a FEMA-defined 100-year flood zone, now requires elevation of the lowest occupied floor of new or renovated buildings within A-zones[7] in the 100-year floodplain above the base flood elevation, BFE (the 100-year flood height, including waves), with lower enclosed spaces used only for parking, storage, or building access. Utilities must lie above the BFE or be designed to exclude water. New or renovated buildings within V-zones[8] must sit on pilings or have an open foundation below the BFE.

[7]A-Zones within the 100-year floodplain exposed to wave heights under 1.5 ft (0.46 m).

[8]V-Zones within the 100-year floodplain exposed to wave heights 3 ft (0.91 m) or higher.

Importantly, stricter city standards require one and two-family houses to provide ~0.6 m of extra freeboard protection above the 100-year flood level; most other buildings must have at least ~0.30 m. This allows an extra margin of safety against flooding, higher storm surges, and sea level rise due to climate change. The 2013 Flood Resilience Zoning Text Amendment eases some regulatory barriers within FEMA-designated flood zones in order to expedite reconstruction of storm-damaged houses and facilitate compliance with higher FEMA-required building flood elevations and new City building codes.

Other structural flood-resiliency enhancements incorporate building flood-proofing (New York City 2014a). Examples of risk reduction strategies include site protection with temporary or permanent barriers, e.g., a floodwall or berm, incorporation of flood-resistant materials into construction or retrofitting of buildings, and raising mechanical and electrical systems from basements. Critical infrastructure are being protected by lifting air vent gratings over subways to prevent water from entering, elevating track switches and electrical equipment, increasing pumping capacity in underground rail and tunnel systems, and building new electric power substations beyond the flood zone. Streets subject to frequent nuisance flooding can be raised, as for example in Averne by the Sea, on Rockaway Peninsula (NYC 2014b).

Programs to strengthen coastal defenses are tailored to specific neighborhood needs (SIRR 2013; New York City 2014a, b). These include raising bulkheads, seawalls, replenishing beaches, building local levees and storm surge barriers, and restoring wetlands and beach dunes. Permanent or deployable floodwalls can be placed at critical points before an impending severe flood storm event. Several smaller, strategically-situated local storm surge barriers are proposed for flood-prone neighborhoods such as Gowanus, Newtown Creek, and Coney Island, Brooklyn (SIRR 2013).

Sea level rise is likely to exacerbate coastal erosion, especially during severe storms like Sandy. Coastal erosion and flooding can severely damage structures, and if unchecked, can undermine foundations, ultimately leading to building collapse. Continual erosion of the city's sandy beaches currently requires periodic nourishment with sand dredged from offshore by the U.S. Army Corps of Engineers. Exposed, ocean-facing shorelines face high beach erosion risks during coastal storms. In particular, three "erosion hotspot" neighborhoods (south shore Staten Island, Coney Island, and Rockaway peninsula) are designated as Coastal Erosion Hazard Areas (CEHA), which require written approval for changes in land use.

Many engineered "hard" structures (such as seawalls, revetments, and groins) can be strengthened to prevent erosion and mitigate flooding. In addition, the City pursues "soft" options, such as re-creation of native wetlands, and constructing Living Shorelines with a mix of plants, soils, and rocks to stabilize the shoreline and develop new habitat (Fig. 13.3; NYC 2014b). Brooklyn Bridge Park located along a former industrial and shipping waterfront on the East River, is a model example of a Living Shorelines approach to coastal defense. The park's shoreline consists of a mix of bulkheads, riprap, newly-planted salt marsh wetlands, salt-tolerant vegetation, and refurbished piers for multi-purpose recreational use.

Fig. 13.3 Example of a living shoreline. Creating a soft edge shoreline, Brooklyn Bridge Park, New York City. *Source* Department of City Planning, City of New York City, 2011

Coastal wetlands in Jamaica Bay provide important ecological services: wildlife habitat, recreation opportunities, and water pollution filtration. They likely also help mitigate flooding by reducing waves (Orton et al. 2015b). However, the salt marshes have greatly deteriorated over the past century (Hartig et al. 2002; Boger et al. 2013, 2016) due in part to historic sea level rise, but particularly to past anthropogenic activities, including extensive dredging; landfill operations; bulk-headed, hardened shorelines; curtailment of sediment supply; and excess nitrogen loading from nearby wastewater treatment plants. An ongoing wetlands restoration program in Jamaica Bay aims to preserve habitat and attenuate storm-driven waves, in partnership with the New York City Department of Environmental Protection, NYC Department of Parks, New York State, and U.S. Army Corps of Engineers. Restoration activities entail replanting of marsh grass, recreating ribbed mussel habitat, and installing pilot wave attenuators[9] to lessen shoreline loss and stimulate sediment accretion, improve management of stormwater overflow, harvest excess sea lettuce (Ulva), and evaluate a pilot oyster bed restoration project (NYC DEP 2015a, b).

The New York City Department of City Planning (DCP) has also initiated a Resilient Neighborhoods program, to help communities in the flood zone adapt to

[9]An anchored island of floating material intended to deflect or minimize wave energy.

and reduce flood risks, coordinate rebuilding efforts with appropriate land use planning, and tailor neighborhood-specific development (NYC DCP 2016; Rebuild by Design 2016). Construction will begin in 2017 on a section of the Dryline project (formerly, the Big U), which consists of a protective ribbon of berms and waterfront parks designed to shield lower Manhattan neighborhoods against future floods and higher sea levels, while also providing outdoor recreation opportunities (LafargeHolcim Foundation 2016).

Much of New York City's resiliency efforts described above respond to current or near-term climate risks. However, the NPCC, in close collaboration with City stakeholders, lays the scientific foundation for stronger measures to help mitigate impacts of elevated sea levels and enhanced coastal flooding. As mandated by City legislation,[10] the NPCC will periodically update its climate change projections, incorporating the latest scientific information. Therefore, ongoing fundamental research to address remaining climate change uncertainties plays an essential role in adaptation planning in New York City. The NPCC, working closely with the City, addresses key knowledge gaps, such as potential ice sheet instabilities, changes in storm characteristics, and surge height interactions with sea level rise. Current and ongoing research focuses on the evolution of extreme events, including coastal and inland flooding, and their changes in intensities and frequencies. Other related research efforts include developing a high-resolution, interactive dynamic mapping tool to illustrate evolving coastal flood hazards, creating indicators to monitor and assess physical and societal climate change risks, investigating ways to reduce societal vulnerabilities and improve resiliency capabilities at neighborhood scales. The city, informed by NPCC research, plans to continue to strengthen its coastal defenses and adjust its adaptation strategies in view of anticipated climate change-induced impacts on coastal areas.

Conclusions

Using an innovative localized approach, the NPCC finds that sea level at the Battery could climb to 46–99 cm (mid-range, 2080s), and 1.9 m by 2100 as an upper bound. Assuming unchanged storm behavior, annual likelihoods for the 100-year flood would increase from 1 to 2.0–5.4%, with an upper limit of 12.7% by the 2080s, and significant expansion of the flood zone. Based on NPCC (2013, 2015) findings, New York City has initiated a program to strengthen coastal defenses, employing approaches tailored to specific neighborhood needs. NPCC continues its collaboration with the city to investigate vulnerability to climate change.

Coastal urban centers worldwide are increasingly aware of the growing risk of coastal flooding and sea level rise and are taking steps toward climate change adaptation, as summarized in publications of the Urban Climate Change Research

[10]Local Law 42.

Network (UCCRN) (Rosenzweig et al. 2011; Dawson et al. 2017). New York City's approach of using the best-available local climate-change data is now officially a Federal policy via Executive Order 11988 (U.S. White House 2015) and can be adapted for use in other cities. The NPCC sea level rise projections are included in a U.S. interagency planning tool devised for the New York City area (USACE 2014). New York City's proposed strategies for enhancing coastal urban resiliency stand as a model for other urban coastal centers to prepare for climate change (Rosenzweig and Solecki 2014; Horton et al. 2016).

References

Bender MA, Knutson TR, Tuleya RE, Sirutis JJ, Vecchi GA, Garner ST, Held IM (2010) Modeled impact of anthropogenic warming on the frequency of intense Atlantic hurricanes. Science 327:454–458

Boger R, Essrog J, Christiano M (2013, 2016) Shoreline changes in Jamaica Bay, Gateway National Recreation Area, 1924–2006: implications for shoreline restoration, PARK Science 20(3), Integrating research and research and resource management in the national parks. http://www.nature.nps.gov/parkscience/index.cfm?ArticleID=648 (published online 5 Apr 2016)

Buonaiuto F, Patrick L, Gornitz V, Hartig E, Leichenko R, Stedinger J, Tanski J, Vancura P, Waldman J (2011) Coastal zones. In: Rosenzweig C, Solecki W, DeGaetano A, O'Grady M, Hassol S, Grabhorn P (eds) Responding to climate change in New York State: the ClimAID integrated assessment for effective climate change adaptation in New York State, vol 1244. Annals of the New York Academy of Sciences, pp 121–162

Coch N (2015) Unique vulnerability of the New York–New Jersey metropolitan area to hurricane destruction. J Coastal Res 31(1):196–212

Colle BA, Booth JF, Change EKM (2015) A review of historical and future changes of extratropical cyclones and associated impacts along the U.S. East Coast Curr Clim Change Rep 1:125–143

Dawson R, Shah Alam Khan M, Gornitz V, Lemos MF et al (2017) Coastal zones in urban areas. In: Rosenzweig C, Solecki W, Romero-Lankao P, Mehrotra S, Dhakal S, Ali Ibrahim S (eds) Climate change and cities: second assessment report of the Urban climate change research network. Cambridge University Press, Cambridge, UK and New York

Elsner JB, Kossin JP, Jagger TH (2008) The increasing intensity of the strongest tropical cyclones. Nature 455:92–95

Engelhart SE, Horton BP, Kemp AC (2011) Holocene relative sea-level changes along the United States Atlantic coast. Oceanography 24:70–79

Estrada F, Botzen WJ, Tol RSJ (2015) Economic losses from US hurricanes consistent with an influence from climate change. Nat Geosci 8:880–884

Federal Emergency Management Agency (FEMA) (2013) Flood insurance study, City of New York, New York: Bronx County, Kings County, New York County, Queens County, Richmond County, Flood Insurance Study Number 360497V000B, ver. 1.0.0.0. U.S.: FEMA. http://www.nyc.gov/html/sirr/downloads/pdf/360497NY-New-York-City.pdf

Gehrels WR, Woodworth PL (2013) When did modern rates of sea-level start? Glob Planet Change 100:263–277

Gornitz V, Couch S, Hartig EK (2002) Impacts of sea level rise in the New York City metropolitan area. Glob Planet Change 32(1):61–88

Hallegatte S, Green C, Nicholls RJ, Corfee-Morlot J (2013) Future flood losses in major coastal cities. Nat Clim Change 3(9):802–806

Hanson S, Nicholls R et al (2011) A global ranking of port cities with high exposures to climate extremes. Clim Change 104:89–111

Hartig EK, Gornitz V, Kolker A, Mushacke F, Fallon D (2002) Anthropogenic and climate change impacts on salt marsh morphology in Jamaica Bay, vol 22. Wetlands, New York City, pp 71–89

Hay C, Morrow E, Kopp RE, Mitrovica JX (2015) Probabalistic reanalysis of twentieth century sea-level rise. Nature 517:481–484

Horton R, Liu J (2014) Beyond hurricane sandy: what might the future hold for tropical cyclones in the North Atlantic. J Extreme Events 1(1): 450007 (30p). doi:10.1142/S2345737614500079

Horton R, Rosenzweig C (2010) Climate risk information. In: Rosenzweig C, Solecki W (eds) Climate change adaptation in New York City: building a risk management response, vol 1196. Annals of the New York Academy of Sciences, New York, pp 149–228

Horton R, Rosenzweig C, Solecki W, Bader B, Sohl L (2016) Climate science for decision-making in the New York metropolitan region. In: Parris AS, Garfin GF, Dow K, Meyer R, Close SL (eds) Climate in context: science and society partnering for adaptation. Wiley, Chichester, pp 51–74

IPCC (2013) Summary for policymakers. In: Alexander L, Allen S, Bindoff L, Church J et al (eds) Climatic change: the physical basis. Contribution of working group I to the fifth assessment report of the intergovernmental panel on climate change. Cambridge University Press, Cambridge

Knutson TR, McBride JL, Chan J, Emanuel K et al (2010) Tropical cyclones and climate change. Nat Geosci 3:157–163

Kopp RE, Horton RE, Little CM, Mitrovica JX, Oppenheimer M et al (2014) Probabalistic 21st and 22nd century sea-level projections at a global network of tide gauge sites. Earth's Future 2:383–406

Kopp RE, Kemp AC, Bittermann K, Horton BP et al (2016) Temperature-driven global sea-level variability in the Common Era. Proc Natl Acad Sci 113(11):E1434–E1441

Kossin JP, Emanuel KA, Vecchi GA (2014) The poleward migration of the location of tropical cyclone maximum intensity. Nature 509:349–352

LafargeHolcim Foundation (2016) The dryline. News release 7 Mar 2017. http://lafargeholcim-foundation.org/projects/the-dryline. Accessed 7 June 2016

Mendelsohn R, Emanuel K, Chonabayashi S, Bakkensen L (2012) The impact of climate change on global tropical cyclone damage. Nat Clim Change 2:205–209

Neumann JE, Emanuel K, Ravela S, Ludwig L, Kirshen P, Bosma K, Martinich J (2015) Joint effects of storm surge and sea-level rise on U.S. coasts: new economic estimates of impacts, adaptation, and benefits of mitigation policy. Clim Change 2015 129:337–349

New York City (2014a) NYC's risk landscape: a guide to hazard mitigation. flooding, Chapter 4.3. http://www1.nyc.gov/assets/em/downloads/pdf/hazard_mitigation/nycs_risk_landscape_a_guide_to_hazard_mitigation_final_highres.pdf. Accessed 7 Mar 2016

New York City (2014b) NYC's risk landscape: a guide to hazard mitigation. coastal erosion, Chapter 4.2. http://www1.nyc.gov/assets/em/downloads/pdf/hazard_mitigation/nycs_risk_landscape_chapter_4.2_coastalerosion.pdf. Accessed 29 Feb 2016

New York City (2016) Department of City Planning (DCP) New York City Planning, Resilient Neighborhoods. http://www1.nyc.gov/site/planning/plans/resilient-neighborhoods.page. Accessed 8 Mar 2016

New York City Department of Environmental Protection (2015a) Wetland protection measures launched in the rockaways. News release 15–82, 17 Sep 2015. http://www.nyc.gov/html/dep/html/press_releases/15-082pr.shtml#.WL8v0HpcjXM. Accessed 7 Mar 2017

New York City Department of Environmental Protection (2015b) Department of Environmental Protection completes $455 million comprehensive upgrade of Paerdegat Basin area. News release 15–90, 26 Oct 2016. http://www.nyc.gov/html/dep/html/press_releases/15-090pr.shtml#.WL8wrXpcjXM. Accessed 7 June 2016

New York City Hazard Mitigation Plan (2014) New York City office of emergency management. http://www.nyc.gov/html/oem/html/planning_response/planning_hazard_mitigation_2014.shtml

New York City, PlaNYC (2014) Progress report 2014: Sustainability & Resiliency. http://www.nyc.gov/html/planyc2030/downloads/pdf/140422_PlaNYCP-Report_FINAL_Web.pdf. Accessed 18 Sep 2015

New York City, PlaNYC (2015) A stronger, more resilient New York: east side coastal resilience project. http://www.nyc.gov/html/planyc/downloads/pdf/150319_ESCR_FINAL.pdf. Accessed 18 Sep 2015

NOAA (2015) Tides & currents. http://www.tidesandcurrents/noaa.gov. (December 2015)

NPCC (2010) New York City panel on climate change 2010: building a risk management response. In: Rosenzweig C, Solecki W (eds) Annals of the New York Academy of Sciences 1196, 354 pp

NPCC (2013) New York City panel on climate change climate risk information 2013. In: Rosenzweig C, Solecki W (eds) Climate risk information: observations, climate change projections, and maps. Prepared for the city of New York special initiative on rebuilding and resiliency. http://www.nyc.gov/html/planyc2030/downloads/pdf/npcc_climate_risk_information_2013_report.pdf. Accessed 21 Apr 2016

NPCC (2015) New York City panel on climate change 2015: building the knowledge base for climate resiliency. In: Rosenzweig C, Solecki W (eds) Annals of the New York Academy of Sciences, vol 1336, 150 pp

Orton P, Vinogradov S, Georgas N, Blumberg A, Lin N, Gornitz V, Little C, Jacob K, Horton R (2015a) New York City panel on climate change 2015, Chapter 4: dynamic coastal flood modeling. Ann N Y Acad Sci 1336(1):56–66

Orton PM, Talke SA, Jay DA, Yin L, Blumberg AF, Georgas N, Zhao H, Roberts HJ, MacManus K (2015b) Channel shallowing as mitigation of coastal flooding. J Mar Sci Eng 3 (3):654–673. doi:10.3390/jmse3030654

Orton PM, Hall TM, Talke S, Blumberg AF, Georgas N, Vinogradov S (2016) A validated tropical-extratropical flood hazard assessment for New York Harbor. J Geophys Res 121. doi:10.1002/2016JC011679

Peduzzi P, Chatenoux B, Dao H, De Bono A, Herold C et al (2012) Global trends in tropical cyclone risk. Nat Clim Change 2:289–294

Perrette M, Landerer F, Riva R et al (2013) A scaling approach to project regional sea level rise and its uncertainties. Earth Syst Dyn 4:11–29

Rebuild by Design (2016). http://www.rebuildbydesign.org/. Accessed 1 Apr 2016

Rosenzweig C, Solecki W (2014) Hurricane Sandy and adaptation pathways in New York: lessons from a first-responder city. Glob Environ Change 28:395–408

Rosenzweig C, Solecki W, Hammer S, Mehrotra S (eds) (2011) Climate change and cities: First Assessment Report of the Urban Climate Change Research Network. Cambridge University Press, Cambridge

Rosenzweig C, Solecki W, DeGaetano A, O'Grady M, Hassol S, Grabhorn P (eds) (2012) Responding to climate change in New York State: the ClimAID integrated assessment for effective climate change adaptation in New York State, vol 1244. Annals of the New York Academy of Sciences

SIRR (2013) A stronger, more resilient New York, PlaNYC, The City of New York, Mayor Michael R. Bloomberg. http://www.nyc.gov/html/sirr/html/report/report.shtml/. Accessed 19 May 2016

Solecki W et al (2015). Climate change and infrastructure adaptation in coastal New York City. In: Glavovic BC, Kelly M, Kay R, Travers A (eds) Climate change and the coast: building resilient communities. CRC Press, Taylor & Francis Group, Boca Raton, pp 125–146

Strauss BH, Kopp RE, Sweet WV, Bittermann K (2016) Unnatural coastal floods: sea level rise and the human fingerprint on U.S. floods since 1950. Clim Cent Res Rep, 1–16 pp. http://sealevel.climatecentral.org/uploads/research/Unnatural-Coastal-Floods-2016.pdf

Sweet W, Park J (2014) From the extreme to the mean: Acceleration and tipping points of coastal inundation from sea level rise. Earth's Future 2:579–600

Sweet W, Park J, Marra J, Zervas C, Gill S (2014) Sea level rise and nuisance flood frequency changes around the United States, NOAA Technical Report NOS CO-OPS 073

Talke S, Orton P, Jay D (2014) Increasing storm tides at New York City, 1844–2013. Geophys Res Lett 41:3149–3155. doi:10.1002/2014GL059574

USACE (2014) Response to climate change: planning for changing sea levels. http://www.corpsclimate.us/sandy/curvesNJNY2_simple.asp. Accessed 20 May 2016

U.S. White House (2015). Executive order establishing a federal flood risk management standard and a process for soliciting and considering stakeholder input. https://obamawhitehouse.archives.gov/the-pressoffice/2015/01/30/executive-order-establishing-federal-flood-risk-management-standard-and-a-process-for-soliciting-and-considering-stakeholder-input. Accessed 30 Jan 2015

Zimmerman R, Farris C (2010) Infrastructure impacts and adaptation challenges. Ann N Y Acad Sci 1196:63–85

Chapter 14
Building an Adaptation Tool for Visualizing the Coastal Impacts of Climate Change on Prince Edward Island, Canada

Adam Fenech, Alex Chen, Andrew Clark and Nick Hedley

Abstract A quantitative risk assessment of Prince Edward Island's coastal residences (homes, cottages), safety and security infrastructure (roads, bridges, water treatment plants, hospitals, fire departments, etc.) and heritage (churches, graveyards, lighthouses, archaeological sites, parks, etc.) was conducted for every one meter of coastline by estimating the possible future coastlines for Prince Edward Island. Over 1000 residences (houses and cottages) and 17 lighthouses were shown to be vulnerable to coastal erosion. Such scientific results were significant but were threatened to sit on a shelf in a scientific report unless communicated sufficiently to the organizations and communities of Prince Edward Island. A geovisual interface, known as the CoastaL Impacts Visualization Environment (CLIVE), was developed enabling citizens to interactively navigate and view a 3-dimensional (3D) virtual environment of the province of Prince Edward Island constructed from accurate historical spatial data and recent LiDAR surveys of topography. A public engagement tour was held across Prince Edward Island during the month of July 2014 at eight communities. This paper introduces the results of the coastal erosion study, the development of the CLIVE tool, and the results of a public consultation.

Keywords Coastal erosion · Climate change adaptation · Visualization · Prince Edward Island · Canada

A. Fenech (✉) · A. Chen · A. Clark
Climate Research Lab, University of Prince Edward Island,
550 University Avenue, Charlottetown, PE C1A 4P3, Canada
e-mail: afenech@upei.ca

N. Hedley
Spatial Interface Research Lab, Simon Fraser University,
8888 University Drive, Burnaby, BC V5A 1S6, Canada

W. Leal Filho and J.M. Keenan (eds.), *Climate Change Adaptation in North America*, Climate Change Management,
DOI 10.1007/978-3-319-53742-9_14

Introduction

The impacts of climate change have been articulated primarily as increases in temperatures, changes in precipitation, or increases in sea level rise. The implications of these climate changes are not readily accessible or do not resonate with individuals or communities as they grasp with the challenges presented by a changing climate. Climate data as well as climate model results require an avenue to show individuals and communities how they will be affected by future climate change in a manner that allows for acceleration of the slow, insidious nature of 'creeping' climate change. This paper outlines how a visualization tool can provide the results from a scientific study in a manner that can be embraced by individuals and communities in their quest for understanding the impacts of climate change, and motivating them towards adaptation actions.

A recent assessment of the science of Canada's Marine Coasts in a Changing Climate (Lemmen et al. 2016) have documented climate change impacts on the East Coast of Canada including that sea levels will rise, sea ice will decrease, and future coastal-erosion rates will likely increase in most areas. Just like the high latitudes of the northern Arctic environment, sensitive small islands such as Prince Edward Island, Canada are at the forefront of the impacts of climate change, and have often been described as the 'canaries-in-a-coalmine'—those areas whose sensitivity makes it a useful early indicator of climate change. The primary challenge that climate change presents to Prince Edward Island is the impact of coastal erosion through storm surges and high water levels. The sensitive sand and sandstone shorelines across Prince Edward Island often experience a wearing away by water, waves, ice and wind. Sea level rise has been measured at Charlottetown, Prince Edward Island to have increased by 36 cm over the past 100 years (1911–2011 from Daigle 2012), and is anticipated to increase by a further 100 cm over the next 100 years (IPCC 2013). At the same time, crustal movement of the Earth's surface at Prince Edward Island is lowering at 10–20 cm/century (Daigle 2012) depending on the side of the Island—not very much but certainly moving in the wrong direction if concerned about coastal erosion. In terms of damaging storms, the Intergovernmental Panel on Climate Change (IPCC), the global community's scientific authority on climate matters, concluded that they were "virtually certain" that there had been an increase in intense tropical cyclone activity in the North Atlantic since the 1970s; and "more likely than not" that these intense tropical cyclones would increase in the North Atlantic in the late 21st century (IPCC 2013). As a result of these anticipated ocean, geological and storm changes, coastal erosion is expected to continue and likely become more severe, threatening public and private infrastructure at great economic cost to the 3295 km of coastline on Prince Edward (Personal communication with Evan MacDonald, University of Prince Edward Island, 2016).

Previous studies have shown that Prince Edward Island has experienced significant coastal erosion (see Davies and MacDonald 2009; Forbes et al. 2004; Genest and Joseph 1989; Giles 2002; Mathew et al. 2010; McCulloch et al. 2002;

O'Carroll 2010; Ollerhead et al. 2013) of as much as 0.80 m/year. A recent study examined the rates of coastal erosion for every metre of coastline on Prince Edward Island (Webster and Brydon 2012) by interpreting aerial photographs of P.E.I.'s coastline for the years 1968 and 2010 using orthorectification and coastline delineation techniques to identify the coastal change (erosion and accretion) during this time period. Using this technique, the overall average rate of coastal erosion on Prince Edward Island from 1968 to 2010 was calculated as 0.28 m/year. An extreme example is Lot 436,485 near Savage Harbour, Prince Edward Island (see Fig. 14.1), three acres of land purchased in 1959 for a total of CAD165. A cottage was built on the land protected by a large sand dune (15 m high) until successive storms eroded the sand dune and over 200 m of shoreline were lost to the sea. The owners were forced to move their cottage, and continue to pay taxes on the land even though it remains underwater (Personal communication with owner, 2012).

While Webster and Brydon (2012) found that the overall average rate of coastal erosion on Prince Edward Island from 1968 to 2010 was 0.28 m/year, erosion and accretion rates of specific areas of Prince Edward Island vary making it unclear as to what coastal infrastructure is at risk. What was needed for a clearer understanding of Prince Edward Island's coastal infrastructure at risk from coastal erosion was a quantitative risk assessment of coastal residences (homes, cottages), safety and security infrastructure (roads, bridges, water treatment plants, hospitals, fire

Fig. 14.1 Loss of property to sea as a result of combined coastal erosion and sea level rise near Savage Harbour, Prince Edward Island (own figure)

departments, etc.) and heritage (churches, graveyards, lighthouses, archaeological sites, parks, etc.). Fenech et al. (Submitted) took the current annual coastal change rates for Prince Edward Island from 1968 to 2010 (both erosion and accretion) as determined by Webster and Brydon (2012), and multiplied by 30, 60 and 90 years to determine possible future coastlines of a high (30-year), medium (60-year), or low (90-year) risk designation. Using these future coastlines, the amount of coastal infrastructure that is at risk to coastal erosion was quantified.

The quantitative analysis of the impacts of coastal erosion on Prince Edward Island infrastructure conducted by Fenech et al. (Submitted) is presented below in Table 14.1. Over 1000 residences (houses and cottages), over 40 garages, 8 barns, and almost 450 outbuildings (buildings such as baby barns and other larger buildings that were clearly not a garage, residence or barn) were shown to be vulnerable to coastal erosion. Even 17 lighthouses, those maritime cultural icons, were deemed to be at risk. It is important to note that the risk to infrastructure increases over time. Of particular concern is the number of residences (houses and cottages) at risk to coastal erosion totalling over one thousand (1004). According to the Province of Prince Edward Island's Thirty-Ninth Annual Statistical Review 2012, which uses data from the 2011 Census of Canada, the average price of a Prince Edward Island residence in 2006 was CAD 144,404 (P.E.I. Statistics Bureau 2013). Using this average price number, the total value of residences at risk to coastal erosion over the next ninety years can be estimated at almost CAD145 million dollars. Although five sewage treatment settling ponds are quantified as being at risk in this study, it is the one on Lennox Island that is at high risk to coastline recession (Fig. 14.2) and is of immediate concern. The study concludes that almost 45 km of roads are also vulnerable to coastal erosion on Prince Edward Island over the coming 90 years under current rates of erosion. Such scientific

Table 14.1 Type and total of buildings and structures at risk to future projected coastal erosion (own figure)

Type	High risk (2040)	Medium risk (2070)	Low risk (2100)	Total
Barn	0	2	6	8
Bridge	124	0	2	126
Commercial	50	49	47	146
Foot bridge	9	0	1	10
Garage	8	14	20	42
Gazebo	5	1	1	7
Lighthouse	6	5	6	17
Outbuildings	90	165	191	446
Residential	148	380	476	1004
Settling pond	1	1	2	5
Tower	0	0	1	1
Windmill	0	1	0	1
Total	441	618	753	1812

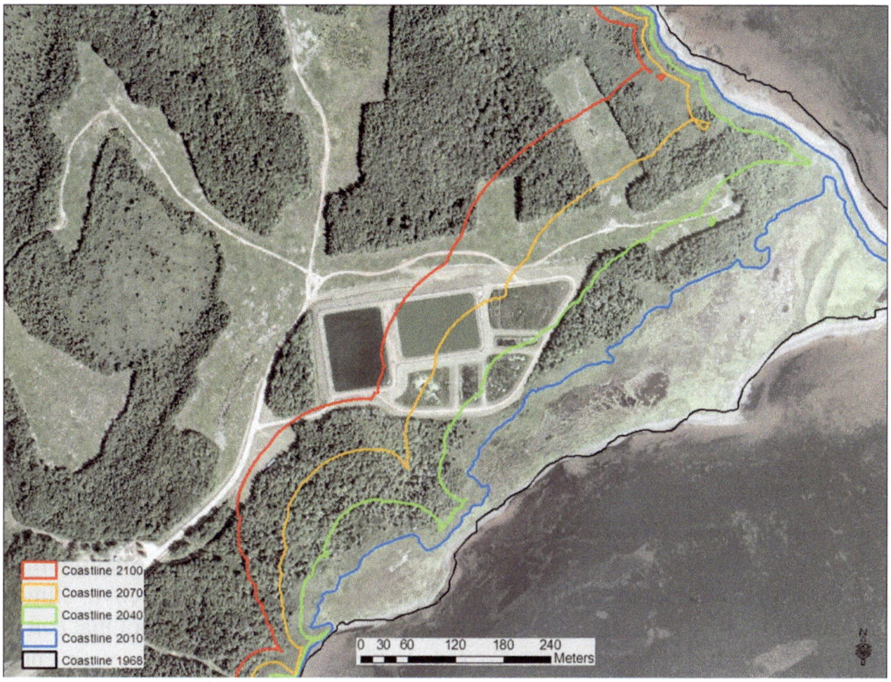

Fig. 14.2 Settling Ponds on Lennox Island depicting areas at high, medium, and low risk to coastline recession. *Background image* is 2010 Prince Edward Island orthophoto (own figure)

results were significant but were threatened to sit on a shelf in a scientific report unless communicated sufficiently to the organizations and communities of Prince Edward Island. But how best to do this?

Visualizing Coastal Impacts of Climate Change

The application of three dimensional (3D) visualization offers a unique learning opportunity by providing the potential to enhance the capacity of PEI communities to strengthen their coastal erosion risk management practices (Burch 2010). An emerging approach to enhancing participation and awareness-building at the local level is the use of 3D landscape visualisation to depict past and future community scenarios. Various forms of imagery including GIS-based tools, 3D modeling and photo-manipulation have been explored to investigate landscape change and management (Al-Kodmany 1999; Tress 2003; Lewis 2006) including some early research on the potential to visualize climate change futures (Dockerty 2005; Nicholson-Cole 2005; Sheppard 2007). These highlight the potential for visualization to influence individuals' perceptions of landscapes, floods, and a changing climate, which in turn may influence cognitive and affective (or emotive) understanding and influence individual and collective behaviour to respond appropriately to risks. Sheppard (2005) has summarized how visualization may contribute as an effective tool for communities in building their capacity to address the impacts of climate change. These include:

1. By integrating the analytical (including predictive) capabilities of GIS-based software with the emotionally-rich and intuitive media of photo-realistic software;
2. By representing recognizable places and local information in a realistic manner (as opposed to more abstract representation) that increases personal relevance;
3. By presenting both past and alternative futures (allowing for choice) to assist with decision making; and
4. By using computer visualization techniques that allow for modification and user-feedback in a participatory manner for refinement and analysis.

Thus the CoastaL Impacts Visualization Environment (CLIVE) was born. CLIVE is an analytical geovisualization tool created by researchers at the University of Prince Edward Island (UPEI) Climate Research Lab and Simon Fraser University's (SFU) Spatial Interface Research Lab. CLIVE is a geovisual interface that combines available coastal data, historical records and predictive climate change models and translates them into a 3D geovisual information tool that can be explored and queried by non-scientist stakeholders. It allows citizens of an entire Canadian province to explore past environmental change, and how climate change may impact coastal communities due to sea-level rise at various scales. By using 3D geo-visualization on two dimensional (2D) displays, CLIVE creates depth cues and high levels of detail not seen by 2D maps. In addition, by using 3D visualization techniques that are closest to the normal human perspective (as recommended by Meng 2002), more geographic variables are able to be shown and understood while keeping the human cognitive load low to the user.

CLIVE combines data from numerous sources, including an extensive province-wide archive of aerial photographs documenting coastline erosion as far back historically as 1968, and the latest high-resolution digital elevation data derived from laser surveys known as LiDAR, a remote sensing technology that measures distance by illuminating a target with a laser and analyzing the reflected light. These historical data and IPCC model projections of future sea-level rise (IPCC 2013) are used to develop analytical visualizations of coastal erosion regimes and potential future sea-level rise scenarios. These geovisual outputs are then delivered using a 3D game engine technology adapted to serious scientific communication.

CLIVE enables citizens to interactively navigate and view a 3-dimensional (3D) virtual environment of the province of Prince Edward Island (PEI) constructed from accurate historical spatial data and recent LiDAR surveys of topography. Users can view this 3D environment from a distance, by flying around it overhead for an overview. They can also explore the data and virtual landscape from first-person-on-the-ground perspectives, to inspect detailed local-scale historical environmental change, and projected future impacts. While navigating CLIVE PEI from any perspective, at any scale, users are able to select and manipulate multi-variate overlays of historical data and projected models through time.

By allowing citizens to view scientific data and explore climate change projections at any scale in their own neighborhood, the aim is to help them understand these often abstract phenomena at local, human scales. By delivering this science and its implications for real coastal communities using agile non-technical game engine technology (rather than specialized and expensive geographical information systems) closes the gap between expert science and citizens. Reconnecting abstract expert science to geographic spaces at risk, with a public information tool, using an inclusive public engagement approach, is a way to connect all stakeholders to this mutual problem. By educating citizens and raising awareness, the CLIVE project aims to encourage engagement, support dialogue and collaborative problem-solving at all scales of society and government.

CLIVE can also be used for the rapid assessment of the vulnerability of coastal areas across Prince Edward Island. By flying over areas of concern such as coastal residences or other coastal infrastructure, sea levels can be raised and lowered to assess the immediate risk due to their low lying topography, and proximity to the sea. Geographic information system (GIS) layers of coastal erosion rates can be toggled on-and-off on the screen to provide a similar assessment of the risk of coastal changes. These rapid assessments are not necessarily used to be quantitative in nature, but to provide a quick view of the vulnerability of coastal areas on Prince Edward Island.

Building the CoastaL Impacts Visualization Environment (CLIVE) Tool

The objective of the Coastal Impact Visualization Environment (CLIVE) 3-D tool was to create a way for users to control flyovers showing the landscapes and coastlines of Prince Edward Island. Sea-level rise impact on the Island are visualized using the "raising the bathtub level" approach, with a 'zoom-in' on areas of flooding risk from sea level rise and coastal erosion. Following MacEachren's Cartography Cube (1995), CLIVE is used mainly as a communications tool. However, the computational models behind the communications tool required the investigation of the effects of sea level rise on coastal erosion on Prince Edward Island. Thus, limitations were built into the visualization to limit environmental variables to not go to unrealistic levels.

Unity, a 3D game engine, was chosen to render the visualization due to its interoperability with the high resolution spatial data. The game engine also easily produces executable files that run on all major personal computer platforms, rather than requiring specialized software be installed on all user computers. The interface of the program is designed to be exact and direct and kept as minimalistic as possible. It was designed with only a few annotations to ensure there were fewer distractions in the visualization and to lower the cognitive load of the user. Extraneous graphical and textual information were kept to a minimum so the user could focus. This was a conscious decision to allow the user to be immersed in the visualization.

The title page of the interface (see Fig. 14.3a) was kept simple and clean. A quick background context was given to inform the user on the context of the project. The controls were also displayed on this screen to give the user the instructions on how to use the program. Features were built into the system to follow traditional 2-D cartography principles. A north arrow was devised to always show point the direction in which the camera was facing. A target elevation tool,

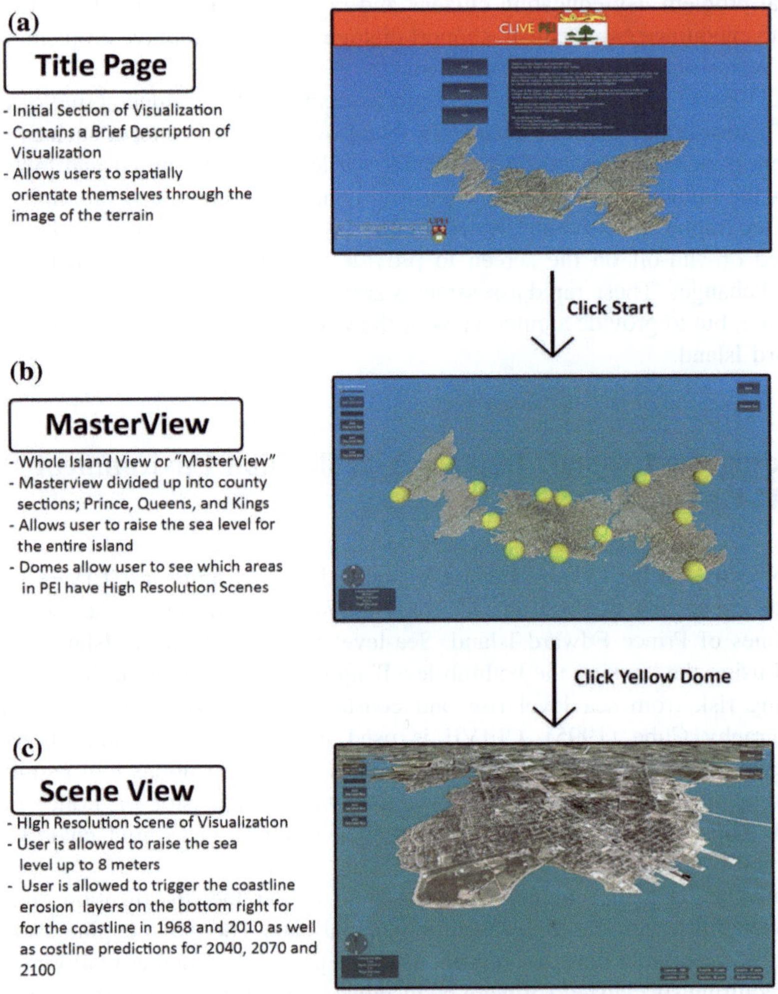

Fig. 14.3 CLIVE tool interface—**a** CLIVE title page; **b** CLIVE master view; and **c** high resolution scene view of Charlottetown, Prince Edward Island, Canada (own figure)

using ray casting techniques, showed the target elevation of a chosen point of terrain. A distance tool was devised, again using ray casting techniques, computes the distance between 2 points or more points, replacing the traditional scale bar. The dynamic part of the visualization is the horizontal slider. The slider controls a plane that changes the seal level, titled the "Sea Level Rise Slider"; the plane as a metaphor for the sea. This was done to allow the user to dynamically change the sea level to a maximum of 8 m.

A flyover camera type was chosen for the visualizations. Flythrough is defined as the functionality for the user to move through the visualization to give the illusion of flying. The cameras were chosen to give the user an effective way to move through the virtual landscape and pick the spatial area in which they wanted to view the sea level rise. The controls are taken straight from the keys of modern day first person viewer video games and are as follows; W and S on the keyboard to pan up and down A and S to pan Left and Right, the mouse changes the rotation of the camera and the key "e" to pause the camera. The "e" is necessary for the user to pause the camera to move the sea level rise slider. Various human computer interfaces were also implemented for CLIVE 3-D. A controller interface was devised to increase the intuitiveness and flexibility of the visualization.

The Master view visualization is the first level of CLIVE. The Master view visualization contains the entire island and a series of yellow "domes" (see Fig. 14.3b). The terrain is separated by 3 different terrain sheets, divided up by the counties of Prince Edward Island—Prince, Queens and Kings. It is a low resolution scene with the purpose of orienting the user and displaying the effects of sea level rise for the entire island. This view serves as the platform to enter the specific domes, which contain the high resolution scenes of the visualization. A user clicks a dome and the camera follows a pre-determined path towards the dome while the visualization loads the high resolution scene. This is to allow the user to understand the direction the camera is going, and maintain their respective spatial orientation. In total, there are 13 high resolution scenes: Alberton, Borden, Cavendish, Charlottetown (see Fig. 14.3c), Greenwich, Lennox Island, Montague, North Rustico, Souris, Summerside, Victoria, West Point and Wood Islands. Spatial data such as coastlines in 1968 to 2010, as well as coastline futures in 30 year intervals: 2040, 2070 and 2100, were put into the high resolution scenes for the user to see the erosion and accretion effects on the coastline. These were derived from creating 3-D models from their respective spatial shape file layers created in the geographic information system used to study the coastal change from 1968 to 2010 (Webster and Brydon 2012).

As an example of a high resolution scene in CLIVE, Lennox Island (see Fig. 14.4a) is located in Malpeque Bay on the north shore of the Island and is home to the Lennox Island First Nation of the Mi'kmaq People. The island is very low lying with a maximum elevation of eight (8) meters above sea level. Through the visualization, it can be seen that Lennox Island is extremely vulnerable to sea level rise, coastal erosion and storm surges. One of the main vulnerabilities is the groundwater contamination from saltwater intrusion as all of the Prince Edward Island's water supply is groundwater, and saltwater from rising seas is

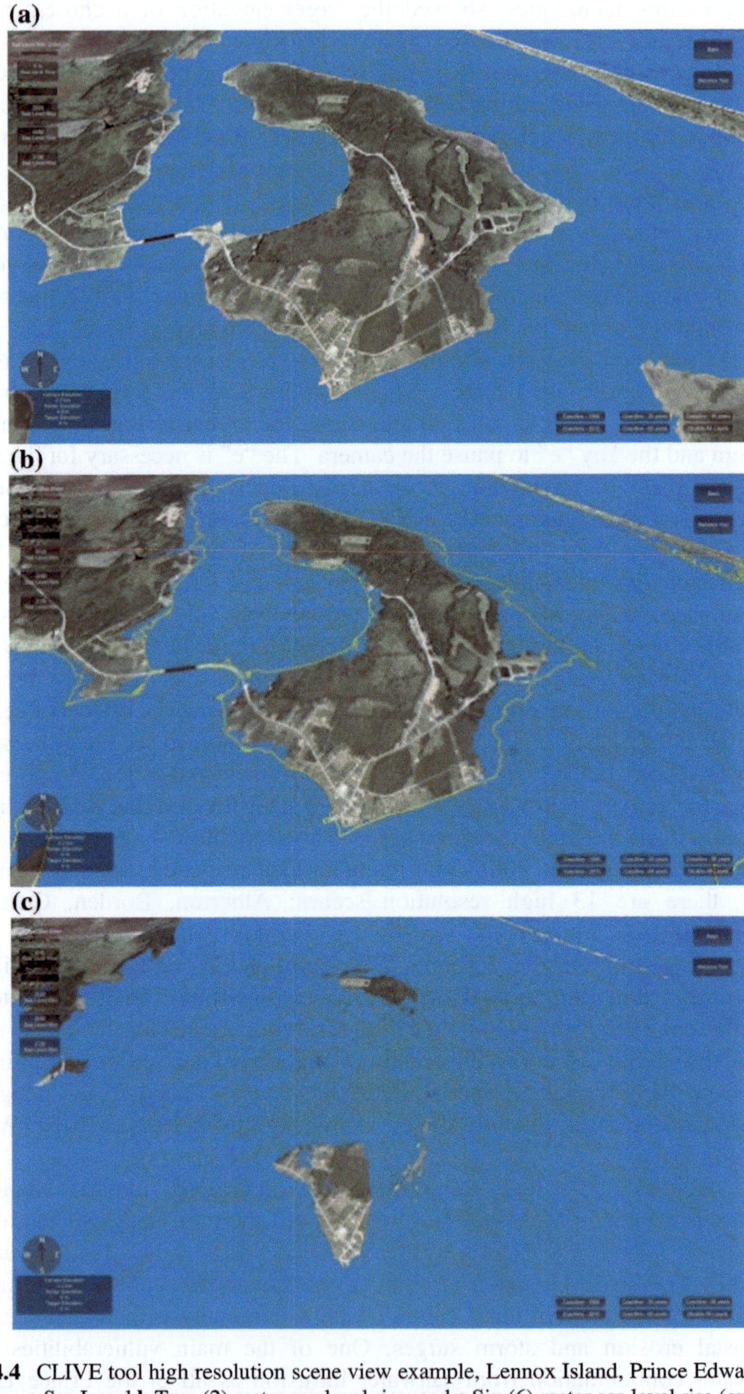

Fig. 14.4 CLIVE tool high resolution scene view example, Lennox Island, Prince Edward Island, Canada—**a** Sea Level **b** Two (2) meter sea level rise; and **c** Six (6) meter sea level rise (own figure)

contaminating the drinking water supply (Hansen 2012). Sea level rise is threatening the archaeological sites around the Lennox Island area, where rising water levels are starting to intrude onto the archaeological sites. With a two (2) meter sea level rise (see Fig. 14.4b), the sewage treatment pond on the east side of the island will be inundated, and could potentially contaminate a multi-million dollar oyster fishery in Malpeque Bay. With a six (6) meter high level water surge (see Fig. 14.4c), most of the island is inundated with the majority of infrastructure on Lennox Island being affected.

The CLIVE tool is limited by many factors including geographic resolution as well as the date of imagery sets. The geographic resolution of the tool is limited by the resolution of the imagery used—in this case, the digital elevation model was resolved at 1.5 m, and the aerial photos at 0.5 m. The two sets of imagery—the digital elevation model and the aerial photos—were also collected in different years, the former in 2008 and the latter in 2010, which may lead to some recent coastal features not being represented.

Introducing the CLIVE Tool to the World

A public engagement tour was held across Prince Edward Island during the month of July 2014 at the communities of Victoria, Souris, Abram-Village, Montague, North Rustico, Charlottetown, Summerside and Alberton (see Fig. 14.5). Each session consisted of a brief powerpoint presentation on coastal erosion and sea level rise followed by a CLIVE demonstration of local areas of concern. The CLIVE community consultation meetings were vibrant, passionate affairs. Each session presented an introduction to PEI's vulnerability to coastal erosion and sea level rise, introduced CLIVE, examined the vulnerability of local communities and answered questions. Each session was also preceded and concluded with a written survey to gauge attendee's knowledge, concern and willingness to adapt to coastal erosion and sea level rise. The concern for coastal erosion of each participant was high (presumably that was the motivation for attending the session), and, in most cases, increased after being introduced to CLIVE. One notable exception where the concern dropped considerably was likely because their property was not affected by the CLIVE visualizations of future sea level rise and coastal erosion. Most importantly, these sessions motivated coastal home or cottage owners to respond to their vulnerability by increasing their resilience to the anticipated sea level rise and coastal erosion.

The leading-edge visualization of climate change impacts (simulation of sea level rise and coastal erosion) known as CLIVE (CoastaL Impacts Visualization Environment) has garnered national and international attention including national text journalism from the Globe and Mail (19 February 2014); national Canadian broadcast coverage from the Canadian Broadcasting Corporation (World Report radio on 11 February 2014), international journal coverage (National Geographic, 16 December 2015) and international television coverage from Al Jazeera media.

Fig. 14.5 CLIVE roadshow locations across Prince Edward Island in 2014—*1* Victoria—Tuesday, July 8; *2* Souris—Wednesday, July 9; *3* Abram's Village—Tuesday, July 15; *4* Montague—Thursday, July 17; *5* North Rustico—Tuesday, July 22; *6* Charlottetown—Wednesday, July 23; *7* Summerside—Thursday, July 24; and *8* Alberton—Wednesday, July 30 (own figure)

The team that created CLIVE won an international award in 2014 from the Massachusetts Institute of Technology for Communicating Coastal Risk and Resilience. As a result of the CLIVE Roadshow to eight PEI communities, the PEI Association of Planners named Dr. Fenech the winner of the 2014 Murray Pinchuk Community Builder Award recognizing the highest standard of community building in the public and private realms, and efforts to develop a coastal erosion visualization tool and work to share ideas on how best to adapt to coastal erosion and sea level rise. A collaborative project is being developed with the University of Southern California and the Southern California Coastal Water Research Project to apply the CLIVE technology to the City of Los Angeles, California, while the Graham Sustainability Institute at the University of Michigan funded the development of a proposal to apply CLIVE to Collingwood, Ontario to address its concerns with lowering levels in the Great Lakes. CLIVE has already been applied to Colchester County in the Canadian province of Nova Scotia.

One of the major obstacles in developing CLIVE for other regions is access to high resolution imagery which can be overcome through the employment of small unmanned aerial vehicles (sUAV). The sUAVs, colloquially known as drones, are a data collection system that provides geo-referenced imagery used to develop 3D visual models such as CLIVE.

Conclusions

The primary challenge that climate change presents to Prince Edward Island is the impact of coastal erosion through storm surges and high water levels. Results of numerous studies have shown significant vulnerability of coastal infrastructure including homes, lighthouses, roads and even wind turbines. Such scientific results were significant but were threatened to sit on a shelf in a scientific report unless communicated sufficiently to the organizations and communities of Prince Edward Island. An emerging approach to enhancing participation and awareness-building at the local level is the use of 3D landscape visualisation to depict past and future community scenarios. A computer tool for visualizing coastal impacts and sea level rise due to climate change was programmed in Unity software by combining fine-resolution aerial images together with a fine resolution digital elevation model. The result was the award winning CoastaL Impacts Visualization Environment, or CLIVE, whose technology is now being applied to places around the world.

This tool has allowed coastal communities across Prince Edward Island, as well as individual home owners, to understand the vulnerability of their coastal infrastructure to rising sea levels and coastal erosion, and motivated communities to adapt to these risks through retreat from the coastlines or, more commonly, the protection of the coastlines through armouring. In this way, the CLIVE tool has helped communities across Prince Edward Island become more resilient to the current and future impacts of climate change.

This tool can be applied anywhere around the world where the digital elevation model and aerial photo imagery is available. The authors have applied the CLIVE approach to other areas of the world including Long Beach in the City of Los Angeles, as well as Colchester County in Nova Scotia, Canada. Where the detailed imagery does not exist to build CLIVE, the application of drone technology to capture this imagery has been deployed for smaller areas interested in risks from coastal flooding, sea level rise and coastal erosion.

References

Al-Kodmany K (1999) Using visualization techniques for enhancing public participation in planning and design: process, implementation, and evaluation. Landscape Urban Plann 45:37–45

Burch SS (2010) Planning for climate change in a flood-prone community: municipal barriers to policy action and the use of visualizations as decision-support tools. Clim Impacts Adapt Sci 1:93–121

Daigle R (2012) Sea level and storm surge projections for PEI, Fenech, Houston, Taylor and Jardine. Prince Edward Island's Changing Climate: Results of the Atlantic Regional Adaptation Collaborative on Climate Change. University of Prince Edward Island

Davies M, MacDonald NJ (2009) Souris Shoreline Erosion Study. Coldwater Consulting Ltd.

Dockerty TL (2005) Visualizing the potential impacts of climate change on rural landscapes. Comput Environ Urban Syst 29:297–320

Fenech A, Brydon C, Taylor E, Angus R, Dickie S, Jardine D (submitted) Coastal vulnerability of Prince Edward Island Residences, Roads and Bridges

Forbes DL, Parkes GS, Manson GK, Ketch LA (2004) Storms and shoreline retreat in the southern Gulf of St. Lawrence. Mar Geol 210(1):169–204

Genest C, Joseph MC (1989) 88 centimeters of coastal erosion per year: the case of Kildare (Alberton), Prince Edward Island, Canada. GeoJournal 18(3):297–303

Giles PT (2002) Historical coastline adjustment at MacVanes Pond Inlet, eastern Prince Edward Island. The Canadian Geographer (Le Géographe Canadien) 46(1):6–16

Hansen BA (2012) Masters Thesis: simulating the effects of climate change on a coastal aquifer, Summerside, Prince Edward Island. St. Francis Xavier University, Anitgonish, Nova Scotia, Canada, 105 p

IPCC (Intergovernmental Panel on Climate Change) (2013) Climate change 2013: the physical science basis. Contribution of Working Group I to the Fifth Assessment Report of the Intergovernmental Panel on Climate Change [Stocker TF, Qin D, Plattner G-K, Tignor M, Allen SK, Boschung J, Nauels A, Xia Y, Bex V, Midgley PM (eds)]. Cambridge University Press, Cambridge, 1535 pp

Lemmen DS, Warren FJ, James TS, Mercer Clarke CSL (eds) (2016) Canada's Marine Coasts in a Changing Climate. Government of Canada, Ottawa, 274 p

Lewis J (2006) Culture and communication: can landscape visualization improve forest management consultation with indigenous communities? Landscape Urban Plann 3:291–313

MacEachern A (1995) How maps work: representation, visualization, and design. The Guilford Press, New York

Mathew S, Davidson-Arnott RG, Ollerhead J (2010) Evolution of a beach-dune system following a catastrophic storm overwash event: Greenwich Dunes, Prince Edward Island, 1936–2005. Can J Earth Sci 47(3):273–290

McCulloch MM, Forbes DL, Shaw RW (2002) Coastal impacts of climate change and sea level rise on Prince Edward Island. GSC Ottawa

Meng L (2002) How Can 3D geovisualization please users eyes better?. Geoinformatics-Magazine for Geo-IT Professionals, Emmeloord

Nicholson-Cole S (2005) Representing climate change futures: a critique on the use of images for visual communication. Comput Environ Urban Syst 29:255–273

O'Carroll S (2010) Coastal Erosion and Shoreline Classification Project, Town of Stratford (Prince Edward Island). Final Report submitted to the Town of Stratford, Prince Edward Island, by Géo Littoral Consultants

Ollerhead J, Davidson-Arnott R, Walker IJ, Mathew S (2013) Annual to decadal morphodynamics of the foredune system at Greenwich Dunes, Prince Edward Island, Canada. Earth Surf Proc Land 38(3):284–298

P.E.I. Statistics Bureau (2013) Province of Prince Edward Island Thirty-Ninth Annual Statistical Review 2012. Department of Finance, Energy and Municipal Affairs, 106 p

Sheppard S (2005) Landscape visualization and climate change: the potential for influencing perceptions and behaviour. Environ Sci Policy 8:637–654

Sheppard S (2007) Future visioning of local climate change scenarios: connecting the dots and painting pictures to aid earth system governance, Human Dimension of Global Environmental Change. Vrije Universiteit, Amsterdam

Tress B (2003) Scenario visualization for participatory landscape planning—a study from Denmark. Landscape Urban Plann 64:161–178

Webster T, Brydon C (2012) Evaluating long-term coastal change on Prince Edward Island: GIS-based spatial analysis of coastline orthophotography, Fenech, Houston, Taylor and Jardine. Prince Edward Island's changing climate: results of the Atlantic regional adaptation collaborative on climate change

Part IV
Information, Communication, Education and Training on Climate Change

Chapter 15
Effective Public Service Communication Networks for Climate Change Adaptation

Rae Zimmerman

Abstract Climate change adaptation is fundamentally linked to the suitability and adaptability of communication networks to confront consequences of climate change and climate-related extreme events. Communications often receive less attention yet are critical to managing adverse outcomes of climate change by emphasizing human and technological interfaces. Several cases of communication failures and ultimate successes in extreme events are diagnosed uniquely using network concepts. In these cases, communications influenced the magnitude and direction of consequences and illustrate successful learning efforts and best practices across disciplines to address communication deficiencies. The overarching research question is how two critical components of risk communication—messaging and the adequacy of communication technology—can be shaped to support adaptations to climate change to reduce its adverse consequences. Messaging and technologies act together at human-technology interfaces. One case involved the interplay of communication technology, human capacity for understanding warnings, and mortality in a record mid-West tornado, and improvements were recommended at human and technological interfaces for more effective communication. A second case involved communications among weather scientists, government operators, system users, and other groups about transit services in a severe New York City snowstorm, and learning from those experiences contributed to reduced human exposures and addressed operational changes for transit services in subsequent storms. A third case addressed communication failures in getting services to New York City's populations in record-breaking storm-related flooding of the New York City subway system, which resulted in recommendations about information access and accuracy. The methodological approach follows case-based research techniques supported by network concepts. These failures and the lessons learned provide key, transferrable strategies and methodologies for climate change

R. Zimmerman (✉)
Robert F. Wagner Graduate School of Public Service,
Institute for Civil Infrastructure Systems, New York University,
295 Lafayette Street—2nd Floor, New York, NY 10012, USA
e-mail: rae.zimmerman@nyu.edu
URL: http://www.nyu.edu/icis; http://wagner.nyu.edu/zimmerman

W. Leal Filho and J.M. Keenan (eds.), *Climate Change Adaptation in North America*, Climate Change Management,
DOI 10.1007/978-3-319-53742-9_15

adaptation in terms of a multi-disciplinary, network approaches for communication. Although many of the analyses emphasize technology for communications improvements, these are embedded in a larger framework of societal and inter-organizational networks that reflect human-technological interfaces.

Keywords Risk communication · Climate change · Extreme events · Infrastructure · Adaptation

Introduction: Rationale and Aims

Communication networks play a central role in developing and supporting adaptations to climate change and extreme events, and can affect how society responds to these conditions. Communications are often hidden, receiving minimal or indirect attention yet are critical to managing adverse outcomes by emphasizing human and technological interfaces. The lack of emphasis upon communications is related to the intermediary role that communications often play in disaster management. The communication context focuses here on the risks associated with climate change and extreme events, that is, the likelihood and magnitude of the occurrence and consequences of these conditions. In light of this focus, the general concept of communication is narrowed to risk communication. The relationships between climate change and extreme events have been highlighted (Herring et al. 2015; The National Academies 2016) as well as similarities in their outcomes. This supports an emphasis upon understanding communications in extreme events as a guide to communications for climate change.

Communication is multi-dimensional and complex, encompassing technical and social aspects. Knowledge of the processes and components originates in numerous, diverse disciplines. The spatial reach of communication in any given event can be enormous, that is, the impact, need, components and participants are far reaching, often well beyond the specific physical location of a particular event. Communications occur among individuals and among organizations and groups from local to global scales. Social-psychological as well as technological dimensions often form the basis for and shape communications and these relationships have been extensively studied, benefiting from a network perspective.

Risk Perception and Risk Communication

An important foundation for understanding the way risk communications are received and acted on is risk perception. Such insights include (Slovic 2000): people have strongly held beliefs (for example based on past experience) that are often difficult to change limiting adaptation to new situations; trusted sources of information tend to be believed to a greater extent; biases exist in the way people process information, for example, under some circumstances people overestimate small risks and underestimate large ones (Slovic et al. 2000); and the form of information can affect how people process information, for example, numerical or quantitative information (numeracy) can be more difficult for some people than others (Peters et al. 2010; Zimmerman 2013:188). Wachinger et al. (2013: 1053) have noted that trust and other factors such as experience with a risk can vary in their effects on protective action, particularly for floods and climate change, where increased perception can either reduce or heighten preparedness depending on the circumstances. Spence et al. (2012) have analyzed public perceptions of climate change as an underlying factor that contributes to the gap in climate communication given the global nature of climate phenomena and its uncertainty. For this reason, Spence et al. (2012: 970) have noted that the translation of the risk perception literature into risk communication strategies has been limited, yet they underscore the value of making these connections.

Risk Communication and Climate Change

An extensive literature on risk communication has provided direction for identifying and analyzing a number of attributes of effective communication (Höppner et al. 2010), one of which pertains to the need to align communication strategies with communication goals. Communications often have numerous and diverse goals or purposes, namely, to disclose, inform, warn, educate, understand and reduce uncertainty, promote and encourage involvement, participation, arbitration, mediation, and negotiation to resolve differences, guide behavior, and build public trust and confidence. Orr et al. (2015: 7) and Höppner et al. (2010: 17) have also identified these goals and others. The Center for Research on Environmental Decisions (2014) has applied risk communication to climate change, taking into account those aspects of climate and extreme events that pose challenges for communication and has suggested how communications can be shaped to convey climate change information.

A number of problems in the communication of climate change risks have been identified in the risk communication literature pertaining to the general public and managers responsible for protecting the public from adverse consequences. Surveys have pointed to the consistently low priority given to climate change relative to other issues though its importance has been growing over time, and Pew surveys show climate change still ranking second to the lowest among other priority areas (Doherty et al. 2015: 2). Communicating climate change risks is particularly challenging given climate change attributes, namely its global nature, uncertainty, and less immediacy than other kinds of problems (Moser 2010: 31; The National Academies 2016: 19). On the other hand, attributes of extreme events, which have been linked to climate change or otherwise share similar consequences, often involve the opposite attributes, namely, they are geographically specific or localized, often have identifiable and certain onsets at least within a range of temporal and spatial ranges, and are immediate making risk communication relatively easier. This supports the fact that the climate change and extreme event linkage (Herring et al. 2015) justifies using extreme event communication to inform communication of climate change, thus filling a gap in ways to communicate climate change. Evidence exists that extreme events and climate change are in fact linked in the scientific community (Walsh et al. 2014; Herring et al. 2015) and surveys identify these linkages in the minds of people to a limited extent, varying by regions of the world (Stokes et al. 2015: 7, 21). Thus, in order to bridge the gap in climate change communications, communications in extreme event cases are used.

Approach and Methodology

Several cases of risk communication failures and successes in extreme events are portrayed and evaluated here using network concepts (Newman 2010) to characterize the existence, type and direction of communication flows among nodes and across links and proximity (even virtually) of senders and recipients. The network lens provides a unique way of relating communications to technological, institutional and social dimensions of the cases, and understanding how communications often act as control points that can change the outcomes of disasters. The method of analysis borrows from research protocols for case analysis for example by Yin (2014) using an analysis of public documents within the case context. The cases generally meet the characteristics Yin (2014: 24) sets forth for case studies in reflecting real world situations, having current relevance, and where the events (phenomena) are not always clearly distinguishable from the context. Source materials used here included government documents, news media interpretations, reports of oversight agencies, and where available, analyses of the cases in the peer reviewed literature. These cases illustrate how communications dramatically altered the magnitude and direction of consequences and how successful learning efforts and best practices across disciplines emerged to correct communication deficiencies.

First a generic framework for risk communications is presented and adapted to consequences of the risks of extreme events and climate change. The overarching research question is how two critical components of risk communication—messaging and the adequacy of communication technology—affect consequences and can be shaped to reduce adverse consequences of climate change.

Second, the networks of interrelationships pertinent to major public services and the communications that accompanied them are presented for key representative cases that exemplify some of the conditions expected from climate change and extreme events.

Finally, conclusions are presented in terms of ways in which consequences were and could have been reduced by virtue of changes in the nature of communications.

Results

Generic Communications Framework

Risk communication frameworks are traditionally portrayed as networks comprised of an initiating source or messenger, the communication or message, a route or channel through which the communication travels, and the recipient or receiver who is potentially in a position to act on the communication (Mayhorn and McLaughlin 2014). Other characteristics have been added to the framework, such as characteristics of the hazards and social contexts (Orr et al. 2015: 5; Höppner et al. 2010: 46) and behavioral modifications (Lindell and Perry 2004). Risk communications are seen as dynamic and interactive, implying multiple directions (Orr et al. 2015: 5 citing Höppner al. 2010: 7; Renn 2005: 55) and having the ability to undertake changes according to stages in the event (Höppner et al. 2010: 18).

The generic communications framework is applicable across a variety of events consisting of two interrelated dimensions or components of the framework. First is messaging including the network structure that defines flows between messengers and recipients in terms of direction, rate and content of communications. The second component is communication technology and its deployment and how it influences messaging and can in turn shape consequences.

Messaging. Hierarchical structures of messages are a useful first step in analyzing messages and focusing on their deficiencies. The first step in the messaging hierarchy is whether messages are sent or not. Second, where a message has been sent and has been heard, the reactions of recipients can be simplified in terms of a number of major steps (Zimmerman 1988). The recipients or target populations follow the message with or without some modifications, try to get the message verified (Zimmerman and Sherman 2011; National Oceanic and Atmospheric Administration (NOAA) 2011) due to lack of trust in messengers (Slovic 1993: 678), ignore the message e.g. due to disagreement or denial (Drabek 1999), or misinterpret the message due to biases in perception (Slovic et al. 2000). Thus, the impacts of communications range from no effect to substantial effects on recipients' knowledge, understanding, attitudes or behavior. Message effects appear across levels ranging from individuals to various social groupings. The form of messages broadly aligns with the human senses individually and in combination, such as visual, audible, and tactile stimuli which have advantages and disadvantages depending on the design and context (Haas and van Erp 2014). Of particular relevance to the cases reviewed is that audible sirens as warning messages alert people of an emergency but do not provide information on event cause and severity thus requiring a combination of messages. In terms of institutional trust, surveys tend to support people's trust in scientific and religious institutions over government and industry, and trust can also depend on organizational characteristics (Slovic 1993).

Technology options and deployment. Communication technologies provide essential services that support communication activities. The purposes technologies support include ongoing information transmission about hazards in general and for actions in times of emergencies. The number and type or variety of such technologies have exploded and are constantly changing (Hilbert and Lopez 2011). Cell phone sites alone have been increasing exponentially (Cellular Telecommunications and Internet Association (CTIA) 2015) and the concentration of cell sites has also been increasing (Zimmerman 2012: 227). Communication technologies in extreme events can initially fail but then improve during recovery stages. These lessons are critical to coping with the consequences of climate change.

How people interact with communication technology depends on what people are used to relying upon. Warning systems are a common form of communication technology that played a critical part in one of the cases but was criticized for its shortcomings. Surveys indicate that people favor television networks for information (Saad 2013). In terms of the individuals people trust to deliver information, people tend to rely on scientists among others (Carsey Institute 2009).

The resilience of information technologies under disaster conditions is a critical issue for their role in communication. Gauges to measure water levels during a storm, for example, are often destroyed as they were during Hurricane Sandy by flooding or wind damage (Gillis 2014). Communication services such as cell phones can become inoperative due to overloading or electric power failures. The means to harden this equipment exist to overcome many of these limitations.

Messaging and communication technologies are not independent entities, and work hand in hand in supporting human-technology interfaces.

Cases of Communication Disruption and Adaptation

Case 1. Tornadoes: Joplin, MO, May 22, 2011

On May 22, 2011, a tornado reaching one of the highest Enhanced Fujita scale categories, EF-5 at its maximum points, descended on Joplin, MO with winds exceeding 200 mph, along a 22 mile length (6 miles of which was considered the damage path) killing 158 people and on record at the time as one of the costliest U.S. tornadoes (NOAA 2011: 1). In 2011 Joplin, MO population was about 50,000 and population density was about 1500 people per square mile (NOAA 2011: ii).

The two aspects of communication are explored for the Joplin case—the first pertaining to the immediate effects of and attempts to deal with physical breaches in communications during response and restoration that affected messaging communication technology for warning systems used for populations at risk and the second having to do with how warning system technologies perform in meeting the needs of populations at risk.

Messaging and communication ruptures in services and implications for response and recovery. Figure 15.1 illustrates the relationships among a number of events that occurred influencing the robustness and performance of communication

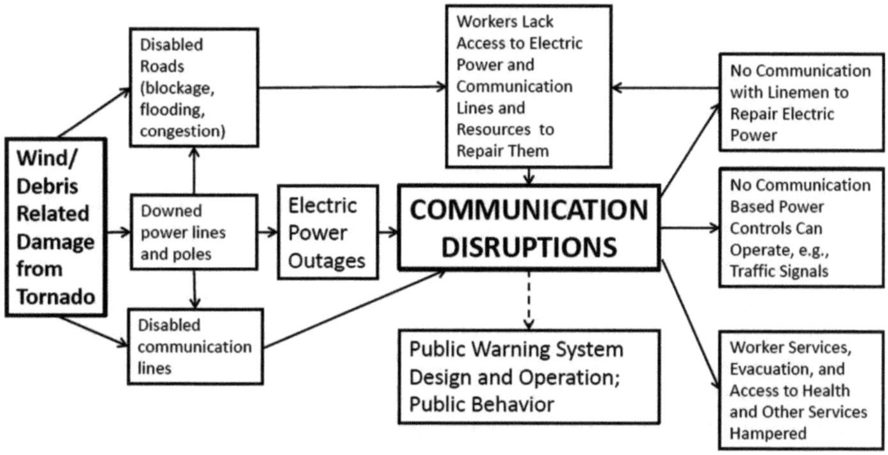

Fig. 15.1 Communication disruptions from the Joplin, MO Tornado of May 22 2011. *Source* Based on information from Penning (2012). *Notes* The figure only shows communication linkages that existed at the time of the event, not added afterwards. The *dashed line* indicates a second communication dimension related to warning system technology not shown in detail on the figure but is described in detail in the text in the public warning system technology section

systems and the ability of those systems to be restored in a timely manner to support effective messaging. According to the accounts by NOAA (2011) and Kuligowski et al. (2014), communication lines were directly damaged by the force of wind and debris as electric power lines were, resulting in not only physical damage to the communication lines but also functional breaches in communications due to electric power outages. These weather conditions in turn affected the ability of workers to access and repair the damaged infrastructure. Workers attempting to repair both electric power and communication lines also faced transportation impediments since roads were blocked by debris as shown in Fig. 15.1. Thus, other infrastructures such as transportation and electric power were critical nodes affecting the ability of communication infrastructure to function.

Public warning system technology. A key objective of the warning systems for the exposed population was to enable people to seek shelter, which was a voluntary action. Such warning systems are critical nodes linking weather forecasting agencies such as the National Weather Service (NWS), emergency response agencies and the population at large. In Joplin, MO, siren warnings were structured to be activated either by a dispatch center on the basis of weather information from the NWS or other more local sources and guidance for wind speed, decibel ratings, and duration from the Federal Emergency Management Agency (FEMA) (NOAA 2011: 5). Kuligowski et al. (2014: 243–246) and National Oceanic and Atmospheric Administration (NOAA) (2011) identified design and operational characteristics of warning systems and their effects on believability and behavior specifically associated with the Joplin tornadoes:

- Emergency communication plans existed prior to the tornadoes in the form of sirens for alerts but were not capable of voice communication.
- Sirens sounded simultaneously throughout the city, and were not interconnected between the City and the county.
- Joplin's sirens were used for both excessive wind events and tornadoes (NOAA 2011: 5) and the sounds were the same for both events creating confusion and uncertainty as to the severity of the event the warnings were communicating.
- Siren sounds differed by community which potentially created confusion especially for travelers moving among areas.
- Joplin also had a reverse 911 system where subscribers received alerts in the form of recorded messages, but volume could be a problem leading to delays so the system was not used on May 22, 2011 Kuligowski et al. (2014: 6).
- Web based systems existed but only for subscribers.
- Most communication systems required people to be proactive in seeking them out.

Technology and messaging operate together. Ripberger et al. (2015) explored perceptions in general of the accuracy of tornado warnings and response in terms protective action. They found that people's perception of the inaccuracy of warnings increased with false alarms and events that go unreported, which in turn contributed to less trust in the NWS, and resulted in their taking less protective action (Ripberger et al. 2015: 54). They identified some demographic characteristics that correlated with trust in warnings, for example, the elderly were more trusting and minorities and less educated people were less trusting. The Ripberger et al. (2015) study for Joplin emphasized as did the NOAA (2011) and the National Institute of Standard and Technology (NIST) (Kuligowski et al. 2014) investigations that reliance on warning devices did not take into account that devices served multiple purposes, for example, high wind warnings and the more dangerous tornadoes, producing unclear messages and thus reducing the effectiveness of the warnings for tornadoes.

Lessons learned from these experiences pointed to the need for the design of communication technologies to be unambiguous in order to support an effective human-technology interface.

Case 2. Snowstorm: New York City Rail Transit, December 26, 2010

Snowstorms are a type of extreme precipitation weather event. Over the weekend of December 26, 2010, New York City experienced a massive snowstorm that at the time ranked sixth in terms of snowfall accumulations in Central Park since the late 19th century, and an estimated 650 buses and 500 people in subway cars were stranded (Kluger 2011: 1). This case focuses on communication relationships in terms of messaging and technologies used to support messages, and draws upon some of the findings in Zimmerman (2013: 189-190). The NYC rail transit portion of the NYS Metropolitan Transportation Authority (MTA) system (the subways) is emphasized here, though the MTA system extends beyond the City north and east (including Metro North and the Long Island Railroad), and communication issues that arose in those areas are not discussed here.

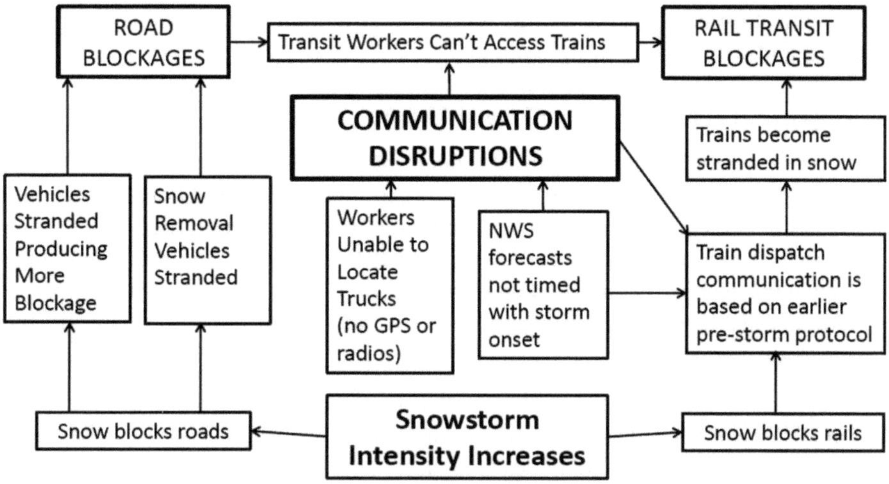

Fig. 15.2 Communication disruptions from Snowstorm, NYC, Rail Transit, December 26, 2010. *Sources* Based on Kluger (2011), Weinstein and Funk (2011), and the New York City Council (January 10 and January 14 2011a, b). *Abbreviations*: *NWS* National Weather Service, National Oceanic and Atmospheric Administration; *GPS* Global Positioning System

Messaging networks were identified between the NWS and the MTA, internally within the MTA and between the MTA and the general population of the City. These relationships are illustrated in Fig. 15.2. One critical messaging system involved train dispatch. As reported by Kluger (2011), the MTA train dispatch system followed a protocol that on weekends, orders were followed on the Friday before and not updated. Therefore, Friday orders not weekend updates were used to dispatch subway trains and buses, that is, operators continued to dispatch them in spite of the storm.

Zimmerman (2013: 190) summarized the communication deficiencies in terms of the absence of a timely emergency declaration and worker communications vital to the snow removal activity. Evaluations of the snowstorm indicated that operating agencies relied on NWS communications that had predicted a slower magnitude and rate of snow accumulation than what actually occurred in part contributing to the delays in emergency messaging. According to NYC Council Hearings (January 10, 2011a), the sequence of communications about the extent of the snowfall by NWS were as follows. An initial notification occurred on December 25, 2010 of 1-3″, which was below the 6″ threshold for a full alert by the NYC Office of Emergency Management (OEM). The next afternoon NWS issued a notification of 6-8″ justifying a full alert by NYC OEM, however, soon afterwards, the NWS prediction was 16″ with 55 mph winds, which was actually exceeded. The extent of the snowfall combined with a holiday weekend reduced the effectiveness of messages between NWS and agencies responsible for action. During the 2010 snowstorm communication linkages were not apparent between the City and travelers to discourage them from using road vehicles and to prevent them from getting stranded and blocking roads for snow removal and emergency vehicles.

Communication between the NWS and the MTA did not prevent trains from being released and ultimately getting stranded.

Technology played a considerable role in the failed communications. According to Kluger (2011) and the NYC Council Hearings (2011a, b):

- Workers operating snow removal vehicles did not have operable and sufficient radio communication devices and locational units based on Global Positioning Systems (GPS). These included sanitation trucks that also performed snow removal functions. Thus workers were unable to receive or send communications about areas requiring snow removal.
- Communication technologies applicable to the general public appeared to rely on news media.

Many lessons were learned from the snowstorm of December 26, 2010. When NYC experienced more storms a month or so later, many of these deficiencies in communication links and nodes had been addressed. The MTA took actions almost immediately to avoid future risk communication problems: Communication technologies were put in place at least for the New York City Department of Sanitation snow removal equipment (Zimmerman 2013). In addition, a street priority system has been implemented for snow removal, with snow removal equipment positioned closer to where workers live, and smaller snow removal equipment is being used to access narrower streets. The City in fact experienced severe snowstorms in the weeks following the December 26, 2010 storm and a number of preventive actions were in place by then.

Case 3. Flash Flooding: New York City Rail Transit, August 8, 2007

Flash floods and intense precipitation events wreak havoc on transportation systems and their users due to the suddenness, unexpectedness, and large volume of water occurring in a short period of time. They are difficult to predict due to their sudden onset. The potential impact of flooding events on the NYC subway system is reflected in the size of its ridership which the MTA (2016) reported for 2015 as reaching the 1948 record for both average weekday ridership (5.7 million) and annual ridership (1.763 billion), ranking seventh in the world for annual ridership. Three climate-related conditions associated with flooding are excess precipitation (rain and snow) exceeding the capacity of drainage systems, storm surge, and sea level rise. These conditions provide the core of communications to enable emergency planning and response and to connect with affected publics.

- Excess Precipitation. In New York City, both average and extreme precipitation are expected to increase during the 21st century (New York Panel on Climate Change (NPCC) 2015: 31) and this is not unlike similar projections for the northeastern U.S. (Walsh et al. 2014). Historically, the MTA (2007) reported that the subway system experienced numerous flooding episodes during the first decade of the 21st century alone that ranged from almost 2 to 7 in. within a short duration of about an hour or so shutting down anywhere from 7 to 20 of its 23 lines. The August 8, 2007 storm alone was considered one of the most severe heavy precipitation events at that time, and many of the subway lines were closed for about 12 h. The significance of increased experiences with flooding events for communication is that they tend to heighten people's perception of the danger though the actual effect depends on what the experience actually is (Wachinger et al. 2013: 1052). Perceptions can be a strong predictor of whether or not people will take protective action.

- Storm Surge. Storm surge accompanying hurricanes and other storms is another extreme event impact, which was a record event for Hurricane Sandy and was a major source of flooding of the subway lines through the tunnels and street openings.
- Sea Level Rise. Sea level rise is also expected to occur along coastal areas on the east and southern coasts of the U.S. (Walsh et al. 2014). Thus, flood related communications are critical to saving lives in the short term during flash flooding as well as in the long-term related to flood plain location. The Regional Plan Association (2016) estimated that many critical services and developments are located in areas potentially affected by sea level rise flooding in the New York Region by 2050, including about two-thirds of energy capacity and close to half of the wastewater treatment facilities among others. Other studies have reinforced the vulnerability of NYC structures to flooding (Zimmerman and Cusker 2001; Zimmerman et al. 2015). Many of the subway system components, in particular, its stations are in flood-prone and storm surge locations. For the subway system alone, Zimmerman (2013–2014) estimated that based on Sea, Lake, and Overland Surges from Hurricanes (SLOSH) models about a third of the NYC subway stations are in one of the four risk zones and under ten percent are in FEMA flood zones (cited from Zimmerman et al. 2015: 65). The City identified about 31 stations vulnerable to flooding (MTA 2007: 49). Unlike communications for extreme events, communicating the risks of sea level rise associated with climate change has proved challenging, given the greater length of time over which the impacts occur.

Figure 15.3 illustrates some of the gaps in communications that the MTA identified during the August 8, 2007 storm event diagrammed from the MTA (2007) report.

Messaging. The network for messaging among the actors in the flash flooding event parallels some of the generic structure that Höppner et al. (2010: 16) outlines beginning with the weather and flood forecasters, then moving to the warners, responders and ultimately to the affected publics. The MTA identified a number of communication problems associated with that one event, attributed primarily to the unexpectedness and rapidity of the event at the outset (MTA 2007: 6). These communication gaps occurred both internally between operations and employees such as field employees and externally between MTA and the NWS (MTA 2007: 11), the MTA and the general public to alert them to the conditions (MTA 2007: 19), and various agencies on the other hand. The communication issues arose with respect to message information accuracy, timeliness, and accessibility (MTA 2007: 7).

According to the MTA, field operations apparently did not communicate with operations personnel within the MTA regarding the extent and location of the flooding, the MTA communications with the general public were weak, and were

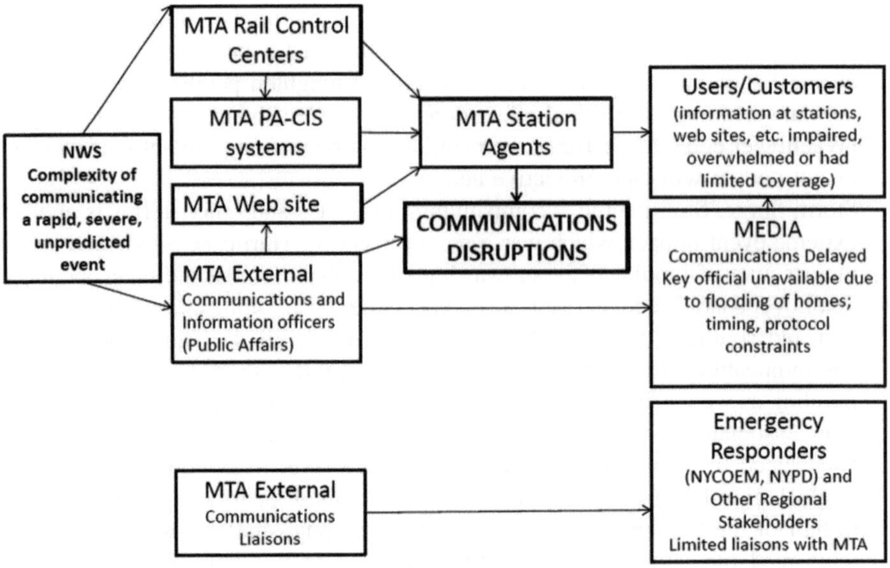

Fig. 15.3 Communication disruptions from flash flooding, NYC, Rail Transit, August 8, 2007. *Sources* Based on information from MTA (2007)

also limited in providing accurate information access and prompt information for the media to disseminate regarding the extent of the disruptions (MTA 2007: 6–7). The specific gaps identified primarily pertained to the lack of coverage of communication protocols and supporting devices and other deficiencies between links in the system to support messages:

- The Station Command Center in the Rail Control Center (RCC) and station field agent connectivity: relied on a "mass call" system covered only 40 out of the 900 booths for simultaneous communication (MTA 2007: 60–61).
- Field personnel ("in signal towers, local dispatchers and station agents") and customer connectivity: various combinations of communications cover three quarters or 348 stations with 120 unserved (MTA 2007: 61–62).
- Train supervisor and personnel connectivity: supported by the "Emergency Booth Communication System".
- Station Agents and RCC supervisor connectivity: mass call capacity problems had to be supplemented by the Emergency Booth Communication System (MTA 2007: 61).
- Media and customer connectivity: Communications with the media were delayed which in turn delayed communications to the general public, and direct communication with the public via public address systems and the capacity to communicate were deficient (MTA 2007: 19).

Considerable lessons were learned with respect to messaging. NYC and the MTA have implemented a number of operational mechanisms to respond and recover from flooding to reduce the problem in both the short-term and long-term, and these have been accompanied by changes in messaging. In terms of the immediate response, NYC preemptively shut down the transit system with warnings to the public within the timeframe of warnings they receive from forecasters as well as the time it takes to shut the system down. They extensively communicated closures, and did so for Hurricanes Irene and Sandy (Zimmerman 2014).

Technology. Many of the deficiencies in messaging were related to technological deficiencies. Communication equipment to connect field personnel to central office managers was not as prevalent. The ability of the agency to communicate with the public was impeded by the lack of adequate or widespread public address systems in both trains and stations and customer information screens (MTA 2007: 60–61). Another set of technologies indirectly affecting communications pertained to barriers to protect communication systems and the electric power infrastructure they depended on from flooding.

A number of lessons learned emerged from this event with respect to both messaging and communication technology. The MTA's self-study and other assessments developed numerous recommendations to address the communication problems that were primarily technological—public address systems, signage, etc.

- For the field personnel and operations link within the MTA, technological solutions were suggested used at the time such as Personal Data or Digital Assistants (PDAs) and blackberries (MTA 2007: 7) which since that time would probably be updated with smart phones. For both the MTA-media link and the MTA-user link, technological approaches and outreach strategies were identified, though these differed for each type of link.
- The technologies for the MTA-user link included visual (including computer based) and audible communications in the form of public address systems, electronic visual displays ("customer information screens"), more effective wireless access for customers (MTA 2007: 7–8), and additional ones in service included customer service lines.
- In 2012, the MTA expanded the visual and audio alert systems, piloting the "Help Point" program, and at that time it planned for installations at 102 stations that customers could use for assistance by connecting directly to the RCC (MTA September 21, 2012) rather than entirely through the intermediary of the MTA Station Agent. They are strengthening the RCC nodes and the liaison nodes with regional stakeholders.
- In 2015, the Federal Transit Administration awarded NYS a $52.4 million grant to strengthen communication systems at subway stations and RCCs to enhance communication links between those to nodes and between them and customers directly (New York State Office of the Governor, September 3, 2015).

With respect to flood prevention and water removal technologies, the City has expanded its use of pumps as well as relying on a network of providers of such equipment in the short-term. Numerous recommendations were made for flood prevention involving functions related to a number of city agencies and organizations (MTA 2007: 17–9). In the longer term, however, the City has had to employ operational controls, such as rerouting riders to make long term repairs to tunnels and equipment damaged during storms, e.g., the "Fix&Fortify" program in response to damages from Hurricane Sandy. Passenger rerouting becomes a problem for areas served primarily by a single subway line which is the subject of a separate communication protocol. Disruption of the R and the L train service to make repairs possible often over a year or two, for example, was the subject of intense public communication.

Communications Lessons from Industrial Accident Cases

Industrial accidents are not directly related to climate change, however, adverse consequences experienced in these circumstances particularly with respect to technology and human-technological interfaces are often analogous to those arising in climate-related and extreme events also, and thus, reflect other potentially useful experiences for climate communications. For example, detection equipment may not necessarily be designed to reflect communication needs. Sensors for pressure, temperature and volume in industrial storage and containment structures have not detected the presence of substances in those vessels or the conditions that affect their release in some historical accidents (Zimmerman 1988). As temperatures increase from climate change, this type of technology will be critical for communications. Communication equipment is often inoperative in disasters. This occurred in some oil/gas pipeline rupture accidents (Jennings et al. 2014). Computerized data systems are only as good as their coverage of anticipated impacts. In some industrial accidents substances released were not programmed in computers, ultimately leaked, and the leakage was therefore not detectable at the control center (Zimmerman 1988).

Conclusions

Communication among individuals, social groupings, and formal organizations will continue to be an essential aspect of the reduction and avoidance of the adverse consequences of climate change and extreme events. These conditions are estimated to continue and become more severe throughout the 21st century (Walsh et al. 2014). This paper has filled the gap in the literature on climate change communications by examining communications for the risks of extreme events given the implications of extreme events for climate change.

The three cases presented use network concepts to portray relationships among communication components, and illustrate some of the communication issues and opportunities to potentially resolve shortcomings. Adaptations and protective measures that occurred subsequent to the events covered are identified that reduced future adverse consequences. Many improvements in the communication systems adopted are strengthening pre-existing communication nodes and links such as creating bypasses and reconfiguring linkages to speed up communications. Many of the measures rely primarily on expanding and updating technologies, however some explicitly address the human-technical system interface for effective messaging, which is vital for users of the technologies. These users are very diverse and include weather services, system operators, system users, the general public, the media, and numerous public and private organizations. Workers in particular are affected who in the cases covered were attempting to access affected areas to restore services (Penning 2012; MTA 2007:12; Kluger 2011). In addition, Moser (2010: 32) has suggested for the penetration of climate change-related issues, expanding the audiences reached and diversifying forums, communication channels, messengers, and framings. These approaches can provide the basis for designing effective communication to potentially avoid adverse effects of climate related phenomena.

Although the suddenness of the phenomena in the three cases impeded communication and the timeliness and direction of the response, valuable lessons were learned many of which were used in subsequent events. The many dimensions of risk communication significant for climate change and extreme events are reflected in the two components addressed here messaging and technology. Messaging needs to reflect interconnections among the various roles, and adapt to the type of hazard, the social and behavioral characteristics of recipients, and the timing of or stages in the event. Technology is similarly interconnected, since the technology and the messaging need to be aligned, different technologies have different social-behavioral consequences, and the design, operation, and maintenance of technologies often reflect human error or innovation.

More specifically, the communication processes that occurred in these three cases exemplify some of the weaknesses but ultimate actual and potential improvements with respect to criteria for effective communication. Höppner et al. (2010: 53) outline the following criteria: the existence of clear goals and a framework for communication, identification of roles or actors and the resources available to them, clear understanding of the recipients (audience) and factors that determine their responses to communication characteristics and content, and the compatibility of the technology or channels with the nature of the message, which in turn reflect audience needs. It is instructive to summarize the characteristics of communication across the three cases in light of uncertainty which applies to each of these criteria.

Uncertainty is a central factor in climate change communication (Morton et al. 2011), and thus, the resolution or at least the explicit recognition of uncertainty is an important goal of risk communication. Uncertainty was pervasive in the three cases that contributed to the suddenness of the phenomena. For Joplin it was where and how severe the tornadoes would be (how long, their size and extent) given the spontaneous nature of tornadoes and the fact that communications in the form of warning sirens did not distinguish between tornadoes and severe wind events. For the NYC snowstorm, small changes in the direction or path of the storm could produce dramatically different locational outcomes. The communication of this uncertainty between the weather predictors and government agencies and weather predictors and the general public was considered deficient. For the NYC flash flooding, location and duration were not immediately certain given the sudden onset of flash floods, and communication modes were not initially adapted for that, however, they were subsequently improved.

Further research is required to expand on the approach taken here. Although the cases contribute valuable perspectives, a further direction is to develop associations between the structure of communication and adequate responses that statistical associations provide. Exploring differences between communication content and channels adopted by service organizations versus users of the service (Ossi and Anttiroiko 2016) are other important directions. Research is also needed to continue to build more cases to enable generalizations to be made across cases, and documents such as situation reports and after action reports would begin to provide such needed guidance. Better understanding of how and why people react to extreme events and climate change need to continually be refined to understand the role of risk communication in protective behaviors, since the literature in this area is generally not decisive (Wachinger et al. 2013). Finally, problems of institutional memory exist that need to be addressed in order to carry forward the lessons from one case to others.

Acknowledgements The following US National Science Foundation (NSF) research grants are acknowledged for their support of this work: NSF grant number 1444755 (Arizona State University lead)—Urban Resilience to Extreme Weather Related Events Sustainability Research Network (UREx SRN); NSF grant number 1441140 Resilient Interdependent Infrastructure Processes and Systems (RIPS) Type 1—A meta-network system framework for resilient analysis and design of modern interdependent critical infrastructures; NSF grant number 1541164 Critical Resilient Interdependent Infrastructure Systems and Processes (CRISP) Type 1—Reductionist and integrative approaches to improve the resiliency of multi-scale interdependent critical infrastructure; and NSF grant number 1316335 RAPID/Collaborative Research: Collection of Perishable Hurricane Sandy Data on Weather-Related Damage to Urban Power and Transit Infrastructure. Disclaimer: Any opinions, findings, and conclusions or recommendations expressed in this material are those of the author and do not necessarily reflect the views of the National Science Foundation.

References

Carsey Institute, University of New Hampshire (2009) America's civic health index. National Conference on Citizenship, Washington, DC

Cellular Telecommunications and Internet Association (CTIA) (2015) The Wireless Association annual wireless industry survey. CTIA, Washington, DC

Center for Research on Environmental Decisions (CRED) and ecoAmerica (2014) Connecting on climate: a guide to effective climate change communication. CRED, New York and Washington. http://www.connectingonclimate.org/

Doherty C, Suls R, Weisel R (2015) Public's policy priorities reflect changing conditions at home and abroad. Pew Research Center. http://www.people-press.org/files/2015/01/01-15-15-Policy-Priorities-Release.pdf

Drabek TE (1999) Understanding disaster warning responses. Soc Sci J 36(3):515–523

Gillis J (2014) Tide gauges needed for research are often victims of storms. The New York Times, http://www.nytimes.com/2014/01/14/science/tide-gauges-needed-for-research-are-often-victims-of-storms.html
Haas EC, van Erp JBF (2014) Multimodal warnings to enhance risk communication and safety. Saf Sci 61:29–35
Herring, SC, Hoerling, MP, Kossin, JP, Peterson, TC, Stott, PA (eds) (2015) Explaining extreme events of 2014 from a climate perspective. Bull Am Meteorol Soc 96(12):S1–S172
Hilbert M, Lopez P (2011) The world's technological capacity to store, communicate and compute information. Science 322(60):60–65
Höppner C, Buchecker M, Bründl M (2010) Risk communication and natural hazards. CapHaz-Net WP5 report. Swiss Federal Research Institute, Birmensdorf, Switzerland
Jennings C, Groner N, Roberts C, Fatica A, Hildebrand M, Noll G, Zimmerman R (2014) Hazardous materials cooperative research program HMCRP 14 guide for communicating emergency response information for natural gas and hazardous liquids pipelines. The National Academies, Transportation Research Board, Washington, DC. HMCRP report 14. Main report available at: http://onlinepubs.trb.org/onlinepubs/hmcrp/hmcrp_rpt_014.pdf. Appendices available at: http://onlinepubs.trb.org/onlinepubs/hmcrp/docs/HM-15_HMCRPReport14_Appendices1-3.pdf
Kluger BL (2011) NYC transit's response to December 2010 blizzard. State of New York, Office of the MTA Inspector General MTA/OIG. MTA, OIG, Albany, NY, Report #2011-07
Kuligowski ED, Lombardo FT, Phan LT, Levitan ML, Jorgensen DP (2014) Final report NIST technical investigation of the May 22, 2011 tornado in Joplin, Missouri. National Institute of Standards and Technology (NIST), Gaithersburg, MD
Lindell MK, Perry RW (2004) Communicating environmental risk in multiethnic communities. Thousand Oaks, CA, Sage
Mayhorn CB, McLaughlin AC (2014) Warning the world of extreme events: a global perspective on risk communication for natural and technological disaster. Saf Sci 61:43–50
Metropolitan Transportation Authority (MTA) (2007) August 8, 2007 storm report. MTA, New York, NY
Metropolitan Transportation Authority (MTA) (2012) Help points coming to 102 subway stations. http://www.mta.info/news/2012/09/21/help-points-coming-102-subway-stations
Metropolitan Transportation Authority (2016) Facts and figures: subways. MTA, New York, NY http://web.mta.info/nyct/facts/ffsubway.htm
Morton TA, Rabinovich A, Marshall D, Bretschneider P (2011) The future that may (or may not) come: how framing changes responses to uncertainty in climate change communications. Glob Environ Change 21(1):103–109
Moser SC (2010) Communicating climate change: history, challenges, process and future directions. Clim Change 1(1):31–53
National Academies of Sciences, Engineering, and Medicine (2016) Attribution of extreme weather events in the context of climate change. The National Academies Press, Washington, DC. doi:10.17226/21852
National Oceanic and Atmospheric Administration (NOAA), National Weather Service (NWS) (2011) NWS Central Region Service Assessment Joplin, Missouri, Tornado, 22 May 2011. NWS, Kansas City, MO
New York City Council (2011a) Report of the divisions of infrastructure and government affairs, oversight: the December blizzard of 2010: evaluating the city's response. The Council, New York, NY
New York City Council (2011b) Briefing paper of the Human Services Division Committee on transportation, oversight: evaluating the MTA' response to the December blizzard of 2010. The Council, New York, NY
New York Panel on Climate Change (NPCC) (2015) Building the knowledge base for climate resiliency: New York City panel on climate change 2015 report. Annals of the New York Academy of Sciences, vol 1336. The New York Academy of Sciences, New York

New York State Office of the Governor (2015) Governor Cuomo announces more than $57 million to improve emergency communication systems in subway stations. https://www.governor.ny.gov/news/governor-cuomo-announces-more-57-million-improve-emergency-communications-systems-subway
Newman MEJ (2010) Networks: an introduction. Oxford University Press, New York, NY, USA
Orr P, Forrest S, Brooks K, Twigger-Ross C (2015) Delivering benefits through evidence: public dialogues on flood risk communication. Literature review—SC120010/R3. Environmental Agency, Bristol, UK
Ossi H, Anttiroiko A (2016) Utility—customer communication: the case of water utilities. Publ Works Manag Policy 21:220–230
Penning M (2012) Linemen rebuild Joplin after twister's destruction. Transmission and distribution world. http://tdworld.com/overhead-distribution/linemen-rebuild-joplin-after-twisters-destruction
Peters E et al (2010) Numeracy skill and the communication, comprehension and use of risk-benefit information (Chapter 20). In: Slovic P (ed) The feeling of risk. Earthscan, London and Washington, DC, pp 345–352
Regional Plan Association (2016) Fourth Regional Plan. Regional Plan Association, Sustainability and Resilience section, New York, NY. http://www.rpa.org/fourth-plan
Renn O (2005) White Paper on risk governance: towards an integrative approach. International Risk Governance Council, Geneva. http://www.irgc.org/IMG/pdf/IRGC_WP_No_1_Risk_Governance__reprinted_version_.pdf
Ripberger JT, Silva CT, Jenkins-Smith HC, Carlson DE, James M, Herron KG (2015) False alarms and missed events: the impact and origins of perceived inaccuracy in tornado warning systems. Risk Anal 35(1):44–56
Saad L (2013) TV Is Americans' Main Source of News. The Gallup Organization, http://www.gallup.com/poll/163412/americans-main-source-news.aspx
Slovic P (1993) Perceived risk, trust and democracy. Risk Anal 13(6):675–682
Slovic P (2000) The perception of risk. Earthscan, UK and US
Slovic P, Fischhoff B, Lichtenstein S (2000) Rating the risks. In: Slovic P (ed) The perception of risk. Earthscan, Sterling, VA and London, UK, pp 104–120
Spence A, Poortinga W, Pidgeon N (2012) The psychological distance of climate change. Risk Anal 32(6):957–972
Stokes B, Wike R, Carle J (2015) Global concern about climate change, broad support for limiting emissions. Pew Research Center, Washington, DC. http://www.pewglobal.org/files/2015/11/Pew-Research-Center-Climate-Change-Report-FINAL-November-5-2015.pdf
Wachinger G, Renn O, Begg C, Kuhlicke C (2013) The risk perception paradox—implications for governance and communication of natural hazards. Risk Anal 33(6):1049–1056
Walsh J et al (2014) Chapter 2: Our changing climate. Climate change impacts. In: Melillo JM, Richmond TC, Yohe GW (eds) The United States: the third national climate assessment. U.S. Global Change Research Program, pp 19–67
Weinstein, E, Funk S (2011) Preliminary review of the city's response to the December 2010 blizzard report and recommendations to Mayor Michael R. Bloomberg. http://www.nyc.gov/html/ops/downloads/pdf/december_2010_blizzard.pdf
Yin RK (2014) Case study research: design and methods, 5th edn. Sage Publications Inc., Thousand Oaks, CA
Zimmerman R (1988) Understanding industrial accidents associated with new technologies: a human resources management approach. Ind Crisis Quart 2(3–4):229–256
Zimmerman R (2012) Transport, the environment and security. Making the connection. Edward Elgar Publishing, Ltd., Cheltenham, UK and Northampton, MA
Zimmerman R (2013) Crisis communications. In: Penuel KB, Statler M, Hagen R (eds) SAGE Encyclopedia of Crisis Management. Sage Publishers, Thousand Oaks, CA, pp 188–193
Zimmerman R (2013–2014) RAPID/collaborative research: collection of perishable hurricane sandy data on weather-related damage to urban power and transit infrastructure. Research

project. National Science Foundation #1316335, University of Washington, and Louisiana State University

Zimmerman R (2014) Planning restoration of vital infrastructure services following hurricane sandy: lessons learned for energy and transportation. J Extreme Events 1(2)

Zimmerman R, Cusker M (2001) Institutional decision-making. In: Rosenzweig C, Solecki WD (eds) Chapter 9 and Appendix 10 in climate change and a global city: the potential consequences of climate variability and change. Metro East Coast. Columbia Earth Institute, New York, NY

Zimmerman R, Sherman M (2011) To leave an area after disaster: how evacuees from the WTC buildings left the WTC area following the attacks. Risk Anal 31(5):787–804

Zimmerman R, Restrepo CE, Sellers J, Amirapu A, Pearson TR, Kates HB (2015) MultiModal transit connectivity for flexibility in extreme events. Transp Res Rec J Transp Res Board 2532:64–73

Chapter 16
Speaking Out or Staying Quiet on Climate Change: Broadcast Meteorologists Influenced by the Need to Be Pithy, Popular and Politically Cautious

Helen Meldrum, David Szymanski, Eric A. Oches and P. Thompson Davis

Abstract Broadcast meteorologists are in a position to convey knowledge about climate change to the public. These media professionals make deliberate choices about what information to omit or present to their audiences. Interviews with broadcast meteorologists revealed a wide range of attitudes about conveying climate science to their viewers. Three significant concerns emerged from conversations with participants: on-air time constraints are non-negotiable, pressure to be a "popular" personality is constant, and there is an enduring apprehension about taking a position associated with a perceived controversial topic or political viewpoint. Discomfort with these constraints affects the content choices of broadcast meteorologists who might otherwise take a stand representing the scientific consensus on climate change.

Keywords Broadcast meteorologists · Attitudes and statements on climate change

Introduction

During 2015, a large number of climate-related news stories were reported, such as U.S. President Obama's Clean Power Plan, Obama's rejection of the Keystone XL pipeline, and 195 countries pledging to cut carbon pollution as part of the United Nations COP21 Paris Agreement. These important events received little attention by the mainstream media, which is failing to adequately inform the public about climate change issues (Kalhoefer et al. 2016). If national reporting is slighting this timely concern, who might step up to provide this vital information? Are the local

H. Meldrum (✉) · D. Szymanski · E.A. Oches · P. Thompson Davis
Department of Natural and Applied Sciences, Bentley University,
175 Forest Street, Waltham, MA 02452, USA
e-mail: hmeldrum@bentley.edu

W. Leal Filho and J.M. Keenan (eds.), *Climate Change Adaptation in North America*, Climate Change Management,
DOI 10.1007/978-3-319-53742-9_16

broadcast meteorologists in media markets across the U.S. a resource that will mobilize to meet this need?

The Intergovernmental Panel on Climate Change (IPCC) is recognized as the definitive source for aggregating and synthesizing scientific information on climate change. The key findings of this organization's most recent major report are unequivocal. Human activities are disrupting the global climate system, and we are risking severe and irreversible impacts (IPCC 2014). Yet there still is doubt and skepticism among various segments of U.S. society regarding how humans have contributed to the problem (Hobson and Niemeyer 2012). Attitudes toward climate change are seen as a barrier to public engagement with the topic. Differences in personal values and political ideologies related to environmental concerns affect how people conceptualize and experience the risks of climate change (Corner et al. 2012). Broadcast meteorologists are also a heterogeneous group with diverse attitudes and political inclinations, and we found that the professionals who participated in our study were no exception.

Broadcast meteorologists themselves may be an important factor in viewer selection of TV channels. The TV weather forecast is the primary reason most viewers choose to tune into a local news station (Wilson 2006). Viewers typically list the weather forecast among the top reasons for watching a particular newscast. TV station executives are well aware of the importance of weather information in the eyes of their viewers, and the competition among stations in the same marketplace can be intense (Davie et al. 2006). Many viewers know the name of their favorite meteorologist (Daniels and Loggins 2010). In addition to live television, social media platforms provide a modern way to share weather-related news. Local television and radio broadcast meteorologists also turn to the Internet and smart phone outlets as mechanisms for expanding their streaming segments (Posegate 2008). Accurate weather reports are meaningful, especially to travelers who depend upon them for safe passage. Broadcast meteorologists take an active role in shaping the daily choices of their viewers and are counted on to attract audiences for advertisers supporting media outlets (Vannini and Mccright 2007).

Broadcast meteorologists are increasingly being called upon to go beyond forecasts and predictions of weather conditions and act as newsroom experts for diverse science topics. This expectation to be both an educator and a broadcaster can cause these professionals to feel pulled in different directions. Negotiating the dual roles of scientist and storyteller can create ambivalence on the part of media weather forecasters, as they feel pressure to distinguish what is important to their audiences and then explain it in understandable terms. Most people form their opinions through consumption of mass media, with television still serving as the primary source of information about climate change. Therefore, it appears that formerly more neutral broadcast meteorologists are less able to maintain their exemption from politics and controversial debate (Thompson 2013). Climate change is a polemical area of discussion, in part because of the potential for verbal missteps associated with this contentious topic.

There may be a hesitance to address climate change because there is a chronic shortage of time, as the average weather segment on a newscast is about three minutes long (Wilson 2009). An example of the increasing time pressure can be found in the history of the local segments featured in the U.S. national programming on The Weather Channel. When this feature was started in the 1980s, the local forecast could last up to 5 min. By 1999 it was down to 2 min, and since 2013 it has been limited to 60 seconds (www.weather.com). At the same time, many people hold the perception that the climate change problem is simply too big and overwhelming to solve (Swim et al. 2009). The public needs assistance in interpreting and responding to often confusing scientific information, especially because they have difficulties in processing and responding effectively to such a long-term, complex challenge (Shome et al. 2009).

Communicating about climate change is not a simple endeavor in the U.S. The Center for Research on Environmental Decisions Report (2014) noted that people often seek information that supports their existing attitudes and reject information that contradicts those beliefs. Political party identification may have a major influence on perception of climate change (Egan and Mullin 2012). Political ideology and individualism also are strongly associated with interpretations of climate data and with perceptions of scientific consensus on global warming (Myers et al. 2013). Agreement or disagreement regarding the human contribution to global warming falls largely along party lines (Hamilton 2011). Conservatives may be more likely than liberals to resolve mental dissonance against climate science because they see more conflict between economic development and government regulation (Scruggs and Benegal 2012).

Broadcast meteorologists are well aware that the public's response to topics related to climate change can be overwhelmingly positive or negative. Many broadcasters may want to avoid being considered "too political." Some have documented their trust of the IPCC data, while others have said the organization is too polarized to be credible and discredit its entire body of evidence (Wilson 2009). Also, Wilson (2012) asserted that approximately two-thirds of TV broadcast meteorologists still think there is significant disagreement among scientists about whether global warming is really happening. It appears also that broadcast meteorologists influence each other in their tight-knit community. For example, we discovered that there was discussion among broadcast meteorology colleagues about whether or not their peers had been targeted by ForecasttheFacts.org (now ClimateTruth.org) activists. The Forecast the Facts campaign was launched specifically to pressure TV meteorologists to inform their viewers about climate change. Perhaps it is not too surprising then that there is an unwillingness to grapple with perceived political questions on television, despite the fact that a well-informed public is essential to promoting effective policy on climate change (Wilson 2000).

What formal or informal rules and norms do media personalities observe when they are on-air or presenting to various public groups? It is widely believed that media weathercasters are chosen more for personal attractiveness and presentation skills, than for scientific acumen (Perryman and Theiss 2014). In the early days of TV broadcasting, broadcast meteorologists provided light banter designed to increase ratings, and they sought to avoid gaffes to preserve their likability. The growth of The Weather Channel on cable television in the U.S. is an example of how "infotainment" can be used as an asset to attract sponsorship. The field of broadcast meteorology is plagued by stereotypes about much larger financial rewards for media talent over working as a scientist who studies atmospheric conditions away from the spotlight. There are many examples of broadcast meteorologists resorting to attention-grabbing actions for increased ratings (Perryman and Theiss 2014). TV weather coverage has swung between silly and serious at several points in the history of broadcasting (Wilson 2006). So how do these conflicts help the broadcast meteorologist to balance demands for being entertaining and authoritative at the same time (Doherty and Barnhurst 2009)? The question is complicated because there is considerable disagreement about what information is experienced as alarming or reassuring to the public (Sandman 1994).

A broadcast meteorologist's credibility is based on audience perception of qualities such as dynamism, extraversion and sociability. Higher levels of involvement and greater feelings of affinity are stimulated by some self-disclosure by a media personality (Perse 1990). Another essential interpersonal quality is required of broadcast meteorologists, namely a sense of humor that can withstand criticism for inaccurate or unwanted forecasts. But this upbeat tone of the televised weather report inevitably will need to shift as extreme weather events become the story in and of themselves (Rosen 1989). Will most broadcast meteorologists ever explicitly and regularly acknowledge that the majority in their audience believe that humans have actually altered the climate, and therefore, weather patterns?

The TV weather segment is still only the few minutes that come before or after the sports scores and traffic report. Research on interaction with TV media personalities has traditionally been conceptualized as an imaginary one-sided friendship and labeled as a "para-social" phenomenon that provides vicarious connections rather than actual relationships. Viewers feel that they know and understand the TV personality in the same intimate way they know and understand flesh and blood friends (Perse and Rubin 1989). If the broadcast meteorologists believe this relationship to be true, would they stay away from a controversial topic because of fear of negative backlash that might alienate their viewers? In addition, in an effort to boost loyalty for the station, these professionals may use a cultivated engagement with viewers in Facebook and Twitter postings so that their audience feels even more personally connected (Stanfield 2014). Given the potential for expanding their roles and additional communication forums available to the broadcast meteorologists we interviewed, we wanted to know if they felt that climate change is simply too complicated and too controversial to discuss in the context of the local forecast.

Prominent scholars in climate change communication acknowledge that more work is required to shift the understanding of the general public from mere awareness to actual engagement (Moser 2016). A recent review of research literature on climate change communication (Wibeck 2014) notes the importance of framing climate issues in ways that make sense to lay audiences. For example, researchers have suggested that discussing climate change in terms of public health concerns may make it more relevant to segments of the public who are currently disengaged or dismissive (Myers et al. 2012). Moser concludes, "If communication researchers want climate communication to be as effective and impactful as it could be, their work must connect more effectively with those who do most of the talking (climate scientists, policy-makers, advocates in all sectors of society, journalists, editors, public intellectuals)" (Moser 2016, p. 356). And yet, the typical citizen does not have much exposure to the list of professionals that Moser identifies as spokespeople. Rather, laypeople with low levels of science literacy frequently tune into media broadcasts for weather reports. Our qualitative research examines weather broadcasters' willingness and ability to address their audiences on climate issues.

The role of local broadcast meteorologists as science educators is an understudied area of inquiry (Wilson 2008). It is clear, however, that TV broadcast meteorologists walk a fine line, projecting authority while also being entertaining enough to ensure adequate ratings for the station's programming (Doherty and Barnhurst 2009). Our research examined whether or not broadcast meteorologists are willing to speak out about specific issues regarding climate change. For example, would they ever address the topic of adaptation, such as behavioral shifts, by urging their audiences to take deliberate actions (e.g. drive less frequently) to manage the consequences of a warmer, more resource-constrained regional climate? Would they be willing to describe a root cause analysis, such as an on-air discussion of greenhouse gas emissions?

Methodology

To augment the knowledge gained from a national quantitative survey on the attitudes of broadcast meteorologists about climate change (Maibach et al. 2010), we conducted an interview-based study with broadcast meteorologists regarding their views on climate change and their perceptions about the impact of their broadcasts on their audiences. The first author (Meldrum) is trained in counseling psychology and elicited detailed responses from participants without using pre-determined, fixed-response categories. This design went beyond conventional quantitative methods and allowed for unanticipated and multifaceted responses.

We conducted a census of 137 broadcast meteorologists at 36 TV and radio stations in New England and eastern New York. The meteorologists' education, significant awards or certifications, number of years working at their current station, as well as the station addresses and national network affiliation (e.g. NBC, ABC, etc.) were collected through internet searches.

The first U.S. postal mailing was sent in September 2013 to 92 potential participants across New England. Two months later we expanded this mailing to include 34 TV broadcast meteorologists in the New York City and Albany metropolitan areas. We also sent invitations to 11 broadcast meteorologists affiliated with radio stations in the New England area. The radio weather broadcasters also all had on-air TV experience. The letter expressed our interest in a 20- to 30-min phone interview, to be scheduled at the broadcaster's convenience. Each recruitment letter was personalized with something distinctive about the broadcaster (e.g., a special award earned) with the aim of boosting cooperation. The letter explained that our faculty team in the Natural and Applied Sciences Department at Bentley University (in Massachusetts) was working collaboratively with meteorology faculty from Plymouth State University (in New Hampshire) to study aspects of informal science education and communication about climate science as part of a National Science Foundation grant.

One month after the postal mailing, a second request was e-mailed to the broadcast meteorologists for whom we had e-mail addresses, about two-thirds of the pool. A third request was sent approximately two months later to those who had yet to respond. Of 137 broadcast meteorologists asked to participate, 97 did not respond, 25 agreed to be interviewed, 13 declined, and two asked to postpone a decision. Twenty-two interviews were recorded in the fall of 2013 and spring of 2014 (about 16% of the total who were contacted). Considering the intense work obligations of these professionals, we felt that 16% was a successful response rate for recruitment. Each participant was thanked via e-mail and invited to a workshop on climate science communication in the media, held at Bentley University in the fall of 2014.

Participants were given open-ended prompts related to their views about their audiences' attention to the topic of climate change, and whether or not they perceived a shifting level of audience awareness. The broadcast meteorologists were sent several general, open-ended questions before the interview in case they wished to make notes. The interviews were digitally recorded with the permission of the broadcast meteorologists and then transcribed. Participants had the opportunity to review the transcripts for accuracy. All interviews were conducted, and copies of the conversations are stored, in accordance with Bentley Institutional Review Board requirements, and all responses are anonymized for publication and other dissemination methods. Ensuring respondents' confidentiality was essential in recruiting interview subjects. We asked the participants questions to assess the depth of analysis that they thought their audience brought to bear on the topic, and whether they felt that their viewers and station management were inclined to take deliberative action in response to climate change. We were interested in how broadcast meteorologists prepared their messages for TV, as well as how they conducted themselves in off-air forums, such as community appearances and social media postings.

Qualitative data analysis was dictated by Braun and Clarke's (2006) schematic for analyzing interview transcripts. This type of inductive analysis emphasizes a progression from basic description and organization towards perception of patterns in content that can be summarized and interpreted. This process allowed the data to dictate the results instead of trying to place the responses into a pre-existing framework. The authors listened to the digital recordings and read the transcripts of the interviews several times, initially noting patterns in the data that suggested potential coding schemes to each of us individually, and then again in conversation with each other. We generated a tentative list of ideas about what was most striking and might be assessed in a meaningful way. Next, we collated the text and sorted quotations into groups representing identified themes. We clarified a set of categories, and asked if refinement of these themes was possible. Some content collapsed into the same descriptive headings, while other content needed to be broken into new sub-categories. Our aim was to categorize the data in a meaningful way while still having identifiable distinctions between subtopics (Braun and Clarke 2006).

We reviewed theme headings that did not seem fully inclusive and considered reworking our categories, creating new ones or finding a fit for excerpts not yet placed in context. We sought to identify the "essence" of what each categorization represented, while taking care not to rely on first impressions. Data analysis was complete when we saw that no new organizational schemes were being generated. We chose not to use a text analysis software package because of our belief that these programs are not yet sensitive to subtle differences in categorizing content.

Our analysis highlights similarities and differences across the participants' comments. The participants' words were taken verbatim from the recorded conversations. Quoted comments have not been edited other than to limit length. The following results summarize the respondents' comments regarding time and role constraints, concerns about alienating viewers, contradicting management edicts, and pushing political boundaries.

Results

In conversation with the broadcast meteorologists, several themes emerged.

Most saw their mission as limited to the immediate weather forecast. They remain very conscious of the opinions held by their viewers and management. However, in contrast, a few did not feel particularly inhibited by the anticipation of a positive or negative reaction from their audiences.

Most of the broadcast meteorologists seemed to blame their on-air time constraints for impeding their ability to speak out on climate change topics. Some remarks reflecting their feelings about being hampered by their allotment of minutes are as follows:

Time is by far the biggest issue.
I can't really do it on television. My time is limited, obviously.
It's about the time commitment to it, not necessarily the actual subject matter.

Some weathercasters were very specific about their exact limits:

...in that three minutes of time, we don't have a whole lot of a chance to get into more detailed discussion.

...it's just that three minutes' time that I have is really forecast driven.......

What I do on a nightly basis in my two or two-and-a-half minutes that I'm given is really not the format for a conversation.....

I try to avoid it because..., you have a 30-second... a 15-second toss to hand it over...

It was also clear that the restrictions were built into the organizational culture at some TV stations:

My boss is like, "No, please don't get into that right now because it's long and it's laborious and it takes up too much time."

In addition to feeling handicapped by short segments, many broadcast meteorologists articulated a feeling that speaking out about climate change was simply beyond the scope of their vocation:

My job is just purely to give you the forecast, what's gonna happen today, what's gonna happen tonight, what's going to happen tomorrow.

I think it's important just to be the authority on the weather at that point, especially with the amount of time I'm given on air.

I basically will stick to the next seven, eight, ten days.

There are things that I just can't do in the position that I'm in.

Some broadcast meteorologists actually went as far as to say that speaking out would be objectionable somehow:

I don't feel like it would really be appropriate... How can you say a sentence about it in the middle of... tomorrow is going to be sunny, the next day it's going to rain...? I don't really think it would work at all.

No way that I am going to get into climate change discussions.

This isn't a place for debate. It's not the type of show that we do.

I'm not up there to give my opinions on what some scientist says or what I think....

On the other hand, a few broadcast meteorologists made the distinction that they will pursue the topic in other settings that do not present the same constraints as the daily forecast:

I definitely could do extra reports....

Any other forum, like when I go to speak to civic organizations... if the topic comes up, I like to....

To obtain a position as an "on air" TV presenter, the broadcast meteorologists have had to learn to cultivate an audience and connect them to their "brand" (i.e. their personality). They need to put the work in and show a competitive attitude in order to obtain an entry-level position at a TV station. And in an era where the Internet regularly makes people infamous through "viral" contagion, fame and personal brand are conflated. These weather broadcasters know that they need to be recognized for providing ongoing value to their viewers. Their comments reflect the fact that most do not want to risk losing the affection of their audiences by making remarks on climate change:

You are better off not giving your opinion whatsoever and that way you're not going to anger anybody.... Being personable on air, but not crossing that line that I'm going to upset anybody; that's how you don't get the e-mails with people saying they hate you.

I would become a very polarizing figure in this market.

I'd be more than happy to voice my opinion, but on air...it's just too heated.

You are putting yourself in a very dangerous spot by speaking out because it is so polarizing.

We don't want people not to like us as broadcast meteorologists. Either way, if you take one side on this, you lose. Any time I see anyone doing it, like another broadcaster, I cringe because I'm like, "Dude, you're probably not doing anything except hurting your career and hurting the station."

You have to show both sides. If you interview someone on this side, you better interview someone on the other side or you're going to get calls and people will be mad.

You're going to polarize some segment of the population.

We don't want to take any sides....

Honestly, I try to play down the middle....

One weather broadcaster let a single viewer know that he would like to say something, but feels that he simply cannot speak out.

I have gotten some e-mails from some people... There's one guy in particular who's very passionate about climate change. He has a nine-year-old son, and he wants me to talk about global warming and that we are the cause.... I finally just kind of gave up and just said, "Man, listen, I agree with you. Between you and me, I agree with you, but...."

A few broadcast meteorologists were particularly blunt, even when referring to off-air contexts:

I pray to God even when I do school talks that somebody doesn't bring it up....

Because 97% of Climatology is marketing...If I tweeted everything that I felt like tweeting I would probably be run out of town.

Perhaps the only people more worried about losing viewers over the topic of climate change are the weather broadcasters' supervisors:

We were pulled aside by management about eight or nine years ago and we were told not to talk about it.

If I were to get up there and say, "Hey, we got this snow storm because of global warming," I'm sure my GM would not be happy with me about it.

Under the two newer managements that we've had, I think that they're actually quite happy that we don't talk about it.

[Speaking about a peer weather broadcaster on the station staff] *Management said that that individual actively does speak out at events and public appearances about the fact that man is not causing this to a large extent, but nothing is done necessarily to temper that. There is just displeasure of the fact that it occurs.*

My agent said to me about three years ago in no uncertain terms, he said, "Do not ever on the radio, on television, in print media, do not ever talk about climate change." No, not at all. It's absolutely too dangerous.

Interestingly, one broadcast meteorologist felt too uninformed to speak out:

There is another deeper motivating reason why I don't touch it and that is because I don't believe I know enough to touch it....and I read the IPCC report....

In contrast, some broadcast meteorologists seem unconcerned about any consequences of sharing their opinions:

As a TV meteorologist, people are looking to you for what they're going to wear, but also insight into the climate. You're one of the experts. You're the go-between between the scientists out in the field and the viewing public

If the anchor would blatantly ask me on air what my thoughts are and I'd give a 15 to 20 second quick piece of my mind as to what my thoughts are. I'm not saying what people should believe but leaving it up to them in providing examples of Tweets or Facebook post or web link are from our website that could let them make the informed decision....

If somebody wants to have a conversation with me off the air, one-on-one, I'm more than happy to engage.

Science is not a deli. You can't just pick and choose what you want. You have to go along with the scientific process, and somehow that got lost in all of this....

This is real and this is something that we need to be concerned about.... I have addressed that plainly, but not in an alarming fashion to the audience in trying to explain the basic physics....

I've never been targeted by somebody who's a hardcore denier. I wouldn't mind if I were. I would be happy to engage with them in conversation and just go head to head with them.

Given the opportunity, if someone says to me, "Come to our school and could you discuss climate change?" Definitely I'm willing to have that conversation with them…. If the subject is too complicated or it's not interesting, I would respect their opinion on that too.

I'm a lot freer going on Twitter: I don't keep my mouth shut on Twitter, although I bite my tongue, but people know how I feel….

I've gotten some nasty e-mails and some letters from people that are just all into this whole climate change. And when I try to tell them that, it's not happening because of what you're doing today…I wouldn't be harsh about it. I would just say…in a lot of the meteorological societies that I'm involved in, there are many people that don't believe in that [IPCC] report…. If you take ten meteorologists in a room, seven out of the ten are not going to think that global warming is a big issue. I strongly believe that.

Forecast the Facts has made inroads and they put up these form letters so everyone gets a form letter from a person and we can tell that they lift it from Forecast the Facts….

We are all aware that political discussions are often frustrating, unproductive, and sometimes hostile. People can assume bad will about anyone who disagrees with them. Many deal with what are perceived as different political values by subjecting others to ridicule and vitriol. Our broadcast meteorologists seemed very conscious of these phenomena:

You learn early on that you don't talk about certain things on television. You don't talk about politics….

To even mention that, ugh, I would just be nervous about the response…. Can't have opinions about politics….

It affects how people view us…. Are we siding on the politicians on this, are we informing or are we taking a side?

I don't get into the politics quite as much….

They kind of spin that back to saying that we're trying to push some kind of political agenda…. You kind of limit how much you talk about…

That's when it starts to get edgy when you start bringing government and politics into it….
I never talk about it on the air because it's such a divisive issue.

On the other hand a few broadcast meteorologists step into the conversation in a way that preserves their authentic beliefs:

Where did I bring up politics? I just told you the level of the ice cap; I just told you the level of the sea. Now I'm getting accused to playing politics.... The ones that do e-mail me, they've got this like cut and paste letter that they get from this outlet called Forecast the Facts...

I try to present more of the scientific facts and maybe kind of lead them towards where I think they should go in terms of policy...
Every once in a while I just tell myself, "Why don't you just stick to the forecast?" Just stick to the forecast. It's so much easier, but then...I feel responsible for bringing the good name of CO_2 back.

Only a small number of our interviewees are willing to address the topic no matter what their own personal views. It seems surprising that so few broadcast meteorologists are stepping up to speak out when at the same time Americans in 2016 are expressing record high levels of belief that global warming is happening (64%), as well as certainty that human activity is causing the changes (65%). This recent article based on a Gallup poll analysis speculated that erratic weather patterns might be more persuasive to the U.S. public than anything they hear from politicians and/or climate-change scientists (Saad and Jones 2016). And yet, most of the people who tell them about their local weather are editing themselves very carefully.

Discussion and Conclusions

Our research supports a growing number of studies assessing the potential value of weather broadcasters as climate educators. As more laypeople come to believe that there is credible scientific consensus on global warming, an important shift occurs that in turn influences a number of other attitudes about engagement towards mitigation or adaption (Cook and Jacobs 2014). In fact, in 2014, broadcast meteorologists from a number of major U.S. cities were invited by President Obama to spend the day at the White House because he hoped to recruit them in spreading the word about climate change and the link to extreme weather events (Davenport 2014).

Typical citizens do not read peer-reviewed climate science research, but rather learn about global warming from people in their everyday lives (Boykoff et al. 2015). It is well documented that media figures are prominent sources for acquiring information about climate change (Van der Linden 2014). Due to the visibility and trustworthiness of weather broadcasters, engaging them as informal science educators on climate change could be a vital method for increasing the literacy of the public (Peters-Burton et al. 2014). One study showed that a series of brief on-air and online educational segments on climate change provided by broadcast meteorologists significantly shifted viewer's comprehension over the course of one year's time (Zhao et al. 2014). Another study suggested that viewing local TV weather forecasts can increase understanding of extreme weather events, which can then expand awareness about the impacts of climate change (Bloodhart et al. 2015).

A journal written specifically for broadcast meteorologists has run a series of articles urging weather broadcasters to embrace their role as station scientists and to augment their duties to regularly address environmental issues (Posegate 2008). The main goal of science education is to encourage individuals to take an informed part in decisions, and to enact appropriate steps to affect their own wellbeing and the welfare of society (Harlen 2010). In light of this definition of what it is to be a science educator, it is easy to conclude that most of the broadcast meteorologists we interviewed do not see themselves as agents in this important endeavor. The majority does not seem to actively identify as serious scientists, perhaps because of their overarching concern of being accepted by their diverse viewing audiences. For most broadcast meteorologists, because their core status is as presenters of practical information for use in the immediate future, it does not appear to cause a conflict for them not to be working as advocates for adaptation to or mitigation of climate change.

Although many of the broadcast meteorologists mentioned time constraints as a barrier, it would actually take only a few seconds to make a brief remark about climate change on a day that related stories were being reported by a desk news anchor. A recent national U.S. conference to discuss how weather broadcasters can communicate climate change topics effectively was convened by the National Center for Atmospheric Research, the National Science Foundation, affiliated organizations and academic researchers (Smith 2016). The report on the conference included a summary of the weather broadcasters' comments on the challenge of communicating climate change science. Similar to our group of respondents, the session participants were concerned about diverse political views held by viewers, station management apprehensions about alienating viewers, and ongoing time constraints. After conducting our interviews, our research team at Bentley University sponsored a comparable workshop on *Science and Storytelling* for weather broadcasters from the Northeastern U.S. (Szymanski et al. 2014). These types of educational programs have great potential to assist broadcast meteorologists struggling with their choice to speak out or stay silent. By providing support, exposure to thought leaders and the chance to share with peers, additional customized workshops could impart a grounding that inspires the necessary confidence for more weather broadcasters to articulate a position of advocacy.

However, without the support of this type of educational outreach, our data indicate that some weather broadcasters still take a strong position that translates as "it is not my job to take a stand" on climate change. A few respondents actually said that they actively avoided any contentious discussion about climate change on air. The comment linking the field of broadcast meteorology and marketing may have been particularly telling. At this point, the comparison to having something to sell (e.g., themselves as personalities) implies that they do not embody their most basic role as science communicators most of the time. The majority of the weather broadcasters that we interviewed are not functioning as science educators, and one of the main reasons is their ongoing concern about popularity. Some mentioned that their management passed along cautions that ratings could be hurt by speaking out on any side of a climate change discussion. Viewers' ratings (and now Facebook "likes") reinforce the fact that television stations do better economically if their weather broadcasters are popular. Many of the broadcast meteorologists seemed complacent about the fact that they are figures in a profit-driven system.

A few weather broadcasters stand out because they appear less concerned about their personal fate and the impressions they convey to their audiences, although it is notable that they are more likely to speak out in forums away from the rolling camera coverage. The comment "science is not a deli" implies an unwillingness to cut out or edit factual statements. Several maintain their interpretation of the journalistic principle advocating "balanced" presentations. However, it would be difficult to give 50-50 coverage to a topic for which such a large majority of climate scientists (97%) are in agreement about causality of global warming (Andregg et al. 2010; Cook et al. 2013). Perhaps then it is easier to avoid the topic of climate change altogether rather than to feel like the portrayal highlights a fringe minority opinion.

A critical area of future research would be to compare the behaviors of U.S. broadcast meteorologists with their peers in other countries. For example, Brazilian citizens show the most concern about climate change, with 85% reporting that climate change is very serious (PEW Research 2010). People in the U.S. (the world's largest per-capita emitter of carbon dioxide) are less troubled. Only 37% of Americans say global warming it is a very serious problem (PEW Research 2010). Perhaps since more citizens in other countries accept the consensus on climate change, there would be less concern about addressing the topic in broadcasts outside of the U.S. Although we are working with the limitation of a small sample size, our results are robust given the depth of the interviews, and especially the distinct groupings in knowledge base and viewpoints expressed by the broadcast meteorologists who participated in our study. Still, replication of our study through outreach to a larger international sample of weather broadcasters might provide more information from which to compare national outlooks.

The topic of climate change may continue to be difficult to discuss with the public because of the complexity of the science, which can engender resistance among viewers who are used to simple rhetoric. There is no easy answer to a societal challenge with such big stakes for everyone involved (Moser and Dilling 2011). Media coverage of global warming is one of the most complicated stories of our time because it involves abstract probabilistic science, unnecessary dueling political viewpoints, conflicting economic models, and a complex interplay of individuals and societies (Wilson 2000). For climate science information to be fully absorbed by media audiences, it must be actively communicated with effective narrative storytelling and delivered by trusted messengers (Shome et al. 2009).

Broadcast meteorologists have unique opportunities to provide compelling examples of climate change, such as how a region might suffer as a result of a rising sea level, which their viewers should see as having salience and legitimacy. Climate change adaptation and mitigation must happen; what continues to be lacking is a sense of public urgency and strong leadership (Leiserowitz 2007). The IPCC urges adjustments to climate change (planned, reactive and anticipatory) to moderate potential damage and to cope with the consequences. It has been shown that well planned, early adaptation strategies can save money and even lives in the short term and the long term. Appropriate mitigation urges the use of new technologies for energy efficiency. The changes that are needed can be as ambitious as innovative urban planning or as simple as redesigning appliances. It is clear that broadcast meteorologists have the potential to play a vital role as climate educators. However, the majority of weather broadcasters who took part in our study cited a variety of reasons why they were unwilling to speak out about mitigation and adaptation topics. U.S. citizens need a basis for action to address the myriad environmental problems that climate change is causing. If weather broadcasters could reframe their thinking about perceived constraints, they could step into the conversation and help shift their viewers from a disengaged or abstract conception of climate change toward a better informed understanding of an important topic in which we all have a personal stake.

Acknowledgements We appreciate that this research was supported by a US National Science Foundation Grant 12-22521 through the Advances in Informal STEM Education program. We also thank our Bentley University undergraduate student Kathryn Foley for assistance in multiple ways throughout our NSF-funded project.

References

Andregg WL, Prall JW, Harold J, Schneider SH (2010) Expert credibility in climate change. Proc Natl Acad Sci 107(27):12107–12109

Bloodhart B, Maibach E, Myers T, Zhao X (2015) Local climate experts: the influence of local TV weather information on climate change perceptions. PLoS ONE 10(11):e0141526

Boykoff MT, McNatt MB, Goodman MK (2015) The cultural politics of climate change news coverage around the world. In: The Routledge Handbook of Environment and Communication, p 221

Braun V, Clarke V (2006) Using thematic analysis in psychology. Qual Res Psychol 3(2):77–101

Center for Research on Environmental Decisions and ecoAmerica (2014) Connecting on climate: a guide to effective climate change communication. The Earth Institute, Columbia University, Copyright © December 2014 by The Trustees of Columbia University in the City of New York and ecoAmerica

Cook J, Jacobs P (2014) Scientists are from Mars, laypeople are from Venus: an evidence-based rationale for communicating the consensus on climate. Reports of the National Center for Science Education, vol 34(6)

Cook J, Nuccitelli D, Green SA, Richardson M, Winkler B, Painting R, Way R, Jacobs P, Skuce A (2013) Quantifying the consensus on anthropogenic global warming in the scientific literature. Environ Res Lett 8:024024

Corner A, Whitmarsh L, Xenias D (2012) Uncertainty, scepticism and attitudes towards climate change: biased assimilation and attitude polarisation. Clim Change 114(3–4):463–478

Daniels GL, Loggins G (2010) Data, Doppler, or depth of knowledge: how do television stations differentiate local weather? Atlantic J Commun 18(1):22–35

Davenport C (2014) Using weathercasters to deliver a climate change message. New York Times, 6 May 2014. http://www.nytimes.com/2014/05/07/us/politics/using-weathercasters-to-deliver-a-climate-change-message.html?_r=0. Accessed 28 Apr 2016
Davie WR, Auter PJ, Dinu LF (2006) Identifying the goals of weather instruction: toward a model approach for broadcast meteorology. Journalism Mass Commun Educ 61(2):149–164
Doherty R, Barnhurst KG (2009) Controlling nature: weathercasts on local television news. J Broadcast Electron Media 53(2):211–226
Egan PJ, Mullin M (2012) Turning personal experience into political attitudes: the effect of local weather on Americans' perceptions about global warming. J Polit 74(03):796–809
Hamilton LC (2011) Do you believe the climate is changing? Answers from new survey research. Carsey School of Public Policy, University of New Hampshire, Durham, USA.
Harlen W (ed) (2010) Principles and big ideas of science education. Association for Science Education, Ashford Colour Press Ltd., UK
Hobson K, Niemeyer S (2012) "What sceptics believe": the effects of information and deliberation on climate change scepticism. Publ Underst Sci. doi:10.1177/0963662511430459
IPCC (2014) Climate change 2014: synthesis report. Contribution of working groups I, II and III to the Fifth Assessment Report of the Intergovernmental Panel on Climate Change [Core Writing Team]. In: Pachauri RK, Meyer, LA (eds) IPCC, Geneva, Switzerland, 151 pp
Kalhoefer K, Robbins D, Seifter A (2016) How broadcast networks covered climate change in 2015: an analysis of nightly news and sunday shows. Media Matters for America, Washington, DC
Leiserowitz A (2007) Communicating the risks of global warming: American risk perceptions, affective images and interpretive communities. In Moser S, Dilling L (eds) Creating a climate for change: communicating climate change and facilitating social change. Cambridge University Press, U.K.
Maibach E, Wilson K, Witte J (2010) A national survey of television meteorologists about climate change: preliminary finds. Center for Climate Change Communication
Moser SC (2016) Reflections on climate change communication research and practice in the second decade of the 21st century: what more is there to say? Wiley Interdisc Rev Clim Change 7(3):345–369
Moser SC, Dilling L (2011) Communicating climate change: closing the science-action gap. In: The Oxford handbook of climate change and society. Oxford University Press, Oxford, pp 161–174
Myers TA, Nisbet MC, Maibach EW, Leiserowitz AA (2012) A public health frame arouses hopeful emotions about climate change. Clim Change 113(3–4):1105–1112
Myers TA, Maibach EW, Roser-Renouf C, Akerlof K, Leiserowitz AA (2013) The relationship between personal experience and belief in the reality of global warming. Nat Clim Change 3 (4):343–347
Perryman N, Theiss S (2014) "Weather girls" on the big screen: stereotypes, sex appeal, and science. Bull Am Meteorol Soc 95(3):347–356
Perse EM (1990) Media involvement and local news effects. J Broadcast Electron Media 34(1):17–36. doi:10.1080/08838159009386723
Perse EM, Rubin RB (1989) Attribution in social and parasocial relationships. Commun Res 16 (1):59–77
Peters-Burton E, Schweizer V, Cobb S, Maibach E (2014) Weathercaster views on informal climate education: similarities and differences according to climate change attitudes. J Geosci Educ 62(3):431–444
PEW Research Center (2010) Obama more popular abroad than at home: global image of U.S. continues to benefit. PEW Research Center: PEW Global Attitudes Project. From: http://pewresearch.org/pubs/1630/obama-more-popular-abroad-global-american-imagebenefit-22-nation-global-survey
Posegate A (2008) Station scientists: beyond the daily forecast. Weatherwise 61(6):20–25
Rosen J (1989) Don't need a weatherman? Harper's Magazine, 35, April, pp 35–36

Saad L, Jones JM (2016) U.S. concern about global warming at eight-year high. Retrieved from Gallop website http://www.gallup.com/poll/190010/concern-global-warming-eight-year-high.aspx?g_source=us%20concern%20about%20global%20warming%20at%208%20year%20high&g_medium=search&g_campaign=tiles

Sandman PM (1994) Mass media and environmental risk: seven principles. Risk Health Saf Environ 5, 251

Scruggs L, Benegal S (2012) Declining public concern about climate change: can we blame the great recession? Glob Environ Change 22(2):505–515

Shome D, Marx S, Appelt K, Arora P, Balstad R, Broad K, Leiserowitz A (2009) The psychology of climate change communication: a guide for scientists, journalists, educators, political aides, and the interested public. CRED Columbia University, New York, p 48

Smith J (2016) Weathercasters hone climate change reporting skills [Press release]. Retrieved from https://www2.ucar.edu. Accessed 28 Apr 2016

Stanfield K (2014) Why should they comment? Understanding local television news viewers motivation for commenting on stations' Facebook pages. Unpublished doctoral dissertation, University of Missouri—Columbia, Columbia, Missouri

Swim J, Clayton S, Doherty T, Gifford R, Howard G, Reser J, Weber E (2009) Psychology and global climate change: addressing a multi-faceted phenomenon and set of challenges. A report by the American Psychological Association's task force on the interface between psychology and global climate change. American Psychological Association, Washington, DC

Szymanski DW, Meldrum H, Davis PT, Oches EA, Foley K, Doner LA (2014) Views of broadcast meteorologists on climate change communication with their audiences. Geological Society of America Abstracts with Programs, vol 46, no 6

Thompson VK (2013) Ethical ambivalence in local television weathercasting: a Rossian analysis. Western Michigan University, Kalamazoo, Michigan, p 178

Van der Linden S (2014) Towards a new model for communicating climate change. In Cohen S, Higham J, Peeters P, Gössling S (eds) Understanding and governing sustainable tourism mobility: psychological and behavioural approaches. Routledge, Taylor and Francis Group, Florence, Kentucky, pp 243–275

Vannini P, Mccright AM (2007) Technologies of the sky: a socio-semiotic and critical analysis of televised weather discourse. Crit Discourse Stud 4(1):49–74

Wibeck V (2014) Enhancing learning, communication and public engagement about climate change—some lessons from recent literature. Environ Educ Res 20(3):387–411

Wilson KM (2000) Communicating climate change through the media. Environ Risks Media, 201–217

Wilson K (2006) Seals of (dis) approval: television broadcast meteorologists debate the value of voluntary credentials. National Weather Dig 30:100–107

Wilson K (2008) Television weathercasters as potentially prominent science communicators. Publ Underst Sci 17(1):73–87

Wilson K (2009) Opportunities and obstacles for television broadcast meteorologists to report on climate change. Bull Am Meteorol Soc 90(10):1457–1465

Wilson KM (2012) Ideology trumps meteorology: why many television broadcast meteorologists remain unconvinced of human-caused global warming. Electronic News 6(4):208–228

Zhao X, Maibach E, Gandy J, Witte J, Cullen H, Klinger BA, Pyle A (2014) Climate change education through TV weathercasts: results of a field experiment. Bull Am Meteorol Soc 95 (1):117–130

Part V
Climate Change, Planning and Health and Examples from Other Regions

Chapter 17
Urban Planning to Prevent Mosquito-Borne Diseases in the Caribbean

Derek W. Willis

Abstract Climate change could lead to a resurgence of malaria in the Caribbean. A major initiative is currently underway to eliminate malaria in Hispaniola. In order to achieve this goal, and prevent a subsequent resurgence of malaria, the anti-malaria intervention policies implemented in Hispaniola will need to account for the impact of climate change. Lessons learned from the policies implemented in Hispaniola over the next 10–20 years could be highly informative to prevent resurgences of malaria in other parts of the Caribbean that are largely free of malaria today. Current malaria funding to control malaria vectors in Hispaniola is being allocated almost entirely to the implementation of long-lasting insecticidal nets and indoor residual spraying. Urban anti-malaria planning is urban planning that includes the suppression of potential larval habitats of malaria vectors as one of its goals. While long-lasting insecticidal nets and indoor residual spraying may be cost-effective in rural areas, it is unlikely that they will always be more cost-effective than urban anti-malaria planning in urban and urbanizing malarious communities. One explanation for the lack of funding to implement urban anti-malaria planning in Hispaniola, as well as other malarious regions, is that policy makers are unaware of the historical evidence of its use to suppress malaria. This chapter presents a case study of the use of urban anti-malaria planning in the town of Cité Duvalier, Haiti in the early 1970s. Using a multi-disciplinary methodology, the cost and impact of urban anti-malaria planning in Cité Duvalier is compared to the cost and impact of indoor residual spraying, mass drug administration and larviciding. This methodology can be used in future work to continue to examine the extensive historical evidence of urban anti-malaria planning interventions. The results of these future studies will enable policy makers to understand when urban anti-malaria planning should be implemented instead of, or in addition to, other anti-malaria interventions.

D.W. Willis (✉)
Center for Research on Environmental Decisions,
Columbia University, New York, USA
e-mail: derekwwillis@gmail.com

W. Leal Filho and J.M. Keenan (eds.), *Climate Change Adaptation in North America*, Climate Change Management,
DOI 10.1007/978-3-319-53742-9_17

Keywords Malaria · Urban planning · Environmental management · Larval source management · Climate change · Adaptation · Mosquito-borne diseases · Vector-borne diseases

Introduction

An ambitious goal has been established to eradicate malaria by the year 2040 (Newby et al. 2016). In order to achieve this goal, policy makers must account for the potential impact of climate change on malaria transmission. Climate change may increase malaria transmission in sub-Saharan Africa where malaria has decreased over the past decade, including areas at higher elevations (Siraj et al. 2014). In addition, climate change could lead to a resurgence of malaria in areas where indigenous cases of malaria have been eliminated.

While malaria has been eliminated throughout most of the Caribbean, malaria is still an important public health problem on the island of Hispaniola. The United States Centers for Disease Control and Prevention is leading an initiative to eliminate malaria by the year 2020 in Hispaniola (PAHO 2015). Hispaniola, as well as the Caribbean in general, is particularly susceptible to the increased likelihood of extreme events (e.g., storms, hurricanes and floods) due to climate change (IPCC 2014). These extreme events are also associated with an increase in the risk of malaria (Hales et al. 2003).

There are two types of interventions for preventing malaria vectors (i.e., mosquitoes that transmit malaria) from transmitting malaria that could be implemented on a large scale by the public sector in order to achieve this goal of eradicating malaria.

The first type is insecticide-based interventions that kill malaria vectors or prevent malaria vectors from biting individuals. The two interventions of this type that are currently being funded and implemented on a large scale by the public sector are long-lasting insecticidal nets (LLINs) and indoor residual spraying (IRS).

The second type of interventions is generally referred to as larval source management (LSM). These interventions target the larval stage of malaria vectors by either eliminating breeding sites for these mosquitoes or preventing the larvae in these sites from developing into adult mosquitoes. Two approaches to implementing larval source management are larviciding and environmental management. Larviciding is the application of an oil or toxin to a larval habitat to prevent larvae of malaria vectors from developing into adult mosquitoes. Environmental management (also referred to as environmental modification or environmental manipulation) involves temporarily or permanently reducing the number of larval habitats of malaria vectors (Keiser et al. 2005). An example of an environmental management intervention would be the construction of drains in a community to reduce the number of larval habitats of malaria vectors.

Urban planning policies incorporate many aspects of environmental management. When sufficient resources are available, an objective of urban planning can include developing drainage systems that prevent or minimize flooding. The increased probability of extreme events from climate change has led to a renewed focus on urban planning as a critical climate change adaptation strategy, especially in low lying coastal areas (Elsharouny 2016; Carter et al. 2015).

In this study, we use the term 'urban anti-malaria planning' to refer to the implementation of environmental management interventions in an urban community with an objective that includes the suppression of malaria transmission.

There is extensive evidence of urban anti-malaria planning being used to successfully suppress malaria, with one of the first cases of its use occurring in the early 1900s (Watson 1921). Despite this evidence, international anti-malaria funding for vector control interventions in urban communities has been allocated almost entirely to insecticide-based interventions (WHO 2016).

One possible explanation for this disconnect between the evidence regarding urban anti-malaria planning and how malaria funding is allocated is the lack of clear guidelines for when urban anti-malaria planning is more cost-effective than insecticide-based interventions. Specifically, policy makers need guidelines that describe how the epidemiological, entomological and socio-economic conditions of a community will affect the potential cost and impact of urban anti-malaria planning.

In order to develop these guidelines for policy makers, the impact and cost of urban anti-malaria planning across a wide range of communities need to be analyzed. Although there is extensive literature on urban anti-malaria planning, the literature needs to be analyzed appropriately in order to develop the necessary guidelines.

This chapter presents a case study of the cost and impact of urban anti-malaria planning in the town of Cité Duvalier, Haiti in the early 1970s. This case study uses data from several articles published in the early 1970s that describe the anti-malaria interventions implemented in Cité Duvalier. A multi-disciplinary methodology is used to compare the impact and cost of urban anti-malaria planning to alternative anti-malaria interventions that are currently favored by international donors for malarious countries.

Methodology

Cité Duvalier was located in an area next to Port-au-Prince and had a population of approximately 16,000 people from 1968 through 1971 (USAID 1972).

Our analysis of the impact and cost of urban anti-malaria planning in Cité Duvalier was based on a dataset we developed from multiple sources (Schliessmann et al. 1973; Carmichael 1972; USAID 1972) for the time period from January of 1968 through July of 1972. The implementation of urban anti-malaria planning in Cité Duvalier began in December of 1969 (Schliessmann et al. 1973). Prior to December of 1969, the following anti-malaria interventions had been implemented in Cité Duvalier: mass drug administration, indoor residual spraying and larviciding (Schliessmann et al. 1973).

We defined the impact of urban anti-malaria planning in Cité Duvalier as having two components. The first component is the impact of urban anti-malaria planning on malaria cases. We estimated its impact on malaria by comparing the number of malaria cases with urban anti-malaria planning in Cité Duvalier to the number of malaria cases in Cité Duvalier prior to its implementation. We also considered whether precipitation or other factors could have played a significant role in the decrease in malaria cases that was observed in Cité Duvalier from 1970 through July of 1972.

The second component of the impact of urban anti-malaria planning in Cité Duvalier is the non-malaria impact. We defined the non-malaria impact of urban anti-malaria planning as the impact, if any, it would have had in Cité Duvalier from December of 1969 through July of 1972 if there had been no malaria transmission in Cité Duvalier during that period.

Finally, we estimated the cost of implementing urban anti-malaria planning in Cité Duvalier and compared it to the cost of the anti-malaria interventions that were implemented prior to December of 1969.

Results

In this section, the impact and cost of the urban anti-malaria planning strategy implemented in the community of Cité Duvalier are presented.

The first step in estimating the impact of urban anti-malaria planning on malaria in Cité Duvalier was determining if there is a causal relationship between its implementation and the change in the monthly number of malaria cases as other factors could have been primarily responsible for the decrease in malaria cases that occurred after urban anti-malaria planning was implemented.

The full implementation of urban anti-malaria planning began in December of 1969. Before December of 1969, there was a strong relationship between rainfall and malaria cases in Cité Duvalier as spikes in rainfall led to an increase in malaria cases (Schliessmann et al. 1973). In contrast, similar increases in rainfall after 1969 did not lead to an increase in malaria cases (Schliessmann et al. 1973). This suggests that the sustained decrease in malaria cases after 1969 cannot be attributed to rainfall.

Although monthly estimates of the total population in Cité Duvalier are not available, there is no evidence in the literature that there was a decrease in the community's population from January of 1968 through July of 1972. In addition, there is no evidence that there was a change in the process that was used to identify and record malaria cases in Cité Duvalier during this period.

It is, therefore, highly likely that the decrease in malaria cases in Cité Duvalier after December of 1969 can be attributed primarily to the implementation of urban anti-malaria planning. In order to estimate its impact on malaria cases, we examine the relationship between the anti-malaria interventions that were implemented in Cité Duvalier from January of 1968 through July of 1972 and the monthly number of malaria cases over that period.

Indoor residual spraying (IRS) and mass drug administration (MDA) interventions were implemented from January 1968 until the summer of 1969. IRS was terminated after February 1969 due to high levels of insecticide resistance (Schliessmann et al. 1973). Mass drug administration involved the community residents taking drugs in order to decrease their risk of malaria infection. Due to the low willingness of the community to take the drugs, this intervention ended in May of 1969 (Schliessmann et al. 1973). Larviciding was conducted in September of 1968 and throughout most of 1969 (Schliessmann et al. 1973).

We estimated the cost of implementing these interventions (IRS, MDA and larviciding) in Cité Duvalier using relevant cost data from multiple sources (Thevasagayam et al. 1979; PAHO 2010; Schliessmann et al. 1973; Carmichael 1972; USAID 1972).

The urban anti-malaria planning that was implemented in Cité Duvalier consisted of developing ditches to improve drainage using heavy equipment from December 1969 to February 1970 and labor-intensive measures from December 1969 until January 1971 (Schliessmann et al. 1973). The cost of the heavy equipment that was used to create 16,484 ft of ditches was $4568 (1970 US$) (Schliessmann et al. 1973). The cost of the labor-intensive measures was approximately $18,607 (1970 US$) and we estimate that it created approximately 4000 ft of ditches (Schliessmann et al. 1973).

Following the completion of this work, there were only 35 cases of malaria in Cité Duvalier over a 27-month period from May 1970 through July 1972. In contrast, there were approximately 586 malaria cases in January of 1969.

Finally, it is important to recognize the non-malaria impact that urban anti-malaria planning had on Cité Duvalier. Several studies note that the drainage system made the land more suitable for agriculture and acted as a convenient water source for the population for bathing and laundry (Schliessmann et al. 1973; Carmichael 1972). Given that flooding had previously been a problem in Cité Duvalier, it is also highly likely that urban anti-malaria planning led to reduced damage due to floods (Schliessmann et al. 1973).

The construction and maintenance of ditches to improve drainage (i.e., environmental management) represented approximately 80% of the total cost of interventions implemented from December 1969 through July 1972, with the cost of larviciding making up the remaining 20%. Expressed in 2015 US dollars, the total cost of environmental management (i.e. improved drainage) from December 1969 through July 1972 was $259,526 while the cost of larviciding was $67,149.

It is important to note that improved drainage reduced the cost of larviciding by approximately 65% due to the permanent elimination of a significant proportion of larval sites (Schliessmann et al. 1973) (Table 17.1).

Table 17.1 Results—cost, malaria impact, non-malaria impact

		Jan. 1968–Nov. 1969	Dec. 1969–July 1972
Number of months per time period		23 months	32 months
Total Cost (2015 US$)	MDA, IRS, larviciding	$292,566	$0
	Urban anti-malaria planning (environmental management and larviciding)	$0	$326,675
Cost per month (2015 US$)		$12,720	$10,209
Malaria cases		1521	54
Non-malaria impact		None	Facilitated agriculture; reduced flooding

Conclusion

In conclusion, we have presented a case study of the impact and cost of urban anti-malaria planning in Cité Duvalier, Haiti. The results of our analysis indicate that urban anti-malaria planning required less resources to implement per month and led to fewer malaria cases compared to the use of MDA, IRS and larviciding. In addition, urban anti-malaria planning made parts of Cité Duvalier more suitable for agriculture and reduced flooding.

We used a multi-disciplinary methodology, linking the disciplines of epidemiology and economics in particular, to conduct our analysis. A unique aspect of this methodology was including estimates of the non-malaria impact of urban anti-malaria planning. We defined the non-malaria impact of an anti-malaria intervention as the impact the intervention would have had on an identical community where there was no malaria transmission. In contrast to insecticide-based anti-malaria interventions, urban anti-malaria planning can benefit communities by, for example, decreasing flooding and making the land suitable for economic activities. If the non-malaria impact of urban anti-malaria planning is ignored when comparing the impact and cost of anti-malaria interventions, the true impact of urban anti-malaria planning on communities may be underestimated.

We recommend that researchers use this methodology to examine the extensive literature of urban anti-malaria planning. One of the major challenges to identifying the most appropriate anti-malaria interventions to implement in a community is the significant heterogeneities in the impact and cost of interventions over space and time. An important factor that contributes to these heterogeneities over space and time is the malaria vector. There are more than 30 species of Anopheles mosquitoes that can transmit malaria. These species differ in the characteristics of the water habitats where they breed, the time of day they bite and the degree to which they bite indoors versus outdoors. It is, therefore, critical to use this methodology to examine urban anti-malaria planning in communities with a wide range of malaria vectors in order to determine how vector behavior affects the impact and cost of this strategy.

Unlike insecticide-based interventions, urban anti-malaria planning is not an intervention that can typically be implemented solely by a government's health sector. It often requires an intersectoral approach, necessitating collaboration between a country's ministries of public health, planning and transportation. The potential cost of this collaboration across sectors should therefore be considered by policy makers in determining whether or not to implement urban anti-malaria planning today.

Urban anti-malaria planning in Cité Duvalier involved the use of environmental management to improve drainage and reduce the density of larval sites of malaria vectors. However, urban anti-malaria planning could rely on different policies in other contexts. For example, policy makers in an urbanizing community could ensure that trash collection services operate effectively in order to prevent trash from collecting in drains and limiting their effectiveness.

Given that larviciding was implemented concurrently with environmental management, it is not possible to estimate the impact of only environmental management on malaria cases. However, as we noted, environmental management decreased the density of larval sites in Cité Duvalier over time. This resulted in a significant reduction in the cost of implementing larviciding compared to the period prior to the introduction of environmental management. If larviciding had been implemented without environmental management, it could have led to a similar decrease in malaria cases. However, this approach would have likely required larviciding being conducted more frequently and over a larger area compared to the larviciding activities that were carried out after drainage was improved. It is therefore unlikely that implementing only larviciding would have been a more appropriate strategy than implementing it along with environmental management.

This chapter presented a case study of the cost, malaria impact and non-malaria impact of an urban anti-malaria planning policy in Cité Duvalier, Haiti. It is our expectation that this case study will not only serve as a template for applying this methodology in future studies but also highlight the need to consider implementing urban anti-malaria planning in Haiti, as well as other malarious countries.

References

Carmichael GT (1972) Anopheline control through water management. Am J Trop Med Hyg 21 (5):782–786

Carter JG, Cavan G, Connelly A, Guy S, Handley J, Kazmierczak A (2015) Climate change and the city: building capacity for urban adaptation 95(January):1–66

Elsharouny MRMM (2016) Planning coastal areas and waterfronts for adaptation to climate change in developing countries. Procedia Environ Sci 34:348–359

Hales S, Edwards SJ, Kovats RS (2003) Impacts on health of climate extremes. In: Climate change and health: risks and responses. World Health Organization, Geneva

IPCC (2014) Climate change 2014: impacts, adaptation, and vulnerability. In: Field CB, Barros VR, Dokken DJ, Mach KJ, Mastrandrea MD, Bilir TE, Chatterjee M, Ebi KL, Estrada YO, Genova RC, Girma B, Kissel ES, Levy AN, MacCracken S, Mastrandrea PR, White LL (eds) Part A: Global and sectoral aspects. Contribution of Working Group II to the Fifth Assessment Report of the Intergovernmental Panel on Climate Change. Cambridge University Press, Cambridge, 1132 pp

Keiser J, Singer BH, Utzinger J (2005) Reducing the burden of malaria in different eco-epidemiological settings with environmental management: a systematic review. Lancet Infect Dis 5:695–708

Newby G, Bennett A, Larson E et al (2016) The path to eradication: a progress report on the malaria-eliminating countries. The Lancet 387(10029):1775–1784. doi:10.1016/S0140-6736(16)00230-0

PAHO: Malaria Surveillance Indicators. C2010. Regional Malaria Program. Accessed 20 Mar 2017. http://ais.paho.org/phip/viz/malaria_surv_indicators_popup.asp

PAHO (2015) PAHO/WHO supports CDC-led effort to eliminate malaria on Hispaniola by 2020. Pan American Health Organization/World Health Organization. Accessed 20 Mar 2017. http://www.paho.org/hq/index.php?option=com_content&view=article&id=10463%3A201%205-paho-supports-cdc-to-eliminate-malaria&catid=740%3Apress-releases&Itemid=1926&la%20ng=en

Schliessmann DJ, Joseph VR, Solis M et al (1973) Drainage and larviciding for control of a malaria focus in Haiti. Mosquito News 33(3):371–378

Siraj AS, Santos-Vega M, Bouma MJ et al (2014) Altitudinal changes in malaria incidence in highlands of Ethiopia and Colombia. Science 343(6175):1154–1158

Thevasagayam ES, Siong Y, Philip G (1979) Temephos (Abate) as a replacement larvicide for oil for the control of Anopheles maculatus, the main vector of malaria in peninsular Malaysia. WHO

USAID (1972) Evaluation report of malaria eradication program—Haiti. USAID

Watson M (1921) The prevention of malaria in the Federated Malay States: a record of twenty years' progress. J. Murray, London

WHO (2016) World Malaria Report 2015. World Health Organization

Chapter 18
Mental Health Impacts of Droughts: Lessons for the U.S. from Australia

Elisaveta P. Petkova, Anne S. Celovsky, Wei-Yann Tsai and David P. Eisenman

Abstract Drought is a condition already endemic to many regions of the United States and is likely to increase in frequency and/or duration under a changing climate. Drought can have wide ranging health impacts, particularly among vulnerable subpopulations such as rural, agricultural communities. Although poor mental health outcomes are an established public health concern for such populations during and following drought periods, the pathways through which drought can impact the mental health of rural residents are not well understood. In addition, most of the studies to date have been carried out in Australia, a country that has decades of experience with protracted droughts. As regions in the United States struggle with the extended droughts, lessons and evidence from Australia may be particularly useful in the development and implementation of surveillance and mitigation initiatives. This article provides an overview of the existing evidence of drought impacts on the mental health of residents in Australia with particular focus on vulnerability and discusses the application of this knowledge to drought response and planning in the United States.

Keywords Drought · Climate change · Mental health · Adaptation

---

E.P. Petkova (✉)
National Center for Disaster Preparedness, Earth Institute, Columbia University, New York, NY 10027, USA
e-mail: elisaveta.petkova@columbia.edu

A.S. Celovsky
Mailman School of Public Health, Department of Biostatistics, Columbia University, New York, USA

W.-Y. Tsai · D.P. Eisenman
David Geffen School of Medicine at UCLA, Division of General Internal Medicine and Health Services Research, UCLA Center for Public Health and Disasters, UCLA Fielding School of Public Health, Los Angeles, CA, USA

W. Leal Filho and J.M. Keenan (eds.), *Climate Change Adaptation in North America*, Climate Change Management,
DOI 10.1007/978-3-319-53742-9_18

Introduction

Global climate change is projected to have substantial and wide ranging impacts on human health including morbidity and mortality due to extreme weather conditions such as heat, floods, droughts and storms, as well as adverse health impacts due to decreased air quality, water-borne diseases, and other infectious diseases (Costello et al. 2009; Franchini and Mannucci 2015; Haines et al. 2006; IPCC 2014; Luber et al. 2014; Luber and Prudent 2009). Climate change may also present risks to mental health through inducing psychological trauma and migration as a result of exposure to natural disasters, and causing various short- and long term mental health impacts such as stress, anxiety and post-traumatic stress disorder (Clayton et al. 2015; Kinney et al. 2015; McMichael et al. 2006; Rice and McIver 2016). The physical and especially the mental health impacts of sub-acute weather disasters such as prolonged droughts have only recently received more attention in the literature (Berry et al. 2010; Stanke et al. 2013; Vins et al. 2015).

Droughts are generally defined as extended periods of deficient precipitation as compared to the average precipitation for the region resulting in a water shortage causing adverse impacts on vegetation, animals, and/or people (Graham 2000). According to the IPCC, meteorological and agricultural droughts will likely increase in frequency and/or duration in certain regions, including parts of North America (IPCC 2014). Areas dependent on snowmelt for their water supply, such as California, are likely to be additionally impacted as more precipitation is predicted to fall as rain than snow, with snowmelt occurring earlier and less abundantly thereby reducing the water supply available for the drier summer months.

Drought is a condition endemic to many regions of the United States. The CDC estimates that in any year in the past century, 14% of the US has experienced drought (CDC 2010). A study by the Natural Resources Defense Council found that one-third of all counties in the United States will face higher probability of water shortages due to climate change by mid-century (NRDC 2010). Despite strong adaptive capacity to drought in the United States and other developed countries, drought can have wide ranging health impacts, particularly among vulnerable subpopulations. Major health impacts due to drought worldwide include nutritional-related effects, increased particulate matter pollution leading to respiratory diseases, increased vector-borne and water-borne diseases, as well as mental-health impacts (Stanke et al. 2013; Vins et al. 2015). As is the case with other environmental risks, susceptibility to any particular drought risk is dependent on the severity of the drought, the underlying demographics and vulnerability of exposed populations, water infrastructure, and national support to help mitigate or reduce the impacts of droughts on affected populations (Stanke et al. 2013).

At the time of this manuscript, California is battling the most severe drought in its history. The state has experienced an extended and severe drought since 2011, with reduced precipitation and snow-pack levels severely constraining agricultural production and prompting large-scale calls for conservation by water management districts and public utilities. The fallowing of lands and reduced employment place additional stress on farmers and rural communities that could translate into reduced mental health, prompting this review on the existing literature between drought and mental health with an eye toward informing interventions for communities in drought-stressed regions of the United States.

This paper aims to present evidence on the mental health impacts of drought in Australia, a country that has been dealing with protracted droughts for decades and accounts for most of the published literature on drought and human health effects in developed countries. The review fills an important gap in the literature at a time when the U.S. is facing the consequences of a prolonged and severe drought. Despite the wide political and social attention to this ongoing drought, literature on its possible mental health impacts on local populations is scant. Therefore, lessons from Australia may be particularly useful in the development and implementation of new research studies as well as surveillance and mitigation initiatives in the region. The paper starts by examining the existing literature of drought impacts on the mental health of residents in Australia. It provides an extensive review of mental health outcomes discussed in the literature with particular focus on vulnerability. The paper concludes with applications of existing knowledge to the current drought in U.S. and recommendations for population adaptation to future droughts.

Methodology

Between February and April of 2015, a systematic search of the literature was conducted in the SCOPUS, PsychInfo, and PubMed databases for papers examining the relationship between drought and mental health outcomes. The search terms utilized were: "drought" AND "mental health" OR "suicid*" (e.g. suicide, suicidal, suicidality). These searches resulted in 178 articles and book chapters from the three databases (88 in Scopus, 60 in PubMed, 30 in PsychInfo). 68 articles were eliminated as duplicate results. The remaining 110 articles were subjected to a review of titles and abstracts to determine suitability for inclusion in the full review. Papers were excluded if "drought" referred to something other than meteorological drought (i.e. "opioid drought"), if drought was not a key exposure of interest, if mental health/suicidality was not included as a specific outcome of interest, or if the study methodology did not include quantitative estimates of the association between drought and mental health impacts. 11 articles were included in the final systematic review (Fig. 18.1).

Reported Mental Health Outcomes

The 11 articles included in this review along with the major findings are presented in Table 18.1. Major outcomes reported by the studies in the literature review include suicide, general mental health, psychological distress and emotional/ behavioral problems.

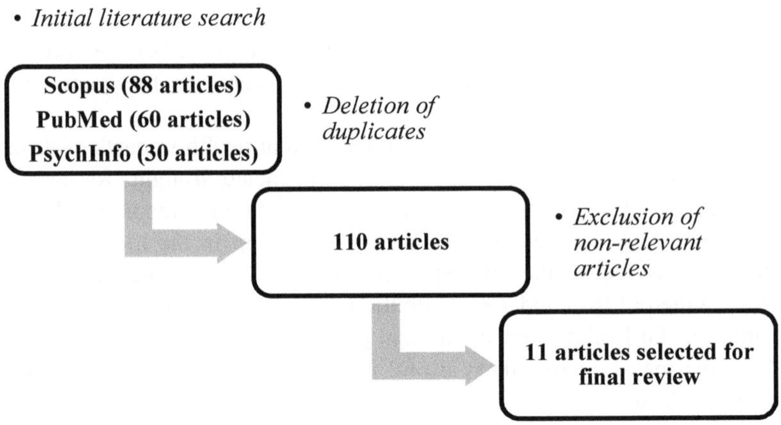

Fig. 18.1 Literature search methodology

Suicide

Two studies of drought and suicide risk were identified for this review. Hanigan et al. (2012) conducted a retrospective review of mortality data in New South Wales between 1970 and 2007 to determine whether suicide rates were associated with drought years, controlling for long-term suicide trends, season, and age-group. Drought was operationalized via the Hutchison drought index which utilizes rainfall data and successfully captures periods of agricultural drought. The study reported a 15% increase in suicide risk for rural males 30–49 years (95% CI: 8–22%) with increased drought duration, while there was a statistically significant decrease in risk among rural females aged 30 years or older. No significant association between drought and suicide was found in urban populations. Guiney (2012) conducted a similar retrospective mortality review in Victoria for 2001–2007 to determine whether suicide was associated with drought among farm workers and reported no association between drought and suicide incidence among farmers. The short study duration (2001–2007), the fact that all of the years included were drought years and small annual number of cases (average = 16), made capturing trends over time challenging and didn't allow for a determination as to whether suicide rates during drought years were significantly different compared to rates during non-drought years.

General Mental Health

Edwards, Gray and Hunter (2015) assessed general mental health using the 5-item Mental Health Inventory from the Medical Outcomes Study 36-item Short-Form Health Survey, an epidemiological tool for determining probable mental illness. The study analyzed an 8000-person survey conducted in agricultural areas in 2007

Table 18.1 Articles included in this review

Citation	Type of study	Outcome/Findings	Population studied
Hanigan et al. (2012)	Retrospective cohort study	• Increased drought duration resulted in increased suicide risk for rural males 10–29 and 30–49 but a decreased risk among rural females 30+ years old. No association was found between drought and suicide risk among urban populations	All deaths/suicides in New South Wales
Guiney (2012)	Retrospective cohort study	• No association between rate of suicides among farmers and drought conditions	All farm suicides in Victoria
Edwards et al. (2015)	Cross-sectional	• 17.8% of farmers in drought regions had MH problems compared to 13.8% of farmers in non-drought regions (but not statistically significant. Difference in MH continuous scores however was significant). 11.9% of farm workers in drought regions had MH problems compared to 5.5% of farm workers in non-drought regions ($p < 0.001$) • Overall, living in a drought-impacted region increased likelihood of experiencing a MH problem with an OR = 1.26 ($p = 0.007$). Farmers who reported that the drought eliminated farm productivity or reduced it to the lowest point ever increased the likelihood of a MH problem by an OR of 8.49 ($p < 0.001$)	Rural residents
Kelly et al. (2011)	Cross-sectional	• Drought and other district-level variables were not significantly associated with psychological distress. Psychological distress was most significantly associated with neuroticism, recent adverse events, personal social support and social capital measures	Rural residents
Stain et al. (2011)	Cross-sectional	• Of those in the high drought exposure group, 28.8% reported high drought worry and 9.8% reported both high drought worry and high psychological distress • High worry about drought was significantly associated with living on a farm or in a very remote area. High neuroticism and low social or community support were associated with increased risk of psychological distress while drought-related worry was associated with measures of socio-economic factors and sense of place/connection to the environment	Adults in drought-exposed rural areas
Gunn et al. (2012)	Cross-sectional	• Sample was found to be significantly more distressed than the broader Australian population (population norms from national survey) • No significant difference in distress by gender. Younger age groups (25–44 and 45–54) reported higher distress compared to an older age group (55–64). Maladaptive coping strategies were significantly associated with increased psychological distress	Farmers

(continued)

Table 18.1 (continued)

Citation	Type of study	Outcome/Findings	Population studied
O'Brien et al. (2014)	Cross-sectional	• Experiencing 20–32 cumulative months of extreme dryness during the 'Big Dry' culminating in recent unbroken dryness of at least a year was associated with increased distress in rural populations but not urban populations (estimated increase of 6.22% in distress compared to other participants)	Nationally representative sample (age 15+)
Stain et al. (2008)	Cross-sectional	• 71.8% of farm workers, 67.6% of farm residents, and 45.8% of non-farmers reported high levels of drought stress • When the farm workers and farm resident categories were collapsed, this group had significantly higher drought stress than the non-farm group as well as a significantly stronger sense of place. Drought stress, however, was not significantly correlated with psychological distress	Rural adults
Friel et al. (2014)	Cross-sectional	• Some types of drought were found to be harmful to mental health because they increase food insecurity, with greater implications for rural areas • Rural people exposed to "constant and long dry" reported moderate levels of distress while those in any other category reported low distress • The type of drought experienced can have positive or negative effects on the association between food insecurity and mental health. For example, increased distress associated with lower-than-average intake of core foods was only significant for people under "zero or moderate drought" • People in "zero to moderate" "very dry" and "constant and long dry" droughts who consumed higher-than-average amounts of discretionary (unhealthy) foods reported significantly more distress than those in other drought categories	Nationally representative sample (age 15+)
Dean and Stain (2007)	Cross-sectional	• Levels of emotional distress as measured by the Strengths and Difficulties Self-Report Questionnaire (SDQ) among rural adolescents ages 11–17 experiencing drought were not significantly different than the Australian norms for that age cohort	Rural adolescents
Dean and Stain (2010)	Cross-sectional	• Emotional and behavioral problems were assessed among rural adolescents ages 11–17 experiencing drought as a follow up of the Dean and Stain (2007) study • Significantly higher levels of distress found in the studied population compared to the Australian norms for their age cohorts. Increased perceived experience of drought significantly correlated with both overall the Strengths and Difficulties Self-Report Questionnaire (SDQ) scores ($p < 0.001$ for ages 11–13 and $p < 0.05$ for ages 14–17) as well as the Emotional Symptoms subscale ($p < 0.01$ for ages 11–13, $p < 0.05$ for ages 14–17)	Rural adolescents

with equal numbers of participants in average, above average and below-average rainfall regions. Drought was measured through a self-report question in the survey, allowing to capture the experience of agricultural or socioeconomic drought. The analysis controlled for demographic variables including age, gender, indigenous status, educational attainment, and geographic area but left income and financial stress uncontrolled as these were hypothesized to lie on the causal pathway. The study reported mental health impacts among farmers and other farm workers while non-agricultural workers and the unemployed were less affected. 17.8% of farmers in drought regions had mental health problems compared to 13.8% of farmers in non-drought regions (but not statistically significant), while the difference in continuous scores was statistically significant ($p < 0.001$). Similarly, 11.9% of farm workers in drought regions had mental health problems compared to 5.5% of farm workers in non-drought regions ($p < 0.001$). Differences for people in non-agricultural employment or the unemployed were inconsistent and depended on scoring of the mental health inventory.

Kelly et al. (2011) characterized general mental health by using an aggregate measure of well-being, combining self-reported overall physical and mental health, levels of distress and related impairment (Kessler-10 LM), as well as functioning and satisfaction with relationships and life. In this cross-sectional study of a random sample of 2462 rural adults in New South Wales, Kelly et al. (2011) examined associations of personal and area-level factors (including drought) that influence mental health and well-being. Drought was operationalized via rainfall data. The authors found that drought, along with other area-level factors was not significantly associated with wellbeing while personal-level factors such as neuroticism, recent life events, and levels of perceived social support were significantly associated with well-being.

Psychological Distress

Most studies identified in this review have used psychological distress scales, such as the Kessler-10 Psychological Distress tool, to assess probable mental health impacts of drought.

Stain et al. (2011) examined the impact of exposure to prolonged drought on psychological distress (Kessler-10), as well as an index of concern or worry about drought, among a randomly selected sample of 302 rural adults. High drought exposure was operationalized as living in an area that had been drought-declared at least 6 months of the past year. Predictors of drought-related stress differed from predictors of psychological distress, indicating that drought-related worry is a separate and distinct phenomena that may not be fully captured by more general measurements such as the Kessler-10 Psychological Distress tool. Drought-related worry was significantly associated with socio-economic factors, employment, personal psychological characteristics and connection with the environment/sense of place. Psychological distress was significantly associated with personal factors, such as higher neuroticism and lower levels of hopefulness, as well as lower perceived social support and connectedness to community.

Gunn et al. (2012) conducted a mixed-methods cross-sectional study of 307 farmers or farm spouses in drought declared areas to determine whether age or gender is associated with risk of psychological distress and/or choice of coping strategy, and which coping strategies were most closely associated with psychological distress experienced during drought. The sample was found to be significantly more distressed than normative Australian data (as measured in the National Health Survey, ($p < 0.001$)), with no significant difference in distress by gender. Younger age groups (25–44 and 45–54) reported significantly higher distress than slightly older farmers and spouses (55–64) but were not significantly different than farmers and spouses 65 and older. In addition, "maladaptive" coping strategies (such as substance abuse, behavioral and mental disengagement, venting, and suppression of competing activities) were significantly associated with increased psychological distress.

The Kelly et al. (2011) cross-sectional analysis of rural mental health in New South Wales (referenced earlier) also analyzed psychological distress alone as an outcome and found no significant effect of drought on distress within rural regions of Australia. Psychological distress was most significantly associated with neuroticism, recent adverse events, personal social support and social capital measures. Furthermore, people living in the "very remote" regions had significantly higher rates of psychological distress than other categories of rurality ($p < 0.001$).

Stain et al. (2008) investigated associations between mental health and measures of community support, social support networks, sense of place, adversity, and perceived problems among rural Australians, with special focus on farmers. 71.8% of farmers or farm workers compared to 56.1% of the overall population reported high levels of drought stress. When the farm workers and farm resident categories were collapsed, this group had significantly higher drought stress than the non-farm group as well as a significantly stronger sense of place. Drought stress, however, was not significantly correlated with psychological distress.

O'Brien et al. (2014) conducted a cross-sectional study of the association between severity of drought exposure and psychological distress within a large nationally representative survey of Australians aged 15 and older, conducted just before the end of the 7-year 'Big Dry' drought. 5012 participants who have resided in the same location during the duration of the drought were selected. Drought exposure was categorized as: "zero to moderate", "very dry", "recent long period", "constant drought", and "constant and recent long dry" based on indices calculated using rainfall data. The study found that experiencing 20–32 cumulative months of extreme dryness during the 'Big Dry' culminating in recent unbroken dryness of at least a year ("constant and recent long dry") was associated with increased distress in rural populations but not urban populations (estimated increase of 6.22%, 95% CI: 1.46–10.98).

Friel et al. (2014) conducted a cross-sectional study on the same nationally-representative sample as O'Brien et al. (2014) to explore the association between food insecurity, drought, and psychological distress. The study utilized pre-established associations between both food insecurity and psychological distress, as well as drought exposure and psychological distress. Informed by the findings of the O'Brien et al. study that rural populations experiencing "constant and recent long dry" drought exposure had significantly increased psychological distress, Friel et al. used this drought exposure category as their reference group in their regression modeling. The study found that exposure to drought acted as a moderator in the association between food insecurity and psychological distress, generally increasing the distress level. For example, People living in "zero to moderate" "very dry" and "constant and long dry" drought condition who consumed higher-than-average amounts of discretionary (unhealthy) foods reported significantly more distress than those in other drought categories.

Emotional and/or Behavioral Problems

Emotional and behavioral problems were assessed in two studies of rural adolescents experiencing drought (Dean and Stain 2007, 2010). Both studies utilized the Emotional Symptom Scale of the Strengths and Difficulties Self-Report Questionnaire (SDQ), a validated self-report scale for detecting emotional and behavioral problems among adolescents that also has data on Australian norms.

In 2004, the authors administered mental health questionnaires to and conducted focus groups with adolescents ages 11–17 in 6 rural New South Wales schools (Dean and Stain 2007). Semi-structured focus groups asked 84 students about their perceptions of the drought, its impacts, and their feelings on living in the country. Themes from the first focus group were used to construct a self-report survey that also included the 5 items from the SDQ. 334 adolescents completed the quantitative survey. The study found that while the ongoing drought was a significant experience in the lives of this population, their levels of emotional distress as measured by this subscale of the SDQ were not significantly different than the Australian norms for that age cohort and concluded that this group as a whole is resilient to the impacts of the drought on their emotional health.

In 2007 the authors returned to the same region in New South Wales and again administered a questionnaire and conducted focus groups to determine if the resiliency observed in the previous study persisted. In this second study the adolescents reported significantly higher levels of distress compared to the Australian norms for their age cohorts (Dean and Stain 2010). Increased perceived experience of drought among these adolescents was significantly correlated with both overall SDQ scores (indicating high Total Difficulties ($p < 0.001$ for ages 11–13 and $p < 0.05$ for ages 14–17)) as well as the Emotional Symptoms subscale ($p < 0.01$ for ages 11–13, $p < 0.05$ for ages 14–17). Unfortunately, the authors were unable to resample the same students who participated in the earlier study and instead conducted a second cross-sectional study of the same population.

Vulnerability Factors

Gender, age, occupation and economic circumstance may influence individual vulnerability during drought periods.

Gender

Hanigan et al. (2012) observed that in drought years between 1970 and 2007, men were at increased risk of suicide while suicide risk for women was lower during drought years than non-drought years. These findings suggest that women may be at less risk compared to men to suffer from mental health impacts during drought periods due to their larger/denser social networks and their increased likelihood of holding non-agricultural jobs.

Age

Gunn et al. (2012) found that younger farmers (25–44 and 44–54 years) reported significantly higher levels of psychological distress as measured by the Kessler-10 than farmers 55–64 years old, but not significantly different from farmers in the oldest age group (65–74 years). Younger rural males aged 30–49 were also at increased risk for suicide according to Hanigan et al. (2012). In addition, adolescents may be at high risk for experiencing drought-related distress according to Dean and Stain (2010).

Occupation

Stain et al. (2008) found that farmers and farm residents reported significantly higher drought stress than non-farming rural residents, though levels of psychological distress were not significantly different between the groups. Similarly, Edwards et al. (2015) reported that farmers and farm workers are most impacted by mental health problems during periods of drought.

Economic Circumstance

Much of the impact of drought on mental health appears to be mediated by the effect of drought on farm production and family incomes. Drought was found to moderate the link between psychological distress and food insecurity, increasing the level of distress felt by individuals who were also experiencing some measure of food insecurity, thus indicating that economic and food insecurity are likely significant vulnerability factors when examining the impact of drought on mental well-being (Friel et al. 2014).

Discussion

Poor mental health outcomes, primarily suicide, are an established public health concern for rural populations who often face several barriers in accessing health care, including mental health care. For instance, suicide rates almost twice the national average were reported among male farmers in the U.S. Midwest during an extended period of drought and economic decline in the 1980s (Gunderson 1991). Other studies have also reported elevated rates of suicide among rural residents, particularly farmers in various regions around the world (Page et al. 2007; Singh and Siahpush 2002). The only study from Australia included in this review that was able to assess suicide incidence during drought versus non-drought periods (Hanigan et al. 2012) suggests an increased risk for suicide during drought periods for rural male residents. Although farmers were not found to be at increased risk for suicide (Guiney 2012), additional research expanding the time period of the study to include more variation in exposure to drought and controlling for long-term trends in suicide rates would help to determine whether the finding of no association between drought and suicide among farmers is in fact correct. One of the critical limitations of the Guiney (2012) study is that the total number of Victorian suicides was compared to the number of farming suicides in the region between 2001 and 2008 without indicating whether population changes in the overall and farming populations were taken into account. Thus, it is difficult to judge if the percentages reported are an adequate measure of the suicide risk among farmers compared to the overall population. Finally, in both the Hanigan et al. (2012) and Guiney (2012) studies, cases were not matched to controls by possible confounders such as age, sex and other demographic and/or health variables. Matched case-controls studies (e.g. matching cases to controls in different regions that have not experienced drought during the same time) in order to control for bias would be especially helpful in further exploring the casual relationship between drought exposure and suicide among farmers.

Drought-induced depression, anxiety, psychological distress and other mental health illnesses can have devastating impacts on individuals, families, and communities. The decimation of local economies as a result of drought-induced farm unemployment and reduced productivity as well as the emotional stress of observing the ecological and social effects of prolonged drought can tax the resilience of rural residents. Many of the studies included in this review confirm that rural populations and farmers in particular are more likely to suffer from mental health issues as a result of drought compared to urban or non-agricultural population. For instance, the nationally-representative study by O'Brien et al. (2014) found that drought is associated with increased levels of psychological distress among rural residents but not among urban residents. This study utilized well defined categories of drought exposure in addition to the Kessler 10 item measure of general psychological distress. Of particular interest is the reported association between increased distress and exposure to prolonged and persistent drought. This finding suggests that while populations may have the capacity to recover from short periods of intense drought or even long periods during which droughts have occurred followed by less dry conditions, exposures to persistent drought lasting over 20 months substantially reduces the adaptive capacity of rural populations. Given the importance of this finding, replicating the study in other geographical settings would be very beneficial for better understanding the impact of drought on mental health.

The analysis of the 8000-person survey by Edwards et al. (2015) presented in this review highlighted the vulnerability of farmers and other farm workers and reported a positive association between the severity of the drought impacts on agricultural production and the impacts on farmers' mental health. This is another well designed study utilizing a large sample that sheds light on the mechanisms by which drought impacts the mental health of rural residents. The finding that the mental health of farmers experiencing substantial decreases in farm productivity is particularly affected during periods of drought provides an important direction for future research. However, since this study utilized a self-reported measure of drought that did not allow assessing drought severity, utilizing additional climate measures of drought would be beneficial in future studies.

Although evidence regarding vulnerability is mixed, farmers and farm workers, particularly in impoverished and isolated communities are likely at higher risk for drought-related mental health impacts (Stain et al. 2008; Friel et al. 2014; Edwards et al. 2015), Younger age groups may also be at increased risk for both suicide (Hanigan et al. 2012) and psychological distress (Gunn et al. 2012). The study by Dean and Stain (2010) that reported significantly higher levels of distress among adolescents exposed to drought compared to national averages raises important questions about the potential serious and long term impacts of prolonged drought in this age group.

The pathways through which drought can impact the mental health of rural residents remain not well understood. Drought can have dramatic impacts on the local economies of agricultural communities as production falls and unemployment rises. These economic consequences may be one driver of mental health impacts on rural residents. Drought-related worry may have an important and distinct influence on farmer's mental health (Stain et al. 2011). In addition to the small sample size of this study, the lack of validation of the Worry About Drought scale is a substantial source of uncertainty. Whether or not high levels of drought worry translate into mental health problems is, unfortunately, beyond the scope of the Stain et al. (2011) paper. In addition, as the Gunn et al. (2012) study reported, "maladaptive" coping strategies such as substance abuse, behavioral and mental disengagement, venting, and suppression of competing activities) are significantly associated with increased psychological distress. However, because the Gunn et al. (2012) study is both cross-sectional and samples only from within populations experiencing drought, the association between drought and psychological distress can't be assessed.

Finally, the various measures of drought used in the literature make capturing of potential mental health impacts as well as comparisons across studies challenging. Four types of drought are commonly identified in the literature: (1) meteorological drought, defined by low rainfall, (2) agricultural drought, defined by dryness in surface soil layers, (3) hydrological drought, defined by prolonged moisture deficits and (4) socio-economic drought, defined by the impacts of meteorological, agricultural or hydrological drought on human well-being (Hennessey et al. 2008). Many studies included in this review utilize a definition of meteorological drought, defining drought as a deviation from annual rainfall measures or compared to rainfall measures during defined periods preceding the studies (Kelly et al. 2011; Friel et al. 2014). Some studies have used self-reported measures of drought to better capture agricultural or socioeconomic impacts (as utilized in the Edwards et al. (2015) study). Since drought is such a slow-rolling and cumulative phenomenon, the agricultural or socio-economic definitions of drought might be able to better assess potential links to mental health outcomes. However, self-reported measures may not be able to adequately capture differences in drought severity. Additional studies, comparing various measures of drought exposure will be critical in deriving accurate, consistent and replicable assessments of the mental health impacts of drought.

Conclusions

This review examined the existing literature of drought impacts on the mental health of residents in Australia, a country with extensive experience with droughts and from which most of the current literature on the mental health effects of droughts originates. The review provides a valuable foundation for evaluating the mental health impacts of the ongoing drought in the U.S. Although the evidence on the relationship between experience of drought and various measures of mental health in rural communities in Australia remains mixed, this review demonstrates that drought may impact the general mental health of populations, as well as levels of psychological distress, suicide, and emotional/behavioral problems. In addition, the impact of drought on mental health is likely mediated or moderated by a constellation of vulnerability factors, such as gender, age, occupation and economic circumstance. It is important to point out that the findings from the reviewed Australian studies may not be entirely replicable in the U.S. due to differences in the population characteristics, healthcare systems as well as environmental, social and cultural factors. Therefore, studies and interventions that are designed based on the information presented in this review should be carefully customized for the selected U.S. communities. Based on this review, we recommend several directions for future work in the U.S. that are discussed in greater detail below. First, health impact assessments need to be carried out in vulnerable farming and rural communities currently affected by drought in the U.S. Next, studies modelled after the Australian studies presented in this review should be developed in order to explore the potential causal relationship between drought exposure and applicable mental health outcomes. In addition, establishing a robust and consistent measure of drought exposure, tailored to U.S. conditions will be essential for quantifying and monitoring health outcomes. Finally, potential studies and interventions should take into special consideration mental health surveillance and service provision in affected U.S. communities.

As droughts are expected to increase in both frequency and duration in California and other regions of the US as a result of global climate change, comprehensive adaptation measure will be necessary to address the potential impacts on the mental health of rural and agricultural communities. A critical step towards designing such adaptation measures is a robust health impact assessment, evaluating the epidemiological evidence base for the causal relationships between drought and mental health outcomes. In addition, a plausible mechanism of these relationships and evidence from local communities are necessary for deriving a health impact assessment (Kovats et al. 2003).

Although some of the mental health impacts in the U.S. can be better anticipated after this review of the literature emerging from Australia, there is an urgent need for research quantifying the impacts of drought on mental illness prevalence in rural areas in California and other drought-affected regions, with focus on measures and accounts for the local social, economic, and cultural factors. Some specific research areas highlighted earlier include carrying out matched case-controls studies in order to explore the potential causal relationship between drought exposure and suicide among farmers, and adapting and replicating some of the comprehensive Australian survey studies that have assessed the impacts of drought on various measure of psychological distress. In addition, evidence from Australia about specific vulnerability factors, such as gender, age, occupation and economic circumstance should be incorporated in the assessments of drought impacts of mental health locally.

Another critical component towards a successful assessment of the potential impacts of drought on climate change, and the subsequent design of adaptation strategies, is the establishment of a robust and consistent measure of drought exposure. This review has highlighted the various exposure measures utilized in studies to date. While some of the self-reported measures that are used in the literature may be able to better capture the complex social dimensions of drought, meteorological measures may be more replicable and better able to assess drought severity. Studies dedicated to analyzing and comparing various drought definitions and utilizing both meteorological and self-reported metrics would be especially beneficial for advancing and bringing consistency to the study of drought related population impacts, including but not limited to mental health.

Finally, ensuring optimal mental health surveillance and service provision in rural communities is essential for minimizing the mental health impacts of the ongoing drought on communities in the United States. Unfortunately, as a function of their rural nature, rural communities are often under-served by medical and mental health practitioners, making care difficult to find even for those who seek it.

Also, since impoverished and isolated communities are likely at higher risk for drought-related mental health impacts, such communities in the United States are also likely to be less able to buffer the economic impacts of drought (due to lowered agricultural production and employment) and less able to access supports and services to mitigate drought impacts. The emergency drought relief bills passed by the California legislature in 2014 contained provisions to assist communities that experienced heavy unemployment due to the drought through $46 million in food and housing assistance to individuals who are out of work because their farms have been affected by drought (State of California 2014). Such social supports may help to buffer the impact of the drought on the mental health of rural residents, but further research is needed to assess both the burden of negative mental health outcomes and the efficacy of such safety net services.

References

Berry HL, Bowen K, Kjellstrom T (2010) Climate change and mental health: a causal pathways framework. Int J Public Health 55:123–132

Centers for Disease Control and Prevention (CDC), U.S. Environmental Protection Agency, National Oceanic and Atmospheric Agency, American Water Works Association (2010) When every drop counts: protecting public health during drought conditions—a guide for public health professionals. U.S. Department of Health and Human Services, Atlanta. Available online at: http://www.cdc.gov/nceh/ehs/docs/when_every_drop_counts.pdf. Accessed 7 Feb 2016

Clayton S, Devine-Wright P, Stern PC, Whitmarsh L, Carrico A, Steg L, Swim J, Bonnes M (2015) Psychological research and global climate change. Nat Climate Change 5:640–646

Costello A, Abbas M, Allen A, Ball S, Bell S, Bellamy R, Friel S, Groce N, Johnson A, Kett M et al (2009) Managing the health effects of climate change. The Lancet 373:1693–1733

Dean J, Stain HJ (2007) The impact of drought on the emotional well-being of children and adolescents in rural and remote New South Wales. J Rural Health 23(4):356–364

Dean J, Stain HJ (2010) Mental health impact for adolescents living with prolonged drought. Aust J Rural Health 18(1):32–37

Edwards B, Gray M, Hunter B (2015) The impact of drought on mental health in rural and regional Australia. Soc Indic Res 121(1):177–194

Franchini M, Mannucci PM (2015) Impact on human health of climate changes. Eur J Int Med 26 (1):1–5

Friel S, Berry H, Dinh H, O'Brien L, Walls HL (2014) The impact of drought on the association between food security and mental health in a nationally representative Australian sample. BMC Public Health 14(1):1

Graham S (2000) Drought: the creeping disaster. NASA Earth Observatory. Available online at: http://earthobservatory.nasa.gov/Features/DroughtFacts/. Accessed 7 Feb 2016

Guiney R (2012) Farming suicides during the Victorian drought: 2001–2007. Aust J Rural Health 20(1):11–15

Gunderson PD (1991) An analysis of suicide among farmers in five north central states, 1980–1988. Surgeon General's Conference on Agricultural Safety and Health, 2 May 1991

Gunn K, Kettler LJ, Skaczkowski GL, Turnbull DA (2012) Farmers' stress and coping in a time of drought. Rural Remote Health 12:2071

Haines A, Kovats RS, Campbell-Lendrum D, Corvalán C (2006) Climate change and human health: impacts, vulnerability and public health. Public Health 120:585–596

Hanigan IC, Butler CD, Kokic PN, Hutchinson MF (2012) Suicide and drought in New South Wales, Australia, 1970–2007. Proc Natl Acad Sci 109(35):13950–13955

Hennessey K, Fawcett R, Kironoa D, Mpelasokaa F, Jones D, Bathols J, Whetton P, Stafford SM, Howden M, Mitchell C, Plummer N (2008) An assessment of the impact of climate change on the nature and frequency of exceptional climatic events. Commonwealth of Australia, Australian Bureau of Rural Sciences, Department of Agriculture, Fisheries and Forestry, 37 pp

IPCC (2014) Climate change 2014: impacts, adaptation, and vulnerability. In: Field CB, Barros VR, Dokken DJ, Mach KJ, Mastrandrea MD, Bilir TE, Chatterjee M, Ebi KL, Estrada YO, Genova RC, Girma B, Kissel ES, Levy AN, MacCracken S, Mastrandrea PR, White LL (eds) Part A: global and sectoral aspects. Contribution of Working Group II to the fifth assessment report of the intergovernmental panel on climate change. Cambridge University Press, Cambridge

Kelly BJ, Lewin TJ, Stain HJ, Coleman C, Fitzgerald M, Perkins D, Carr VJ, Fragar L, Fuller J, Lyle D, Beard JR (2011) Determinants of mental health and well-being within rural and remote communities. Soc Psychiatry Psychiatr Epidemiol 46(12):1331–1342

Kinney PL, Matte T, Knowlton K, Madrigano J, Petkova E, Weinberger K, Quinn A, Arend M, Pullen J (2015) New York City panel on climate change 2015 report, Chap. 5: Public health impacts and resiliency. Ann New York Acad Sci 1336(1):67–88

Kovats RS, Menne B, Ahern MJ, Patz JA (2003) National assessments of health impacts of climate change: a review. Climate change and health: risks and responses. World Health Organization, Geneva

Luber G, Knowlton K, Balbus J, Frumkin H, Hayden M, Hess J, McGeehin M, Sheats N, Backer L, Beard CB et al (2014) Climate change impacts in the United States: the third national climate assessment: human health. In: Melillo JM, Richmond TTC, Yohe GW (eds) U.S. Global Change Research Program, Washington, DC, USA

Luber G, Prudent N (2009) Climate change and human health. Trans Am Clin Climatol Assoc 120:113–117

McMichael AJ, Woodruff RE, Hales S (2006) Climate change and human health: present and future risks. The Lancet 367(9513):859–869

Natural Resources Defense Council (NRDC) (2010) Climate change, water, and risk: current water demands are not sustainable. Available online at: http://www.nrdc.org/globalwarming/watersustainability/files/WaterRisk.pdf. Accessed 7 Feb 2016

O'Brien LV, Berry HL, Coleman C, Hanigan IC (2014) Drought as a mental health exposure. Environ Res 131:181–187

Page A, Morrell S, Taylor R, Dudley M, Carter G (2007) Further increases in rural suicide in young Australian adults: secular trends, 1979–2003. Soc Sci Med 65(3):442–453

Rice SM, McIver LJ (2016) Climate change and mental health: rationale for research and intervention planning. Asian J Psychiatry 20:1–2

Singh GK, Siahpush M (2002) Increasing rural-urban gradients in US suicide mortality, 1970–1997. Am J Public Health 92(7):1161–1167

Stain HJ, Kelly B, Carr VJ, Lewin TJ, Fitzgerald M, Fragar L (2011) The psychological impact of chronic environmental adversity: responding to prolonged drought. Soc Sci Med 73(11): 1593–1599

Stain HJ, Kelly B, Lewin TJ, Higginbotham N, Beard JR, Hourihan F (2008) Social networks and mental health among a farming population. Soc Psychiatry Psychiatr Epidemiol 43(10):843–849

Stanke C, Kerac M, Prudhomme C, Medlock J, Murray V (2013) Health effects of drought: a systematic review of the evidence. PLOS Currents Disasters

State of California, Office of the Governor (2014) Governor Brown, Legislative Leaders Announce Emergency Drought Legislation, 19 Feb 2014. http://gov.ca.gov/news.php?id=18415. Accessed 7 Feb 2016

Vins H, Bell J, Saha S, Hess JJ (2015) The mental health outcomes of drought: a systematic review and causal process diagram. Int J Environ Res Public Health 12:13251–13275

Chapter 19
Linking Science and Policy on Climate Change: The Case of Coquimbo Region, Chile

Sonia Salas, Angelo Araya and Andrés Bodini

Abstract The generation of knowledge and its relation with the development of public policies has recently become a matter of concern. This paper explores the interaction, in the public's perception, between scientific knowledge and policies on climate change and water resources, a critical issue when making decisions based on scientific evidence. The study, which employed a qualitative paradigm, generated an interview guide which was administered to a sample of 26 subjects (13 researchers and 13 politicians). The analysis considered four dimensions: evidence, relevance, relationships and "supply and demand". The political sector highlights a lack of information for drawing up policies. Researchers point out that they do not know the best strategies to transfer their results to the political world. In the *relevance* dimension, both groups (politicians and researchers) agree on the characteristics that define this knowledge and note that funding plays a significant role as one of the main elements hindering the science-policy interface. The *relationships* dimension stresses the need to generate fluid communication channels between the two sectors. In short, the process of science-policy interactions has important flaws due to several factors: the political system, the funding platform, the duration of projects, the length of political administrations, and centralism in connection with regional development. In summary, findings revealed that the interaction between science and policy on climate change issues is weak. Its of utmost importance the development of systematic dialogue opportunities associated with institutional commitments and the inclusion of stakeholders while prioritizing specific areas of the science-policy interface on the climate change issue.

S. Salas (✉) · A. Araya · A. Bodini
Universidad de La Serena, La Serena, Chile
e-mail: salassc@yahoo.com

S. Salas
Centro de Estudios en Zonas Áridas, CEAZA, La Serena, Chile

S. Salas
Saint Mary's University, Halifax, Canada

A. Bodini
Unidad de SIG, SECPLAN, Ilustre Municipalidad de La Serena, Chile

W. Leal Filho and J.M. Keenan (eds.), *Climate Change Adaptation in North America*, Climate Change Management,
DOI 10.1007/978-3-319-53742-9_19

Keywords Science-policy interface • Climate change • Drought • Adaptation

Introduction

Global climate change is believed to be manifesting itself through a series of effects such as rising land and ocean temperatures, rising sea levels, decreasing precipitation, and the recurrence of extreme events (IPCC 2013). The current situation is complex due to its associated economic losses, risks to nutritional security (Wheeler and von Braun 2013), and impacts on water resources (Ludwig et al. 2016), all of which are expected to become global-scale scenarios (Dai 2013). In the specific case of Chile, the severe effects of water scarcity and desertification can be observed to be progressing from north to south, affecting the whole country and especially some northern regions such as Coquimbo (CEAZA 2014; La Tercera 2015; Díaz 2016).

In this context, Chile stands out as a country that is highly vulnerable to climate change (Ministerio del Medio Ambiente 2011), given that it combines geographical and climatic weather conditions that increase the likelihood of variations in the level of precipitation, temperature, and presence of extreme events (DGF-Conama 2006; CEPAL 2012a; Fiebig-Wittmaack et al. 2012; IPCC 2013; Santibáñez et al. 2014). In the case of northern Chile, current scenarios forecast a marked decrease in the level of precipitation, which will further compound the severe existing process of desertification, affecting the productive sectors and especially communities that practice subsistence agriculture (Ministerio del Medio Ambiente 2015).

Therefore, it has become necessary to make a joint effort to address climate change, involving all stakeholders: researchers, politicians, businesspeople, and citizens (Díaz and Hurlbert 2014). This "field of interaction" or "contact surface", known as an *interface*, constitutes a key area nowadays, when policymaking and decision-making processes are required to have some degree of coordination. In this context, the science-policy interface has become necessary and provides an opportunity for strengthening and interlinking relationships between scientists and politicians (López-Rodríguez et al. 2015). Macleod et al. (2008) point out that the number of science-policy interactions may have increased due to the complexity of science, the oscillation of natural processes, and the complexity of environmental policies. The task, however, sometimes becomes difficult and dysfunctional (Sutherland et al. 2012) due to the existence of institutional and cultural barriers between scientists and politicians (Díaz and Hurlbert 2014), as well as because of the difficulties involved in accessing scientific data (Larigauderie and Mooney 2010). According to Van Enst et al. (2014) such difficulties result from three meta-problems: the strategic use of knowledge by politicians, the strategic development of knowledge in the sphere of science, and a lack of operational adjustment between supply and demand. In this area, the literature mentions the existence of certain conditions for the science-policy link to be successfully established. Scott et al. (2012) list four conditions: inclusiveness, participation, interaction, and influence. Young et al. (2014) state that it is necessary to make research and politics part of a joint effort by promoting inter- and trans-disciplinary research with mixed work groups and by generating schemes and structures to foster dialogue; meanwhile, Sarkki et al. (2014, 2015) put forward that knowledge must be credible, relevant, and legitimate. At a more global level, Crewe and Young (2002) developed a three-dimensional integrative proposal in order to understand the interface process. This approach comprises the following variables: context (cultural, structural, and institutional dimensions), evidence (credibility and communication), and links (influence and relations).

As can be observed, the issue of the science-policy interface entails multiple dimensions, proposals, variables, results, and discussions; however, it is a concept that is present in current debates on climate change and that has become critical in the vulnerable geographical areas which are suffering the negative effects of global changes. In the case of Chile, only one study was found which describes the interaction between science and policy (Bórquez 2011), and its author notes the presence of more hindering than facilitating factors in the interface. If we consider the high degree of vulnerability of the northern region of the country (area of study) as well as the forecast of negative effects, it is clear to see the significant relevance of exploring the interaction between scientific research on climate change and how such research has influenced policy and decision-making. The country of Chile and the Coquimbo region in particular, need to develop public policies linked to the findings provided by science on climate change to ensure the well-being of the inhabitants who are facing a severe process of desertification.

The present study is part of the research results of a research project as a joint effort by the countries of Argentina, Bolivia, Chile and Canada. Its central aim was to explore the interaction process, or interface, between (social-natural) science and public policy in the Fourth Region of Chile. In order to do this, the three-dimensional approach proposed by Crewe and Young (2002) was used, with the addition of a "supply and demand" dimension.

A Look at Chile's Political System

Chile is defined as a unitary, democratic, and presidential republic comprised by 3 branches: executive, legislative, and judicial. Political administration is in the hands of the President, who is guided by the political constitution established in 1980. In administrative terms, there is a cabinet comprised by 23 ministries that are in charge of managing different areas. Each ministry has a robust hierarchy which includes services, divisions, and departments; as a whole, they shape the national public administration system that generates political and technical instruments, develops bills, supervises the enforcement of policies, and proposes actions.

The country is divided into 15 regions. The administrative head of each region is the *Intendente*, who represents the head of state in the region and is in charge of fulfilling the political guidelines of each government. Regional councilors, a democratically elected political body, form another part of this administrative organization; together with the *Intendente*, they constitute the regional government tasked with the political administration of the territory. In addition, a subunit exists: the *Comuna*, which is administered and directed by a mayor and a group of city councilors (municipality), elected by the citizens. At the level of ministries, there is a regional secretary who is in charge of representing and politically articulating the ministry in each region. This secretary (SEREMI) executes, supervises, and proposes multiple local policies. Even though the country is defined as a decentralized republic, a number of decisions that govern the regions are made centrally, which results in uniformed guidelines that exclude the contextual components of each territory.

The legislative branch is housed in the National Congress, which is comprised by a Chamber of Deputies and the Senate, both democratically elected. The objectives of the legislative branch are to represent the citizens, supervise the government's actions, and generate legal bodies. The 120 deputies are elected every 4 years, while the Senate is made up by 38 members who are elected every 8 years. The Congress mainly works by means of discussion and legislation sessions; in addition, special commissions are created to focus on various issues; both deputies and senators have roles which may differ in some aspects.

Chile's political administration has only recently expanded to focus on the environment: the Ministry of the Environment was created in 2010 to take over the functions of the National Environment Commission (Comisión Nacional de Medio Ambiente, CONAMA). Water resources are administered by the Ministry of Public Works, specifically, the General Water Commission, which is ruled by the 1981 Water Code. Water is considered to be a public good subject to private use, a definition that, according to several stakeholders, has commercialized its use and has led to the creation of a virtual "water market" (Hearne and Donoso 2014).

This study is a contribution to a field of research that has been scarcely explored in Chile as well as the region, in spite of the country's recent political interest in the issues of climate change and its effects. This topic has resulted in actions and reports such as the National Plan for Climate Change Adaptation (Plan Nacional de Adaptación al Cambio Climático) (Ministerio del Medio Ambiente 2015), the signing of international agreements such as COP 2015 (Gobierno de Chile 2015), and a number of reports on climate and water resources (World Bank 2011; CEPAL 2012b; Ministerio de Obras Públicas 2013; Ministerio del Interior y Seguridad Pública 2015). Likewise, the present study conveys the notion that scientists and politicians work in separate fields, but establish points of connection every once in a while. The above initiatives acquire relevance when considering future country-level proposals in connection with the "Climate Change Bill" (Proyecto de Ley de Cambio Climático) or substantial reforms to the Water Code currently in force.

This study intends to describe and analyze the science-policy interface in connection with climate change.

Methodology

This study intends to describe and analyze the science-policy interface in connection with climate change. A qualitative methodology was used through an exploratory design (study case type), based on an interpretative paradigm. The science-policy interface was examined in the sphere of climate change and water resources, focusing on the case of the Fourth Region of Chile. Data were collected via semi-structured interviews and complemented with documentary analysis.

Purposive, snowball-type sampling was used. The sample comprised 26 subjects (13 scientists and 13 politicians) whose background and demographic data are described in Tables 19.1, 19.2, 19.3 and 19.4.

Instruments

Documentary analysis: Used to identify the main research projects and initiatives in the field of climate change and water resources in the region. The analysis of public policy was conducted by examining the political-administrative information associated with the stakeholders and institutions involved, the political guidelines in force, and the technical-political instruments active at the regional and national levels. The inclusion criteria were defined during this stage. Scientists had to be part of projects, publications, or news (either currently or in the past); in addition, they had to belong to a scientific institution (regional, national, or ministry-related research centers or universities). In the case of politicians, the aspects considered were their curriculum, parliamentary bills, participation in activities, media presence, and current or past positions.

Semi-structured interview: In order to generate common analysis criteria and frameworks for all the participating countries, a standardized interview guidelines were produced. The defined theoretical approach considered a thematic nucleus focused on 4 nodes: **supply and demand** (availability of information for politicians and supply of information by scientists) along with the categories introduced by Crewe and Young (2002): **evidence** (relevance of knowledge, credibility, contextualization, and information transfer), **relations** (type of relations, strategies for delivering and requesting information, intermediaries), and **context** (elements that facilitate or hinder the interface). It must be noted that two interview guidelines were generated, one for scientists and another for politicians; even though the

Table 19.1 Sample distribution by group

Scientists	Social Sc.	5
	Natural Sc.	8
Politicians	Regional members of parliament	4
	Regional authorities	4
	Regional technical-political officials (SEREMI)	2
	National technical-political officials (Ministries)	3

Table 19.2 Averages and distribution by age, years of experience, gender, and academic degree (Scientists)

Average age	56.62 years			
Average number of years of research experience	23.15 years			
Distribution by sex	Male	69.3% | n = 9	Female	30.7% | n = 4
Academic degree	Doctor: n = 7	Master of Arts/ Science: n = 5	Bachelor: n = 1	

Table 19.3 Averages and distribution by age, years of experience, and gender (Politicians)

Average age	46.33 years			
Average number of years of political experience	15.38 years			
Distribution by sex	Male	53.8% | n = 7	Female	46.2% | n = 6

Table 19.4 Distribution by experience in the scientific field, university education, and workplace (Politicians)

Experience in the scientific world	Yes: 30.8% | n = 4	No: 69.2% | n = 9
University education	Natural Sciences: 69.2% | n = 9	Social Sciences: 30.8% | n = 4

instruments had the same contents, the approaches used differed depending on the target group. Another noteworthy aspect is that, in Chile, some topics emerged which were not present in the interview guidelines and which were considered in the final report.

The information was analyzed with *NVIVO 10*, using the thematic content analysis technique. The technique known as axial coding yielded 4 hierarchical nodes: supply and demand, evidence, relations, and context; in addition, subnodes were defined and coded in an open manner.

The present study was focused on the region of Coquimbo, and the results observed are therefore limited to the area selected. It must be noted that Chile has a centralized government and that a large percentage of the interviewees (especially politicians) work in the country's capital while usually spending only one week per month in their districts. This centralized situation can be appreciated through their interviews where their perspectives have both a regional and a national dimension.

Results

Supply and Demand

Supply of Information

With regard to the supply of scientific information, it was identified that a number of projects and studies had been conducted about the water resources of Chile and the Fourth Region. In addition, a number of initiatives have been executed in connection with climate change, mainly by natural scientists, whose results are partially known by politicians. On the other hand, scientists acknowledge that the objectives of these projects are not in tune with the requirements of the political world, which is linked with the fact that politicians generally do not request information from them, nor do they attend the activities in which results are disseminated. On the other hand, the research process has other priorities, such as results, and academic productivity, information transfer being only a secondary concern. Therefore, the project's main achievements are associated with the generation of scientific products: technical reports, articles, books, and presentations. A noteworthy element is that the participants consider that project results must be disseminated, but according to an academically-related logic focused on articles and reports which function as a means for validation and positioning within the scientific world, without politicians as their target audience.

From the point of view of politicians, there is a wealth of information sources which are used in the formulation of policies and decision-making, such as the centralized information provided by government agencies or commissioned studies, meetings, the Library of the National Congress, and the information collected on-site by consulting the protagonists themselves, the latter being a substantial and valid indicator for the political sector. It should be noted that politicians can also make use of technical reports and products supplied by research centers.

Demand for Information

Regarding demand, scientists consider that politicians have some difficulties for accessing information due to certain aspects of the interaction processes during or at the end of the execution of projects. Also, they believe that they have no connections due to how much attention the research process demands from them and because of politicians' lack of interest in participating in activities organized by scientists. Difficulties in accessing such information, for scientists, are connected with the contents of the studies: they consider that the concepts used and the results provided are hard to understand.

Politicians (legislators, executives, or technical officials) are aware of the need of having scientific information available, especially in the form of diagnoses and tools that allow them to make decisions or develop technical-financial instruments. Despite a lack of information in the past, several scientific studies are currently ongoing (in the vulnerability and adaptation areas) and information is being produced at internal level. Scientific information is regarded as an important element for the political world, because it is a resource with an additional value. Therefore, having scientific information available may constitute an "extra element" for political processes and debates.

Politicians mention the existence of information voids, which may be real or the result of a lack of knowledge about the studies conducted. In this sphere, it appears that politicians do not commonly turn to scientists to access information. It is important to stress that public workers in charge of generating policies (political-technical role) have the power and the ability to obtain direct, focused, and current information; however, they work with processed and limited materials which are submitted as final products to the politician (e.g. minister). This means that scientific information is generally known by technical personnel before reaching a political authority. Also, this technical personnel tends to engage in constant and contingent dialogue with scientists. Such conversations are, however, informative in nature, because decision-making depends on other factors such as priorities of the executive branch of government, political party influence, and relations with stakeholders.

In terms of future perspectives, over the course of the interviews, some researchers pointed out that one of their interests was to study the topics that the political participants listed as priorities, which reflects some degree of coincidence between supply and demand. This observation highlights the need for better communication management and coordination between both sectors.

Evidence

Relevance

The relevance of knowledge, according to the participating scientists, has two aspects: "intellectual" and "functional", the latter covering various areas. The observations made reveal a tendency in the scientists to privilege "intellectual" knowledge, a choice that is justified in terms of their professional training and the generation of scientific capital. For some academics, relevance is associated with the ability to obtain funding for projects and maintain one's position within the scientific community, whereas social relevance is linked with the all-encompassing nature of knowledge, which involves its application in multiple areas.

The relevance of knowledge, for politicians, resembles the scientific view, because these participants identify two types: intellectual and applied. Politicians value "scientifically backed" knowledge, which they see as being generated in accordance with scientific standards that guarantee and/or certify that it has undergone "quality control" processes.

Credibility

Credibility is defined by scientists with reference to various dimensions that constitute clear and indisputable indicators of scientific quality: validity of the data, methodology/procedures employed, and scientific results. In addition, these participants highlight certain contextual aspects of science: recognition by one's peers, the weight of one's host academic institutions, and one's ability to secure competitive funding. Scientific articles are regarded as "*a measure of worth and a validation platform*", and thus emerge as objects with an additional value (considering elements such as journal quality, type of study, printing house, indexing, and impact factor). These characteristics configure the position of a researcher in the academic environment and therefore constitute a measurable form of verification among peers.

For politicians, scientists' reliability is mainly associated with their trajectory, scientific publications, number of studies carried out, visibility outside of academia, and number of specialization programs completed. Research results are also relevant to politicians, because they can be translated into policies or applied to decision-making; thus, they must be coherent, consistent, and above all clear.

Contextualization of Information

For scientists, contextualizing the information generated is not a priority, because they are influenced by "global trends" that tend to be out of step with national issues. Time management also becomes important, given that some information that is not relevant nowadays may be so in the future. An important point is the language used: scientists, in their everyday work, do not need to make any simplifications; in fact, they must conduct very deep and complex analyses that involve levels of abstraction that differ greatly from those of daily life.

Politicians state that policymaking must consider context and applicability, taking into account a territorial perspective. Given this situation, scientific language acquires importance due to the fact that politicians require information that they can use in quick decision-making processes; therefore, it must be digestible and simple and should be conveyed through clear and understandable language.

Information Transfer

Information transfer is a highly controversial subject, given that scientists and politicians do not agree on who should be in charge of the process. In both communities, the participants mention several individuals or agencies in charge of transferring the results of a study (science journalists, "transfer experts", outreach workers, and institutions). On the other hand, some scientists and politicians state that researchers should have a "dual role": to generate and transfer knowledge; alternatively, they suggest that other stakeholders (universities) should be in charge of knowledge transfer. Even though there is evident disagreement on this issue, we can conclude that the participants perceive a clear need to establish a public or private national agency devoted to knowledge transformation and transfer. This platform could take different forms and include a number of different members; basically, however, it must be robust and have the power to act independently.

Dialogues Between Social and Natural Scientists

According to scientists, interdisciplinarity in research is not a priority for projects nowadays; when it is present, dialogues between the social and the natural sciences, despite not having been wholly conflict-laden, do not denote a communion allowing them to conduct academically and socially valuable joint studies. Some of the reasons for this lack of integration between disciplines include: lack of collaborative work experience, dissimilar work rates and knowledge of analytical frameworks, and different codes and languages.

Politicians tend to value the results of studies which are close to their areas of expertise; so, if a politician has a degree in one of the natural sciences, he/she will ascribe more validity to empirical research because he/she considers that it can yield objective and measurable results. On the other hand, if a politician received an education focused on the social sciences, he/she will emphasize the importance of social information for understanding phenomena from the point of view of the participants themselves.

Relations

Types of Relations

Scientists point out that relations with politicians are rather distant, with encounters occurring in specific situations and due to contingent events; there are no requests for information, nor is there a type of dialogue based on trust or on formal institutional ties. Both groups of participants acknowledge that relations between them are difficult due to the existence of different languages, logics, roles, and priorities that prevent their two worlds from coming together and dialoguing. Even though some changes have already taken place (mainly, the creation of research centers or the provision of State partial support funding for them), the interviewees believe that no information transfer or any significant collaborations can be expected to occur in the future.

The situation as described by politicians has two sides: on the one hand, they emphasize the existence of close and cooperative ties with researchers and therefore assess their relationship as positive; on the other hand, certain barriers that keep their dialogue from being more fluent are identified. Politicians consider that there is a lack of knowledge about the work of scientists, and argue that these professionals form closed and rather hermetic circles that prevent further collaboration.

Strategies and Methods for Requesting and/or Delivering Information.

Scientists do not define clear strategies for managing and/or disseminating the information associated with their studies. The participants' statements reveal the existence of personal dynamics which are centered on the work of each group: researchers tend to include politicians in debates, while politicians invite tenders for projects or contact researchers directly (when necessary).

In the case of legislators, the strategies and means of contact are not uniform and depend on the interest of each group, and involve consultancies, tenders for studies, conferences, and invitations to events.

Intermediate Participants

The limited dissemination of the available scientific information and the lack of time of scientists and politicians, prompts the need to generate interaction opportunities that function as parts of a chain, facilitating the communication process and establishing networks. The creation of platforms in universities and centers, as well as the position of "manager of the center", could emerge as attractive alternatives for strengthening the interaction processes between science and policy.

In the case of Chile, civil society is regarded as an intermediate participant: given that it introduces changes, it constitutes an engine that invigorates discussion and catalyzes several processes. Likewise, social participants constitute vehicles that convey information from scientific to political spheres (and vice versa) in order to generate pressure or set agendas.

Context

Barriers to the Science-Policy Interface Process

Scientists point out that the main hurdles originating in the political world involve political-institutional aspects (centralism), short administrations, insufficient academic training, the public's lack of trust in politics, and the scarce time available to politicians, all of which makes it difficult to access and/or deliver information. Political ideologies (parties) make it more difficult to discuss certain topics; also, shifts in political posts keep deeper dialogues from happening. On the other hand, political agendas and contingent events, along with the great dispersion of the institutions in charge of water resources, tend to generate bureaucracy and distance. Lastly, the funding system for science is considered to be detached from current priorities, given that it involves short and finite projects.

The main hurdles present in the scientific world, according to politicians, are the following: the prioritization of topics which are unrelated to the country's requirements; a lack of university influence in political processes; hermetic scientific groups; national scientific policies which are insufficient and lack a powerful funding structure. In addition, the participants note the lack of platforms that can better organize the available scientific information. Different forms of validation are observed in the two groups (scientists = generation of information/politicians = votes, in the case of authorities who depend on electoral processes). Knowledge dissemination and transfer are not regarded as priorities by scientists.

Facilitating Factors in the Science-Policy Interface

Scientists value government funding (CONICYT [National Commission for Scientific and Technological Research] and Regional Governments) for the research centers which they consider to be facilitators of knowledge dissemination; thus, such centers are viewed positively given that they provide materials for making decisions or produce information about the local and national climate situation. This is the case of the Center for Advanced Studies in Arid Zones (Centro de Estudios Avanzados en Zonas Áridas, CEAZA), in the Coquimbo Region, and the Center for Climate and Resilience Studies (Centro de Estudio del Clima y la Resiliencia, CR2), in Santiago, both of which have emerged as relevant institutions in the fields of information generation and management.

The facilitating factors mentioned by politicians involve their very positive view of the role of scientists and their contribution to the country; in addition, they value the role of international agencies such as the Intergovernmental Panel on Climate Change (IPCC), which are regarded as credible and authoritative platforms. Scientific information constitutes an important material in the decision-making process; also, interacting with scientists is to speak with people who are well-versed in the matter.

Conclusions

The interaction between science and policy on climate change in Chile's Fourth Region is considered to be at a weak level, which is consistent in a certain way with Iyalomhe et al. (2013), who point out that it is hard to implement an interface at a regional scale. On the other hand, it can be observed that the decision-making process is complex and involves selective information (Young et al. 2014). The use of scientific information by politicians is not a linear and unidirectional process (Wesselink et al. 2013); however, according to the present study, this appears to be a controversial topic for scientists. Based on the above information, it could be inferred that one of the main problems associated with the science-policy interface is the establishment of common goals and objectives, an element also reported by Huggel et al. (2015).

With respect to supply and demand, academic production is considered to be in tune with current issues, however, the knowledge derived from that production does not reach politicians (which is consistent with von Winterfeldt 2013). There appears to be an "invisible balance" of supply and demand. There is no contact surface between both and politicians accuse scientists of being "isolated". Regarding the *evidence* dimension, some difficulties for integrating social-natural information are identified in the scientific world (as pointed out by Weichselgartner and Kasperson 2010), where information transfer appears to be a complex issue about which there is no clarity; also, the specificity of scientific language and the complexity of the information doubtlessly poses an obstacle to political participants, a point mentioned in a prior study (Janse 2008). Relationships between scientists and politicians are regarded as rather distant and lacking in formal contact opportunities and clear strategies for both communities to engage with one another. In this regard, a key point emerges: the generation of "mediation" opportunities, which require the creation of institutional, social, or individual platforms that play a mediating and linking role; to a certain extent, "translators" or "intermediaries" are needed to facilitate a more fruitful interaction. A similar point is made in existing studies (Jones et al. 2008; Godfrey et al. 2010; Lidskog 2014; Van Enst et al. 2014) that have stressed the importance of organizations that function as links between science and policy. On the other hand, the interviewees insistently mention stakeholders, who could adopt a critical role as true "mobilizers of knowledge" in order to generate a "*trident*": science, policy, and society. Lastly, the *context* dimension was found to be the most important one in the present study, which is in line with a prior article (Court and Young 2003) describing facilitating and hindering factors. For scientists, the bureaucracy of political processes, which entail long discussion/analysis periods, along with centralism, hinder the scientific sector and prevent the establishment of satisfactory relationships with the political sector; in this regard, Court and Young (2005) note that it is relevant that scientists understand the importance of the political context. It should also be taken into account that the means of validation and incentives are different for scientists (generation of articles) and politicians (participation opportunities); scientists focus on creating research projects, training students/professionals, and publishing scientific articles, which causes a gap that is widened by their lack of knowledge about political processes. It must be pointed out that elements such as scientific language, incentives, the synchronization of the time available to scientists and politicians, and the difficulties encountered by the latter when trying to access research echo the findings of a previous study conducted in Chile (Bórquez 2011).

The transmission of knowledge should be strengthened by the facilitators' actions, the creation of a Ministry of the Environment, the work of climate change agencies, Chile's signing of a number of treaties, and the country's participation in international forums. In terms of contingent events, a salient element is the water crisis affecting the north of the country. This event is believed to have forced the interaction between science and policy mostly in the field of climate change adaptation, which should result in the establishment of a successful science-policy interface (Iyalomhe et al. 2013). Of relevance is the role of regional and national research centers (CEAZA*-CR2*) that have generated an incipient but promising platform for knowledge transfer. This point is exemplified by Adapt-Chile, which has worked with municipalities in implementing plans for climate change adaptation (Adapt-Chile 2013), an effort that has managed to capture the attention of the political world.

With respect to future strategies intended to bridge the gap between science and policy, it is recommendable to generate systematic dialogue opportunities associated with institutional commitments and to include stakeholders (e.g. Proyecto CAS; Krellenberg 2012); likewise, given the water crisis affecting the country, it is advisable to work institutionally with pioneering models in the science-policy interface (Morgan 2014) in order to improve climate change adaptation processes, especially in Chile's arid and semiarid north.

It is necessary to prioritize four areas of the science-policy interface in connection with climate change: (a) the information available to politicians or decision-makers (McConney et al. 2016); (b) the visibility of the available scientific knowledge on climate change, considering information needs or voids; (c) the fostering of new initiatives; and (d) the strengthening of the Chilean scientific system by a government seeking to integrate knowledge production with national necessities in the area of climate change and water resources. The previous statement is consistent with the interviewees' perception that Chilean science is currently undergoing a crisis catalyzed by the State's passive role (Astudillo 2014; CNN Chile 2015; Bajak 2015) regarding these issues. Lastly, the current situation makes a priority for researchers and politicians to "understand" how knowledge works and develops in practice (Nursey-Bray et al. 2014). The results obtained make it necessary to identify the best strategy to ensure that the resources devoted to the development of scientific research on climate change result in public policies that benefit social development and well-being.

By using an approach with 4 well-defined dimensions, the present study has a high degree of potential repeatability at a regional and local level, making it possible to apply this methodology in other sectors with emphasis on climate change and adaptation to water resources. However, one of the study's limitation was that the information collection strategies used were restricted to documentary and interview analysis for financial reasons. It would have been more fruitful to establish direct dialog opportunities between both sectors (scientists and politicians), while including other stakeholders considering that the interface is trident-shaped: science, policy, and stakeholders.

Acknowledgements The present study was part of the project "*Bringing Together Climate Science and Policy: Contributions for Influential Science in Latin America*", which was funded by the IDRC (International Development Research Council) Canadian agency.

References

Adapt-Chile (2013) Diálogos de Ciencia y Política para la Acción Climática a nivel Municipal. [Online]: http://www.adapt-chile.org/web/wpcontent/uploads/2015/04/Dialogos_de_Ciencia_y_Politica1.pdf

Astudillo BP (2014) Chile needs better science governance and support. Nature 511(7510):385

Bajak A (2015, Noviembre) Chile's scientists take to the streets over funding. Nature News. [Online]: http://www.nature.com/news/chile-s-scientists-take-to-the-streets-over-funding-1.18800

Bórquez R (2011) Brecha entre la investigación científica y las políticas públicas en chile, el caso de la investigación relacionada con el cambio climático. Tesis para optar al grado de Magister en Gestión y Políticas Públicas. Universidad de Chile, Santiago, Chile

CEAZA (2014, Octubre) Boletín Climático. [Online]: http://www.ceaza.cl/wp-content/uploads/2014/10/boletin.ceazamet.2014.10.pdf

CEPAL (2012a) La economía del cambio climático en Chile. Santiago, Chile, Naciones Unidas

CEPAL (2012b) Disponibilidad futura de los recursos hídricos frente a escenarios de cambio climático en Chile. Santiago, Chile, Naciones Unidas

CNN Chile (2015, Agosto) Actualidad Central: La crisis de la ciencia, ¿Cuánto se investiga en Chile? [Online]: http://www.cnnchile.com/noticia/2015/08/29/actualidad-central-la-crisis-de-la-ciencia-cuanto-se-investiga-en-chile

Court J, Young J (2003) Bridging research and policy: insights from 50 case studies. ODI working paper #213, ODI, London

Court J, Young J (2005) Bridging research and policy in international development: context, evidence and links. In: Stone D, Maxwell S (eds) Global knowledge networks and international development. Routledge, UK

Crewe E, Young J (2002) Bridging research and policy: context, evidence and links. Overseas Development Institute, London, p 2002

Dai A (2013) Increasing drought under global warming in observations and models. Nat Clim Change 3(1):52–58

DGF-Conama (2006) Estudio de variabilidad climática para el siglo XXI. Informe Final. Departamento de Geofísica. Facultad de Ciencias Físicas y Matemáticas, Universidad de Chile. Comisión Nacional del Medio Ambiente, Santiago, Chile

Díaz P (2016, 22 Febrero) Los desafíos que impone la sequía en la región de Coquimbo. Economía y Negocios (El Mercurio). [Online]: http://www.economiaynegocios.cl/noticias/noticias.asp?id=227995

Díaz H, Hurlbert M (2014) Making science count: climate change and the science/practice interface. In Leal Filho W, Alves F, Caeiro S, Azeiteiro UM (eds) International perspectives on climate change. Springer International Publishing, Berlin, pp 203–215

Fiebig-Wittmaack M, Astudillo O, Wheaton E, Wittrock V, Perez C, Ibacache A (2012) Climatic trends and impact of climate change on agriculture in an arid Andean valley. Clim Change 111 (3–4):819–833

Godfrey L, Funke N, Mbizvo C (2010) Bridging the science-policy interface: a new era for South African research and the role of knowledge brokering. S Afr J Sci 106(5–6):44–51

Gobierno de Chile (2015) Contribución Nacional Tentativa de Chile (indc) Para el Acuerdo Climático París 2015. [Online]: http://www4.unfccc.int/submissions/INDC/Published%20Documents/Chile/1/Chile%20INDC%20FINAL.pdf

Hearne R, Donoso G (2014) Water markets in Chile: are they meeting needs? In: Easter KW, Huang Q (eds) Water markets for the 21st century. Springer, Netherlands, pp 103–126

Huggel C et al (2015) A framework for the science contribution in climate adaptation: experiences from science-policy processes in the Andes. Environ Sci Policy 47:80–94

IPCC (2013) Climate change 2013: the physical science basis. Working Group I Contribution to the IPCC 5th Assessment Report. [Online]: http://www.ipcc.ch/report/ar5/wg1

Iyalomhe F, Jensen A, Critto A, Marcomini A (2013) The science-policy interface for climate change adaptation: the contribution of communities of practice theory. Environ Policy Gov 23 (6):368–380

Janse G (2008) Communication between forest scientists and forest policy-makers in Europe—a survey on both sides of the science/policy interface. For Policy Econ 10(3):183–194

Jones N, Jones H, Walsh C (2008) Political science? Strengthening science-policy dialogue in developing countries. Overseas Development Institute, London

Krellenberg K (2012) A science-policy approach towards local adaptation planning: the case of Santiago de Chile. In: Otto-Zimmermann K (ed) Resilient cities 2. Springer, Netherlands, pp 233–240

La Tercera (2015, Febrero) En la IV Región ya no se habla de sequía sino de desertificación. [Online]: http://diario.latercera.com/2015/02/17/01/contenido/pais/31-183752-9-en-la-iv-region-ya-no-se-habla-de-sequia-sino-de-desertificacion.shtml

Larigauderie A, Mooney HA (2010) The intergovernmental science-policy platform on biodiversity and ecosystem services: moving a step closer to an IPCC-like mechanism for biodiversity. Curr Opin Environ Sustain 2(1):9–14

Lidskog R (2014) Representing and regulating nature: boundary organisations, portable representations, and the science–policy interface. Environ Polit 23(4):670–687

López-Rodríguez MD, Castro AJ, Castro H, Jorreto S, Cabello J (2015) Science–policy interface for addressing environmental problems in arid Spain. Environ Sci Policy 50:1–14

Ludwig F et al (2016) 9. Perspectives on climate change impacts and water security. In Pahl-Wostl C, Bhaduri A, Gupta J (eds) Handbook on water security. Edward Elgar Publishing, UK, pp 139–160

Macleod CK, Blackstock K, Haygarth P (2008) Mechanisms to improve integrative research at the science-policy interface for sustainable catchment management. Ecol Soc 13(2):48

Ministerio del Interior y Seguridad Pública (2015) Política Nacional para los Recursos Hídricos 2015. [Online]: http://www.interior.gob.cl/media/2015/04/recursos_hidricos.pdf

Ministerio del Medio Ambiente (2011) Segunda Comunicación Nacional de Chile ante la Convención Marco de las Naciones Unidas sobre Cambio Climático. Santiago, Chile

Ministerio del Medio Ambiente (2015) Plan Nacional de Adaptación al Cambio Climático. [Online]: http://portal.mma.gob.cl/wp-content/uploads/2016/02/Plan-Nacional-Adaptacion-Cambio-Climatico-version-final.pdf

Ministerio de Obras Públicas (2013) Estrategia Nacional de Recursos Hídricos 2012–2025. [Online]: http://www.mop.cl/Documents/ENRH_2013_OK.pdf

McConney P, Fanning L, Mahon R, Simmons B (2016) A first look at the science-policy interface for ocean governance in the wider caribbean region. Front Mar Sci 2:119

Morgan E (2014) Science in sustainability: a theoretical framework for understanding the science-policy interface in sustainable water resource management. Int J Sustain Policy Pract 9 (2):37–54

Nursey-Bray MJ et al (2014) Science into policy? Discourse, coastal management and knowledge. Environ Sci Policy 38:107–119

Santibáñez F et al (2014) Atlas del cambio climático en las zonas de régimen árido y semiárido. Regiones de Coquimbo Valparaíso y Metropolitana. Centro AGRIMED, Facultad de Ciencias Agronómicas, Universidad de Chile, Santiago

Sarkki S, Niemelä J, Tinch R, van den Hove S, Watt A, Young J (2014) Balancing credibility, relevance and legitimacy: A critical assessment of trade-offs in science-policy interfaces. Sci and Pub Policy 41(2):194–206.

Sarkki S et al (2015) Adding 'iterativity' to the credibility, relevance, legitimacy: a novel scheme to highlight dynamic aspects of science–policy interfaces. Environ Sci Policy 54:505–512

Scott CA, Varady RG, Meza F, Montaña E, de Raga GB, Luckman B, Martius C (2012) Science-policy dialogues for water security: addressing vulnerability and adaptation to global change in the arid Americas. Environ Sci Policy Sustain Dev 54(3):30–42

Sutherland WJ et al (2012) A collaboratively-derived science-policy research agenda. PLoS ONE 7(3):e31824

Van Enst WI, Driessen PP, Runhaar HA (2014) Towards productive science-policy interfaces: a research agenda. J Environ Assess Policy Manage 16(01):1450007

von Winterfeldt D (2013) Bridging the gap between science and decision making. Proc Natl Acad Sci 110(Supplement 3):14055–14061

Weichselgartner J, Kasperson R (2010) Barriers in the science-policy-practice interface: toward a knowledge-action-system in global environmental change research. Glob Environ Change 20 (2):266–277

Wesselink A, Buchanan KS, Georgiadou Y, Turnhout E (2013) Technical knowledge, discursive spaces and politics at the science–policy interface. Environ Sci Policy 30:1–9

Wheeler T, von Braun J (2013) Climate change impacts on global food security. Science 341 (6145):508–513

Wold Bank (2011) Chile—Diagnostico de la gestion de los recursos hídricos. [Online]: http://water.worldbank.org/node/83999

Young JC et al (2014) Improving the science-policy dialogue to meet the challenges of biodiversity conservation: having conversations rather than talking at one-another. Biodivers Conserv 23(2):387–404

Chapter 20
Selection Support Framework Fostering Resilience Based on Neighbourhood Typologies

Laura Kleerekoper, Jeroen Kluck and Andy van den Dobbelsteen

Abstract Selecting climate adaptation measures through a straightforward approach: that is possible with the selection support framework presented in this paper. The framework results from Dutch urban design studies aiming to adapt to climate change. The framework includes adaptation measures for eight neighbourhood typologies on the scale of a street, square or building block. Numerous case studies provided input for the selection support framework. Based on this scientific work the framework provides an easy indication of climate adaptation opportunities, with focus on heat, and can be adapted to fit typical neighbourhoods in any country. The selection support framework enables urban professionals to select a set of climate adaptation solutions tailored to typical characteristics of a neighbourhood typology.

Keywords Climate adaptation · Case studies · Integral design · Urban microclimate · Neighbourhood typologies

Introduction

Climate change predictions pose a new challenge for urban developments. From practice the question rises what can be done. Climate adaptation shows in many cases to be a difficult subject due to uncertainties, multiple stakeholders and complexities in urban areas (Klok and Kluck 2015; Rovers et al. 2014). Microclimate and water solutions need to merge with other urban design challenges

---

L. Kleerekoper (✉) · A. van den Dobbelsteen
Faculty of Architecture, Delft University of Technology,
Julianalaan 134, 2628 BL Delft, The Netherlands
e-mail: l.kleerekoper@hva.nl

L. Kleerekoper · J. Kluck
Amsterdam University of Applied Sciences, Weesperzijde 190,
1097 DZ Amsterdam, The Netherlands

W. Leal Filho and J.M. Keenan (eds.), *Climate Change Adaptation in North America*, Climate Change Management,
DOI 10.1007/978-3-319-53742-9_20

and need to be integrated into policy to realise climate-resilient urban areas. This study aims to achieve a better integration of especially adaptation to increased heat because this is a subject underexposed to many designers (Pijpers-van Esch 2015).

The urban microclimate, and thus heat in cities, can be influenced by the design of buildings and the surrounding public and private areas (Oke 1988; Katzschner 2010). The elements that influence the urban microclimate are vegetation, water, urban geometry, materials and colour (Kleerekoper et al. 2012). Lenzholzer (2015) and Pijpers-van Esch (2015) explained more profoundly the processes of the urban microclimate for temperate climate zones.

For the analysis of the microclimate of a specific location several methods exist, e.g. CFD-based calculation methods (Schrijvers et al. 2015), ENVI-met (Bruse and Fleer 1998), SOLWEIG (Lindberg et al. 2008), Rayman (Matzarakis et al. 2007) and expert knowledge or GIS-informed climate mapping (Lenzholzer 2015; van der Hoeven and Wandl 2013). These methods require (expensive) specialist input data and some make use of complex calculation models or time-consuming interviews.

Possible solutions for adaptation to heat and water are presented by, for example, Pötz and Bleuzé (2012), Voskamp and van de Ven (2015). Also for the implementation of adaptation measures different proven policy instruments are available: financial instruments, regulative instruments, planning instruments, voluntary agreements, informative instruments, monitoring instruments (Stelljes 2015).

However, the actual implementation of measures is not yet part of common design practice, and municipalities do not know if and what action is needed (Helmer and Tijlhuis 2015). Therefore, the aim of this paper is to assist planners and designers in increasing the role of climate adaptation in urban developments. This starts with the hypothesis that planners and designers will benefit from a clear guidance that corresponds with terminology they use in daily practice. Based on expert input, the use of neighbourhood typologies has been selected.

The research question that follows from this is: How can urban areas become climate robust considering the morphology of neighbourhood typologies? To answer the question, first the parameters that differentiate neighbourhoods in relation to their microclimate were indicated. The neighbourhood typologies are the basis for a selection support framework. About 20 case study designs in The Netherlands generated input for this framework. For each case a location-specific design was made to improve the vulnerable areas or to preserve the quality of the non-vulnerable neighbourhoods.

Scale is a determining aspect in the decision-making process regarding climate adaptation measures. Usually the neighbourhood scale is chosen because this can be managed by municipalities or housing corporations. However, the effects of measures in the first place is on the local street scale. The neighbourhood typologies selected have a characteristic building typology and organisation of public space. This enables a general statement of the measures most appropriate for these neighbourhoods. For example, historic urban areas have a completely different starting-point than garden cities: the soil of an historic city centre is almost completely sealed off with pavement, where temperatures can really accumulate, while a garden city can be characterized by large open spaces with a lot of green, where temperatures may remain close to the rural situation.

A first step towards this integration is to specify the general and location-specific measures. Often a measure can be generally applicable for different types of streets and locations. A green roof or facade, for example, can be realized on almost all roofs and facades. But the direct effect of the green on outdoor temperatures is only relevant at a short distance from the wall or roof (Kikegawa et al. 2006). This implies only green on lower buildings is effective. And, are canvas awnings a good shading device for streets? They are especially applicable in streets where there is no option for trees and no traffic running through. Moreover, such a shading device is only feasible with many passers-by. In this study the integral designs link to the physical, social and economic aspects of an urban area.

The selection support framework helps to select the measures that are especially fit for a certain neighbourhood typology. The typology framework allows a first indication of the most appropriate microclimate measures, without the use of extensive data-analysis or computer modelling.

Methodology

Two sub-questions were posed to answer the main question "How can urban areas become climate robust considering the morphology of neighbourhood typologies?". The first sub-question was: "Which neighbourhood typologies can we distinguish that differ in microclimate and in possible measures to improve the microclimate?" An analyses of the physical properties of 20 neighbourhoods indicated vulnerability to heat stress. Hereto, three parameters were analysed: the urban surface characteristics, the height-to-width ratio and shape of the building block. The second sub-question was: "Which design solutions are appropriate for a specific neighbourhood typology?" This question was answered through a qualitative method based on case studies. The case studies are further explained in the following section and illustrated with two examples.

The classification of neighbourhoods by category can be based on many characteristics. The following paragraphs describe the translation of urban typologies to microclimate categories.

Traditionally, urban typologies are classified in relation to their construction period, form of the building block and organisation of public and private space (Wassenberg 1993; Ibelings 1999; Baeten et al. 2004; Lorzing et al. 2008). Beside the common urban typologies, other methods of classification have emerged to support decision makers with the appropriate data about their cities and regions. In the year 1998 the *Woonbehoefte-onderzoek* (WBO) made a classification of residential environments according to the level of urbanity. The result is three main typologies from urban, to suburban, to village & rural environments (ABF 2006). In the beginning of the twentieth century the reflection on a period of fifty years of constructing for housing shortage demanded better insight into the quality of urban areas. The *Rosetta method* redefines the WBO residential areas into five main typologies: highly urban, urban, suburban, village and rural (Prins et al. 2010). Although this recent classification offers insight in the quality and ambiance of an urban area it does not offer a differentiation of the main aspects that are required to improve the microclimate.

For the microclimate there are many parameters of interest such as the quantity of paved surface, type of material, colour, amount and type of vegetation, amount and type of water, height and width of streets and inner courtyards, openness, orientation and built form (Kleerekoper et al. 2012). For cities in the US Stewart and Oke (2012) presented a climate classification according to physical parameters. With Stewart and Oke's method many parameters are needed to come to a classification; some features are overlapping or indicating the same process. In order to improve the accessibility for urban designers and planners working with the urban microclimate, the amount of parameters was brought down from ten to three. The three most important determinants for the microclimate by which Dutch neighbourhoods distinguish themselves are building height, form of footprint, as well as the percentage of green/water in relation to the urban surface; see Table 20.1 for more detail. These three parameters were also selected as indicators by an extensive typological research of Dutch urban types by Berghauser Pont and Haupt (2009). The three parameters enable a classification based on microclimate categories and on traditional urban typologies used by urban designers and planners.

The ratio of paved ground surface and roof and wall surface versus natural (green and water) surface varies per building morphology type and density and is a valuable indicator for the urban microclimate. Heat accumulation in the stony materials and the reflection of radiation between these surfaces are to a large extent responsible for urban heating. In the urban surface analysis walls are an important dimension: the vertical surfaces receive a lot of radiation due to the relative low sun angle in The Netherlands. In American cities wall and roof surfaces in areas of tall, densely positioned buildings exceed the ground surface. Conversely, wall and roof areas of low-density single-family detached houses form only a small proportion of the ground surface area (Ellefsen 1991). In Canada walls are 28–54% of the total surface area in the city (Voogt and Oke 1997). In some neighbourhoods transforming these vertical surfaces can have a significant effect on the local microclimate, and therefore should be visible in the ratio of stony versus natural areas,

Table 20.1 Categorization of urban types (based on Berghauser Pont and Haupt 2009) in relation to the microclimate

Building height	Footprint	Percentage green/water
Low (up to 3 layers)	Strip	Little green (0–10%)
Middle high (4–6 layers)	Open urban block	Moderate green (10–30%)
High (7–10 layers)	Closed urban block	Much green (30–50%)
High-rise (9 and more layers)	Spread buildings	Abundant green (50–100%)

before and after the transformation. Therefore, the surface analyses in this study also included these vertical surfaces.

The parameters given in Table 20.2 were determined using a combination of GIS mapping (ArcMap), aerial images from Google Earth and personal photographs or Google Street views. The source map for the analysis was the TOP10NL, a detailed vector map (Middel 2002). The building height is based on the average height and the footprint on the TOP10NL. The percentage of green and water requires a combination of different sources: the municipal green ground area could simply be calculated from the TOP10NL maps, for green roofs and private gardens Google Earth was used to determine a percentage of the private area that may be counted as green.

In addition to the three parameters given above, also land use, height/width (H/W) ratio, function (residential, industry, city centre, office park, agriculture, sports, recreation, mixed urban functions), density (inhabitants per hectare, dwellings per ha, FSI, GSI) and street trees are important for the urban microclimate. These are not directly part of the categories, but are appointed shortly per category.

From the additional parameters in the paragraph above, especially street trees are important. Municipal green on the ground, green on roofs and private green in gardens were all included in the percentage of green/water. However, street trees were not included in the surface analyses because not all municipalities could provide GIS data with street trees and their properties.

In order to provide a clear image of the characteristics of the microclimate categories Table 20.2 combines common urban typologies and their period of

Table 20.2 Relation between microclimate category and urban typology

Urban typology		Microclimate category		
Typology	Period	Height	Footprint	Green
Historical city block and pre-war city block	Before 1910 and '10–'30	Middle high	Closed urban block	Little green
Garden town	'10–'30	Low	Closed urban block	Moderate to much green
Residential housing	'30–'40	Low	Closed urban block	Little green
Post-war garden city low-rise	'45–'55	Low	Open urban block	Moderate to much green
Post-war garden city high-rise	'50–'60	Middle high/high	Open urban block	Moderate to much green
High-rise city centre	'60–present	High-rise	Spread buildings	Little green
Community neighbourhood	'75–'80	Low	Strips Open urban block	Little to moderate green
Sub-urban expansion —'Vinex'	'90–'05	Low	Strips Closed urban block	Moderate green

origin. The urban typologies that are related to the microclimate categories in this paper are mainly based on the typology description in ‘An urban typology’ (Lorzing et al. 2008). For example: the *historical city blocks constructed before 1910* pertain to the category *middle-high closed urban block with little greenery*, while the *post-war garden city with low-rise,* which dominates the Western part of Amsterdam, pertain to the category *low open urban block with moderate to much greenery*.

Case Studies

The case studies presented in this section provide insight into the type of adaptation measures that are fit for a certain microclimate category. For each category at least two neighbourhoods were analysed and elaborated with a design solution or strategy. Table 20.3 presents an overview of the case studies with the neighbourhood typology and microclimate category, as explained in the previous section. The integral design studies presented in this paper are a follow-up from the designs presented by Kleerekoper et al. (2012). The additional case studies enable the formation of general guidelines per typology.

In each case study design, a different redevelopment program was addressed in combination with opportunities for climate adaptation. This part of the study can be described as ‘typological research’, according to the scheme made by de Jong and van der Voordt (2002), presented in Fig. 20.1. In the case studies the context is variable (different neighbourhoods) and the object is determined (different cases per microclimate category). Each separate case study however was a ‘design study’ in itself, with one specific neighbourhood, with context and variable climate adaptation solutions (objects) thus determined.

Table 20.3 Selection of case study neighbourhoods based on microclimate categories

Typology		Microclimate category			neighbourhood 1	neighbourhood 2	neighbourhood 3
		height	urban block	green			
Historical city block & pre-war city block	before 1910 & '10-'30	middle-high	closed	little green	City Centre, Geertebuurt *Utrecht*	Bergpolder-Zuid *Rotterdam*	Zuidwal *Den Haag*
Garden town	'10-'30	low	closed	moderate to much green	Tuindorp, *Utrecht*	Tuindorp Nieuwendam *Amsterdam*	Tuindorp Oostzaan *Amsterdam*
Workingclass housing	'30-'40	low	closed	little green	Ondiep *Utrecht*	Transvaal *Den Haag*	Rivierenwijk *Utrecht*
Post-war garden city low-rise	'45-'55	low	open	moderate to much green	Slotermeer, Couperusbuurt *Amsterdam*	Watergraafsmeer Jeruzalem *Amsterdam*	
Post-war garden city high-rise	'50-'60	high	open	moderate to much green	Overvecht *Utrecht*	Kanaleneiland *Utrecht*	Schiebroek Zuid *Rotterdam* jaren 60
High-rise city centre	'60-present	high-rise	spread	little green	Station area, *Den Haag*	Lijnbaan *Rotterdam*	
Community neighbourhood	'75-'80	low	strokes	little to moderate green	Lunetten *Utrecht*	Zevenkamp *Rotterdam*	
Sub-urban expansion - Vinex	'90-'05	low	closed	moderate green	Leidsche Rijn, *Utrecht*	Ypenburg *Den Haag*	

Fig. 20.1 Types of design related study by Jong and Voordt (2002)

	OBJECT *Determined*	*Variable*
CONTEXT *Determined*	Design research	Design study
Variable	Typological research	Study by design

The case studies selected are located in the four major cities of The Netherlands: Amsterdam, Rotterdam, The Hague and Utrecht. The municipalities of these four cities were stakeholders within the research project and had a role in the process of providing data, indicating problem areas from their point of view and responding with feedback on design solutions and strategies. In this section the methodology and the selection of case studies is further explained.

In conducting practical case study research, unpredictable factors determining the course of the process are inevitable. As a result, the case studies differ in size, the section of a neighbourhood presented does not always exactly match the area presented in the design solution(s) and there is quite some difference in the elaboration detail of the design solutions. These inequalities are a result of the different ways the design solutions were developed. Most case studies were done by the first author, some by students (MSc, BSc and grammar school) and others by researchers of the Climate Proof Cities programme (Rovers et al. 2014). However, the area chosen for the analyses of the land and urban surfaces always is a representative section with, for example, a representative amount of green at the border and a homogenous building type. An example is the early development of castles and their community and facilities; the occupancy and land use of this type was very characteristic: a formal garden within the castle, dwellings for the castle community outside the castle with shared gardens, surrounded by the facilities (Tummers and Tummers-Zuurmond 1997).

In the following sections, two case studies are presented individually.

Case Couperusbuurt, Amsterdam (Post-war Garden City Low-Rise)

This garden city has a particular urban plan that breaks with the traditional urban block in which the inner courtyard is private property. There are several variations of urban blocks in the low-rise neighbourhoods, such as the L-shaped blocks that form semi-enclosed squares, but also straight strips. Entrances of dwellings are situated at both the outside and inside of the blocks. The strip buildings often have a park along the backside. The streets between the blocks have an average width of 15 m. The inner courtyard has a size of 40 * 60 m and is surrounded by buildings of two layers and a roof; inside the inner courtyard winds a public road. The H/W ratio is thus between 0.6 and 0.3. The open interior without fences and predominantly with grass provides a good ventilation.

The Couperusbuurt is a neighbourhood of the Slotermeer district in the West of Amsterdam. Its garden city roots very well emerge looking at the land surface cover in Fig. 20.2. This shows a 'garden' (green and water) surface covering almost half of the total surface, which is a remarkable achievement, especially taking into account the quite high Floor Space Index (FSI) of 0.6 due to the six layer buildings at the South-East part of the area. The facades form a relatively large part of the stony surface.

The neighbourhood has been listed for redevelopment by the municipality. Dwellings need to be upgraded and, as an extension of the upgrade, also the amount of parking places should increase. There are several ways to achieve this goal, however, the most apparent option is to use the inner courtyards. Figure 20.3 shows an impression of a situation in which the courtyard is used for parking without a decrease in thermal comfort. This can be achieved by planting trees in the middle, using permeable concrete grass tiles and changing the roof colour to light and reflective and the façade colour to a middle-dark tint.

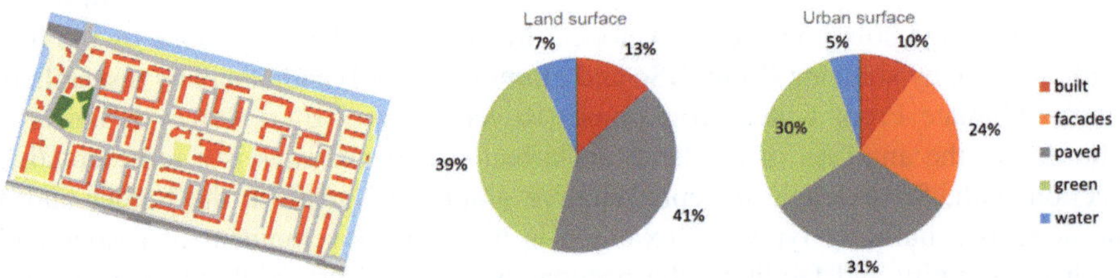

Fig. 20.2 The relation between stony and *green* surfaces in the Couperusbuurt neighbourhood, Amsterdam, with and without the vertical façade surface, map source TOP10NL (Middel 2002)

Fig. 20.3 *Above* Existing semi-public inner courtyard. *Below* Adaptation measures which improve thermal comfort and increase parking space are light roofs, middle-dark facades, additional trees and permeable pavement with grass

An important part of the strategy for the Couperusbuurt is the determination of an appropriate additional user function of the inner courtyard to increase the value for both residents and passers-by. For a more sustainable water system the courtyards could additionally function as a water retention and infiltration area. The small apartments could be transformed into more spacious dwellings without decreasing indoor and outdoor comfort by taking into account solar access. As a result of the transformation from two apartments into one-family houses the need for additional parking space is reduced drastically. As part of the redevelopment strategy energy opportunities such as solar panels or seasonal storage will increase the real estate value and lifespan.

A conclusion from the garden city cases is that there is sufficient green space to cope with climate change. The challenge is to preserve this green. However, when redeveloping these areas, especially this green is in danger of disappearing. In the garden cities most of the green is semi-public and managed by the municipality. The quality of the semi-public green often is low and there are few functions linked to the green. This could be part of the solution: improve the quality of the green and link multiple functions to it.

Case Zuidwal, The Hague (Historical City Block)

In The Hague the Zuidwal neighbourhood is a mix of historical dwellings and some post-war dwellings with many commercial activities. Figure 20.4 presents the relation between green or stony land and urban surface. The neighbourhood is for 95% covered with stony surface. And the share of the façade surface is almost 3/4th of the urban surface. Compared to the other microclimate categories the fraction of stony surfaces is very large. In line with the more dominant façade surface the FSI of 2.8, which is high as well.

In one of the central shopping streets of The Hague, the Grotemarktstraat, the outdoor climate is not regulated to provide a thermally comfortable place for shopping or passing through. However, applying measures is a big challenge: every square metre is occupied, on the ground, the facades, the underground and even in

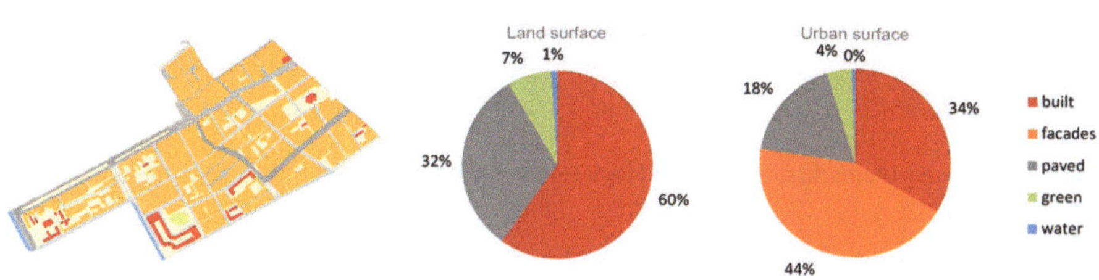

Fig. 20.4 The relation between stony and *green* surfaces in the Zuidwal neighbourhood, The Hague, with and without the vertical façade surface, map source TOP10NL (Middel 2002)

the air. Planting trees is extremely difficult due to the tunnel underneath and the lack of space on the ground. Despite tram cabling high up in the canopy, there is still space available. A support system for Christmas decoration is already available, with some attachment points; this could also hold up canvas awnings to provide shade on hot sunny days, see Fig. 20.5 for an impression.

The canvas sheets are a temporal measure that will be in place during the hottest period of the year. Nevertheless, the street needs sufficient daylight which can be achieved through the use of light colours, some level of translucency and only covering the North side of the street. In any case, a shading device will reduce indoor daylight as well, which might lead to more artificial lighting.

For the analysis of the effect on thermal comfort in the street and surrounding area the canvas shading device was modelled and simulated in ENVI-met. The solution was compared with the existing situation and two other options that could be possible on the location: trees on the North side of the street and light-coloured roofs with an albedo of 0.85 instead of 0.3. The simulation results are presented in Fig. 20.6. The North side and middle of the street receive the most solar radiation, this is visible in receptor points A, D, G (North) and B, E and H (middle). The canvas sheets lower the Physical Equivalent Temperature (PET) with about 20 °C at locations in the sun.

Fig. 20.5 The Grotemarktstraat in The Hague with canvas sheets to improve thermal comfort on hot days

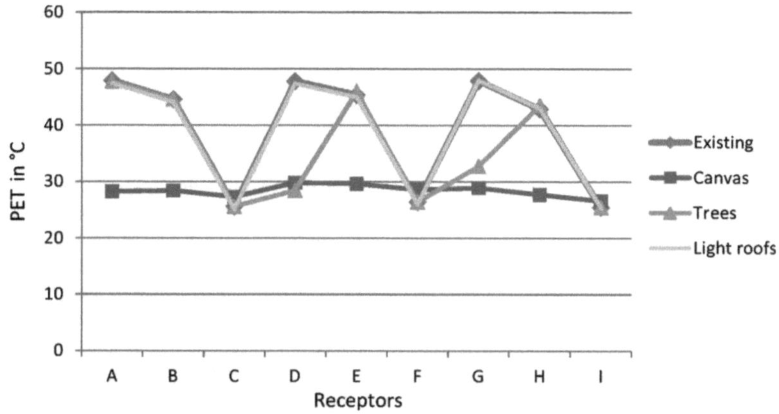

Fig. 20.6 ENVI-met simulation to analyse the effect of three different adaptation measures on thermal comfort (PET) in the Grotemarktstraat at 13:00 h at a height of 1 m on a typical heatwave day

A Selection Support Framework

In the previous sections two examples of case studies are presented. After analysing all of the 20 case studies, the appropriate measures for a specific microclimate category were extracted. The main design solutions are summarized and presented in Table 20.4.

Conclusions

This paper presents a new categorisation of neighbourhoods in order to combine microclimate indicators with traditional urban typologies. This categorisation forms the basis for a selection support framework for urban planners and designers. Without profound knowledge of the urban microclimate the framework enables them to indicate vulnerability, and most important, select from a set of measures that improve or preserve the qualities of the local microclimate. Besides exploring possibilities, the integral design studies also showcased examples.

Instead of using professional simulation programs or extensive GIS mapping, the urban designer or planner can place an area in one of the microclimate categories introduced in this paper. These categories are based on building height, footprint and the percentage of green/water in relation to the urban surface. Note that with the urban surface also the vertical surfaces of façades should be included. The microclimate categories can also be determined for other countries with different building styles and urban morphologies.

Table 20.4 The selection support framework presents design solutions for specific neighbourhood typologies

Historical city centre and pre-war city block (middle-high closed urban block with little green)	• Temporary and flexible measures: canvas sheets above streets and plazas, the spraying of fine water droplets at pedestrian routes, watering the streets to reduce the radiant heat from stony surfaces, seating elements shaded in summer and protected from wind in winter and place elements that can generate ventilation during hot periods • Fixed and robust measures: pergolas covered with deciduous climbers, arcades along south facades, coverings over a part of the street or a walkway, shallow water streams through the street, fountains, white roofs, green roofs, facade vegetation (climbers) or green facade (many individual plants)
Working class neighbourhood (low closed urban block with little green)	• Tiny gardens along the facade and green facades • Type of pavement: semi-pavement, permeable pavement, light colours • Parking solutions combined with a structure for climbing plants • Flat roofs scan contribute to thermal comfort at street level: a white reflective coating, with a sufficiently thick substrate layer • In case there is no possibility to add more green the generation of ventilation is especially important. For example, by more height differences of the buildings or by using the principle of 'hot' and 'cool' places between which the air will be moving (thermal draft) • Find parking solutions: streets can be planted with a double line of trees or one line of trees at the side that receives most solar radiation. In some cases, trees as espaliers can provide a solution when the position of the tree cannot be placed far enough from a building wall
Garden town (low closed urban block with moderate to much green)	• Promoting green in private gardens: inspire people and create awareness about the importance and benefits of green. Another option is to offer trees, hedges or plants for free to the inhabitants. Or, through the water board or council tax, charges can be adjusted to the degree of pavement in gardens, or charge less taxes when rainwater is collected and infiltrated on site

(continued)

Table 20.4 (continued)

	• Add street trees on strategic places
Garden city low-rise (low open urban block with moderate to much green)	• Promote green in private gardens • Increase the value of semi-public green: improving quality of green and attach multiple functions to the green areas. The semi-public inner courtyards have a green lawn of about 30 * 50 m. The functions linked to these should be chosen with care: no nuisance for residents and attractive to non-residents. For example a route to walk the dogs, water storage, butterfly and bee gardens, fruit and nut orchards, etc.
Garden city high-rise (middle-high/high open urban block with moderate to much green)	• Promote green in private gardens • Increase the value of semi-public and public green. The large size (70 * 100 m) of the public inner courtyards are very suitable for special functions, such as water treatment area with reed plants, mixed cropping (urban agriculture or vegetable gardens), a petting farm, water playground, dog training field, lawn or events area with fixed barbecues and permissions for ice cream, snack or Dutch doughnut stand. Throughout the neighbourhood one can think of: walking paths, paved paths for cycling and inline skating, trail for mountain bikers
High-rise city centre (high-rise with little green)	• Offer a diversity of places: sheltered spots out of the wind, shady spots and perhaps places covered from rain • Reduce the extensive amount of pavement, for example, by realizing roof parks • Large surfaces can also be used to transform solar energy into thermal energy or electricity • Reduce the exhaust of anthropogenic heat: car free zones, design buildings that do not need air conditioning and make sure all exhaust air leaves the building at the top
Cauliflower neighbourhood (low strips and open urban blocks with little to moderate green)	• Parking solutions combined with green and shadow elements. Reduce the number of cars is not directly an option here because the neighbourhoods are often at a great distance from the centre. One could think of (electric) car sharing or parking on the edge of several building blocks. Thus when more space becomes available it can be used for additional playgrounds, more trees, rainwater harvesting and infiltration meadows • Green facades

(continued)

Table 20.4 (continued)

	• White roofs
Sub-urban expansion—VINEX (low closed urban block and strips with moderate green)	• Promote green in private gardens, also try to discourage the use of dark anthracite tiles • Parking solutions combined with green and shadow elements

For future analyses, it is recommended to include street trees or tree canopy density. The current rapid development of geographical data registration of all elements in public space does facilitate the inclusion of street trees.

To strengthen the generic conclusions in this paper further, each neighbourhood typology should be complemented with at least three design proposals. The selection support framework is now under further development by using it in actual (re)development plans. The new case studies also address pluvial water while costs and benefits are calculated to further increase practical use. Moreover, these new cases will generate feedback on the value of the selection support framework in practice.

This paper enables categorisation and selection of measures in the first design phase without the necessity of specialised knowledge about the urban microclimate. Nevertheless, the actual implementation and engineering of some solutions does require more expert knowledge.

Acknowledgements As part of the research project of Climate Proof Cities, this study was funded by the governmental institute in The Netherlands, Knowledge for Climate program, supported by stakeholders, among which the cities of Rotterdam, The Hague, Amsterdam and Utrecht.

References

ABF (2006) ABF Woonmilieutypologie. ABF Research BV, Delft

Baeten J, Berg JJ, Betsky A, van Bergeijk H, Ibelings H, Kuipers M, Steenhuis M, de Vletter M, Teunissen M, Stissi V (2004) Gewoon architectuur 1850–2004. NAi, Rotterdam

Berghauser Pont MY, Haupt PA (2009) Spacematrix. Space, density and urban form. Delft University of Technology

Bruse M, Fleer H (1998) Simulating surface-plant-air interactions inside urban environments with a three dimensional numerical model. Environ Model Softw 13(3–4):373–384

de Jong TM, van der Voordt D (2002) Ways to study and research: urban, architectural and technical design. DUP Science, Delft

Ellefsen R (1991) Mapping and measuring buildings in the canopy boundary layer in ten U.S. cities. Energy Build 16(3–4):1025–1049

Helmer M, Tijlhuis N (2015) Voetje voor voetje. Klimaatverbond Nederland

Ibelings H (1999) Nederlandse stedenbouw van de 20ste eeuw. NAi, Rotterdam

Katzschner L (2010) Outdoor thermal comfort under consideration of global climate change and urban development strategies. Adapting to change: new thinking on comfort, Windsor, 9–11 April, Network for comfort and energy use in buildings, London

Kikegawa Y, Genchi Y, Kondo H, Hanaki K (2006) Impacts of city-block-scale countermeasures against urban heat-island phenomena upon a building's energy-consumption for air-conditioning. Appl Energy 83(6):649–668
Kleerekoper L, van Esch M, Salcedo TB (2012) How to make a city climate-proof, addressing the urban heat island effect. Resour Conserv Recycl 64:30–38
Klok L, Kluck J (2015) Reasons to adapt to urban heat (in The Netherlands). In: ICUC9—9th international conference on urban climate jointly with 12th symposium on the urban environment, 20–24 July 2015, Toulouse
Lenzholzer S (2015) Weather in the city. How design shapes the urban climate. nai010, Rotterdam
Lindberg F, Holmer B, Thorsson S (2008) SOLWEIG 1.0–modelling spatial variations of 3D radiant fluxes and mean radiant temperature in complex urban settings. Int J Biometeorol 52 (7):697–713
Lorzing H, Harbers A, Schluchter S (2008) Een stedenbouwkundige typologie. NAi, Rotterdam
Matzarakis A, Rutz F, Mayer H (2007) Modelling radiation fluxes in simple and complex environments—application of the RayMan model. Int J Biometeorol 51(4):323–334
Middel G (2002) TOP 10 data set P1661. Topografische-Dienst, Data Archiving and Networked Services
Oke TR (1988) Street design and urban canopy layer climate. Energy Build 11(1–3):103–113
Pijpers-van Esch MME (2015) Designing the urban microclimate: a framework for a design-decision tool for the dissemination of knowledge on the urban microclimate to the urban design process. Delft University of Technology, TU Delft
Pötz H, Bleuzé P (2012) Urban green-blue grids for sustainable and dynamic cities. Coop for Life, Delft, The Netherlands
Prins E, Bos R, Stam A, van den Berg R, Witteman B, Meijer F, Witsen PP, Fuchs P (2010) Rosetta-methode. Een gemeenschappelijke taal voor woonmilieus in de regio
Rovers V et al (2014) Final report climate proof cities 2010–2014. URL: http://www.knowledgeforclimate.nl/urbanareas/climateproofcities_finalreport. Accessed on 07 Apr 2016
Schrijvers PJC, Jonker HJJ, de Roode SR, Kenjereš S (2015) On the local day time urban heat island of an idealized 2D city. J Appl Meteorol 1–41
Stelljes N (2015) A comparison of local adaptation strategies—results from a qualitative data analysis of nine coastal adaptation strategies. In: European climate change adaptation conference 2015, Copenhagen, Denmark, URL: http://www.ecologic.eu/sites/files/presentation/2015/stelljes_15-5-12_ecca.pdf. Accessed on 21 Apr 2016
Stewart ID, Oke TR (2012) Local climate zones for urban temperature studies. Bull Am Meteorol Soc 93(12):1879–1900
Tummers LJM, Tummers-Zuurmond JM (1997) Het land in de stad. THOTH, Bussum
van der Hoeven F, Wandl A (2013) Amsterwarm. Gebiedstypologie warmte-eiland Amsterdam, TU Delft
Voogt JA, Oke TR (1997) Complete urban surface temperatures. J Appl Meteorol 36(9):1117–1132
Voskamp IM, van de Ven FHM (2015) Planning support system for climate adaptation: composing effective sets of blue-green measures to reduce urban vulnerability to extreme weather events. Build Environ 83:159–167
Wassenberg F (1993) Ideeën voor naoorlogse wijken. Delftse Universitaire Pers, Delft

Part VI
Final Considerations

Chapter 21
A Cross-Disciplinary Perspective on the Factors Shaping North American Adaptation Research

Jesse M. Keenan

Abstract This chapter highlights a number of factors and inquiries shaping the development of adaptation research in North America. While some of these steering factors are unique to the institutions and geography of North America, others are poised to resonate among the global academy. In addition, this chapter provides a narrative review of the trends and trajectories of research in North America that represent unique developments for the emerging field of adaptation. While the production of North American adaptation scholarship has lagged behind the rest of the world, an emergent local academy is wrestling with the growing pains of various unresolved aspects of parallel bodies of research in disaster risk reduction and community resilience. These growing pains offer insight into the struggle to operationalize and make practical, feasible and equitable the knowledge of adaptation science.

Keywords Climate change adaptation · Adaptation science · North America · Community resilience · Disaster risk reduction

Introduction

This book chronicles some of the more notable, recent contributions to the advancement of adaptation scholarship in North America. As an emerging field of study, adaptation draws upon a variety of methods and bodies of knowledge in science, applied science and social science. With greater understanding of the complexity of human and natural systems—within and outside of various domains of ecology and socioeconomics—some of the most critical areas of inquiry are those that intersect the physical, biophysical and social manifestations of climate change. This chapter provides a narrative review of the trends and trajectories of

J.M. Keenan (✉)
Lecturer of Architecture, Graduate School of Design, Harvard University,
407 Gund Hall 48 Quincy Street, Cambridge, MA 02138, USA
e-mail: jkeenan@gsd.harvard.edu

W. Leal Filho and J.M. Keenan (eds.), *Climate Change Adaptation in North America*, Climate Change Management,
DOI 10.1007/978-3-319-53742-9_21

research in North America that are shaping an arguably unique body of scholarship. While the production of North American adaptation scholarship has lagged behind the rest of the world, an emergent continental academy is wrestling with the growing pains of various unresolved aspects of parallel bodies of research in disaster risk reduction and community resilience. These growing pains offer insight into the struggle to operationalize and make practical, feasible and equitable the knowledge of adaptation science.

Of course, whether the multidisciplinary research in adaptation has arisen to a paradigmatic state to qualify as a science is debatable (Kuhn 1970). As Davidson et al. (2016, p. 27) note "[s]uch a paradigm would entail the appearance of a consensus around… its constituent concepts, framing theories, language and metaphysical assumptions—thus permitting agreement on the problems to be solved and the development of… [a] science." By this definition, one could argue that a more focused, mature and well-resourced body of international research is nearing a critical mass on a number of important paradigmatic fronts, including theories of transformation (Park et al. 2012; Revi et al. 2014; Pelling et al. 2015; Manuel-Navarrete and Pelling 2015), measuring adaptive capacity (Lockwood et al. 2015; Williams et al. 2015; Weis et al. 2016) and boundaries and frontiers (Adger et al. 2005; Dupuis and Biesbroek 2013; Preseton et al. 2013; Steffen et al. 2015; Felgenhauer 2015). The extent to which this heterogeneous critical mass represents an emerging consensus across disciplines will be the subject of future reviews. However, as described in this chapter, there are factors in North America—particularly the exporting of the wide umbrella of the 'integrated approach' to resilience thinking (Berkes and Ross 2013)—that are exposing unresolved aspects of adaptation scholarship that present challenges for the global academy.

Geography

North America comprises some of the most diverse human and natural geographies on earth. From volcanic tropical islands in the Caribbean to vast ice sheets in the artic, human settlement is equally varied from isolated fishing villages to some of the largest metropolitan areas on earth. Yet, the political jurisdictions of these selected geographies are unified by close legal, economic and ecological connections. Like much of the world, the people who inhabit North America are facing a variety of social and environmental stressors, including an aging demographic, growing economic inequality, population displacement, overburdened healthcare systems, and increased systematic vulnerability in economic and infrastructural systems. With climate change, these stressors are amplified by the increased probability occurrence and severity of extreme weather events, as well as the incremental environmental changes that are nearly totalizing in their impact (Ford et al. 2010; Dawson 2015).

In the north, permafrost thaw is leading to depopulation of lands that have been continuously inhabited for tens of thousands of years (Ford and Smit 2004; Durkalec 2015). In the west and southwest, droughts are being exacerbated by climate change to challenge the viability of agricultural industries and resource economies that have long sustained the region (Feng et al. 2010; Mann and Gleick 2015). In the east, sea level rise together with greater vulnerabilities and risks of tropical storms is redefining the parameters of coastal development that supports hundreds of millions of people (Scavia et al. 2002; Tebaldi et al. 2012; Neumann et al. 2015). In the Caribbean, warming oceans, drought, vector borne disease, and resource depletion are creating the perfect storm for diminishing not only tourism but also local autonomy (Feld and Galiani 2015; Williams 2016; Beckford and Rhiney 2016). Not only do these risks, vulnerabilities and impacts cross jurisdictional boundaries, they are simultaneously leading to increased transnational migration and economic interdependencies (Agrawala et al. 2011; Nawrotzki et al. 2013; McLeman 2013).

Research Trajectory

The ultimate ends of socioeconomic adaptation have historically been referenced to climate mitigation—or a failure thereof—as a transition to renewable energy (Sims 2004), low impact food systems (Antle and Capalbo 2010) and sustainable urbanization (Hamin and Gurran 2009). In the first generation of adaptation research, North American scholars focused much of their time on understanding the nature of the impacts of climate change and the extent to which matters of physical science can be attributed to the manifestation of environmental change. Often drawing on the varying legacies of sustainability and climate mitigation, recent scholarship has begun to conceptualize, measure and evaluate frames, models, and empirical observations ascribed to either autonomous or planned adaptation outside of the immediate scope of energy and the environment.

A more recent generation of scholarship has tracked along three primary lines of inquiry. The first track has begun to conceptualize normative positions for the design of policies and material interventions. Whether it is the institutional design of public health policies (Keim 2008), the development of operational management protocols for ports (Becker et al. 2012), or the architectural design of buildings (Keenan 2014), this body of research has sought to develop processes, and supporting analysis, that promote some objectively reasoned notion of a robust adaptive capacity and/or transformative adaptation. To this end, adaptation is conceptualized either as a capacity to accommodate the unknown or it is positioned to a known endpoint defined by performance or status goals. This endpoint is generally defined within complimentary conceptualizations of a state of resilience. As will be discussed, various lines of thought in ecological, socio-ecological, urban, disaster and community resilience have complicated the operationalization of these distinct but interrelated concepts (Davidson et al. 2016).

As represented by contributions to this book, the second track has sought to engage local actors and stakeholders in finding a more nuanced set of impacts, interrelationships, and feasible, if not actionable, responses or plans. Growing out of a body of participatory or action planning research, this track has sought to utilize empirical science to inform risk assessment and options analysis for various adaptation pathways. The third track has incorporated many formal methods associated with communications and behavioral science to examine the interaction between science, public policy and private decision making. As well represented in this book, this track covers everything from broadcast journalism to public communications.

Conflating Adaptation, Resilience and Sustainability

What unites these research tracks is the task of developing metrics for robust and successful adaptation. However, defining the nature of success often undermines the consistent conceptualization of adaptation and resilience as processes and not as outcomes. As a general proposition, resilience has not only dominated popular usage in the U.S. thanks to the efforts of well-meaning institutions such as the Rockefeller Foundation (Rodin 2014), this usage has also haphazardly conflated resilience with adaptation (*see generally*, Presidential Directive 2013). To this end, resilience has served political ends more substantively as a metaphor for social capacity to resist, recover and thrive (Dias 2015). It has even been argued that a broader response to global terrorism has helped accelerate the institutionalization of the resilience metaphor following 9/11 (Coaffee and Rogers 2008; Coaffee 2013).

Whether it is climate change, or even global terrorism, the incremental institutionalization of resilience in the U.S. and Canada has been under the conditions of utilizing tailored definitions that are largely normative, if not aspirational, and lack the descriptive specificity required to effectively steer policy and decision making to wide diversity of sectors and actors. As will be discussed, for now, resilience is defined within the context of its limited application to disaster risk reduction and management (Meerow and Newell 2016). While resilience is increasingly recognized as a boundary object between disciplines, its use as a boundary object between concepts (i.e., resilience and adaptation) (Brand and Jax 2007) is questionable given the multi-scalar (Folke et al. 2010), multi-ontological basis for the application of each of these concepts (Olsson et al. 2015).

In this sense, the research and practical challenge is one in the same: how do we operationalize 'resilience thinking?' to applications that require measurement for the purposes of the allocation of limited resources, the design of material interventions and the division of power by and between potentially competing social interests and organizations? This is a particularly unique set of challenges in North America where short-term political biasing—often in denial of climate change—and a well-developed academy, industry and administrative state for disaster management have steered resilience into a narrowly drawn normative set of meanings. Interviews with dozens of regulators conducted by the author have suggested that normative definitions that are disconnected from scientific bodies of work on resilience and adaptation represent significant challenges for measurement science and other aspects of empirical benchmarking that are prerequisites for the stewardship of government budgeting allocations. These findings are consistent with the scholarship that has warned of counter-productive disconnects between transitional legislative intent and the execution of localized resilience for a number of years (Smith and Stirling 2010; Bahadur and Tanner 2014; Mulligan et al. 2016).

Canada and Mexico are much more closely aligned with the formal academics of Europe, Australia and Asia that have in recent years coalesced around an emerging consensus as to the parameters of adaptation science, which is inclusive of resilience frameworks as a related but distinct concept and body of knowledge. The conceptual conflation has led to not only confusion among scholars, it has also led to confusion among the very actors and stakeholders whose behaviors, beliefs, perceptions and actions are the subject of social science research. Increasingly, this conflation is recognized for not only its practical inconsistencies (Keenan et al. 2015; Kythreotis and Bristow 2016), but also for its theoretical inconsistences in social science, as Olsson et al. (2014) noted.

> [W]hereas resilience theory aims to prevent transitions—or rather, hinder the collapse of a productive system—social theory commonly used in sustainability studies—from transition theory to political ecology—aims to locate and analyze multilevel or multiscalar resistance against change while seeking to stimulate social transformation (Geels 2010). This incommensurability is problematic for at least two reasons. First, sustainability research needs to consider both continuity and change while also distinguishing between them (Geels 2011). Second, transformation for the sake of persistence of the system—rather than transformation for profound change—appears counterintuitive to social science thinking (Olsson et al. 2015).

Likewise, scholarship has consistently under-recognized the subjectivities associated with adaptation and resilience. In popular terms, adaptation—like resilience—is normatively defined as an absolute good (Béné et al. 2012; Alexander 2013; Rigg and Oven 2015). As such, much research has failed to critically explore, or recognize, whether a particular goal for adaptation is truly adaptive across a variety of scales, actors and time horizons beyond the intended object or benchmark.

Likewise, these North American research tracks have sought to identify and to understand the nature of autonomous adaptation within the context of limited resources, with the implicit acknowledgment of the transactional and marginal costs associated with scaled adaptation efforts (Mendelsohn 2015; Bonen et al. 2016). Increasingly, adaptation—particularly built environment adaptation research—is looking to a body of research in sustainability that addresses programmatic development within the context of resource constraints (Childers et al. 2015; Johnson et al. 2015). From an institutional perspective, sustainability likely offers many hard earned lessons in terms of organizational mainstreaming, alternative economic utilities and social acceptance. However, the downside to such alignments are that one risks conflating process with intention. Interviews with state and local policy makers in the U.S. and Canada suggests that the public often references adaptation or resilience as the 'new' sustainability. To this end, the challenge for the science-to-policy contingent is to draw the distinctions between sustainability, resilience and adaptation. A failure to do so is likely to thwart legislative intent for future policy development and decision making.

Parallel Lines of Inquiry: Disaster Risk Reduction

An exploration of interdependencies between systems and cascading impacts has also recently defined the scholarship. This focus came from two directions. The first was a desire to identify co-benefits from a single investment in order to maximize the utility of a given intervention that often does not pass an investment threshold under a single set of benefits (Spencer et al. 2016; Rose 2016; Whelchel and Beck 2016). The other trajectory came from developments in impact analysis, particularly in terms of spatialization, that have identified precise relationships between social, economic and environment vulnerabilities within the context of multi-hazard risks (Shepard et al. 2012; Carr et al. 2015; McKee et al. 2015). In this regard, disaster risk reduction scholarship is developing tools that are increasingly recognized as helpful within the context of adaptation to the extent that they are seeking to address the fundamental question: adapt to what?

However, scholarship in disaster risk reduction has also worked to confuse adaptation and resilience in ways that are counter-productive to both academies (Schipper 2009; Ireland 2010). With comparatively robust civil defense systems and institutions, North America has a well-developed academy in disaster risk reduction. Together with the field of emergency management, much of this scholarship is focused on responding to and planning for specific risks that represent rapid, isolated, and irregularly occurring extreme events. In addition, these extreme events fall within a well-defined categories that often have associated probabilistic profiles (Mileti 1999). While these contributions are valuable, particularly with greater occurrences of extreme events with climate change, adaptation research is also concerned with time horizons that are either longer or of an unknown duration. Particularly with climate change, the notation of change is often understood to be incremental in nature. In addition, adaptation is concerned with a variety of risks known and unknown, with the recognition that it is often the indirect impacts to climate change that drive adaptation processes. Disaster risk reduction is primarily concerned with known risks. Although, it should be acknowledged that has made reciprocal advances in understanding systematic vulernability and risks associated with dependent and interdependent systems (Cutter et al. 2008a; Resurreccion and Santos 2012; Haraguchi and Kim 2016).

The more fundamental gap between disaster risk reduction and adaptation is the quality of the status of an object in a post-event or post-change scenario. Disaster risk reduction is primarily concerned with resilience and its elasticity to the relative operations of the status quo. The outcomes of risk reduction are benchmarked to well defined states of relief, recovery and reconstruction (i.e., single equilibrium). By contrast, adaptation is often oriented towards more flexible manifestations of future states that are alternatively defined by actors that are either subject to, or initiating, such adaptation processes (i.e., multi-equilibrium). While the language of transformation is unresolved between resilience and adaptation, disaster risk reduction and socioeconomic adaptation have generally conducted research at specific geographic, organizational and institutional scales that reflect their often divergent constituencies and audiences. However, with time, disaster risk reduction audiences, such as emergency managers and floodplain managers, are increasingly recognizing the necessity of incorporating time horizons and risk profiles consistent with the broad and long-term scope of adaptation. This will likely provide the basis for much adaptation research for the years ahead. Future research inquires may include the utilization of projections within risk management decision making; the development of adaptive management governance mechanisms for incorporating up-to-date data; and, legal and economic frameworks for accommodating uncertainty in the provision of public investment in disaster risk mitigation, including the identification of economic co-benefits between risk mitigation and socioeconomic adaptation.

Disaster risk reduction in North America has developed out of ethical and fiscal necessities, often as a byproduct of risk and vulnerability in dense urban agglomerations that have experienced a variety of disasters in the 20th century. One could argue that the concentration of economic, social and cultural power in urban centers in North America has led to an amplification of disaster risk reduction methods, models and techniques as a coordinated policy response to extreme weather and climate change. Certainly within the U.S., climate change has been increasingly institutionalized within the context of emergency management functions of the government. By contrast, European jurisdictions have historically experienced many few natural disasters, with perhaps the exception of earthquakes in Italy, and, as a consequence, have not institutionalized climate change within the more narrow confines of extreme weather and disaster risk management. Adaptation research has yet to fully explore the nature of potential synergies and conflicts between broad scale public adaptation and the adaptive capacity of institutions of disaster risk management to work under a variety of time horizons, risks, vulnerabilities and stakeholders.

Parallel Lines of Inquiry: Community Resilience

The focus on top-down public policy has left the door wide-open for research that seeks to identify and evaluate the extent to which local or state agencies, private firms and private social organizations are or are not understanding, evaluating or internalizing (i.e., adaptation and/or resilience) their vulnerability relative to multi-hazard risks, including climate change. Some of the chapters of this book have very precisely engaged in areas of inquiry within this research agenda. To connect top-down policy with bottom up execution, Community Resilience has arisen in North America as a series of frameworks that connect social and physical systems within a variety of scales, including the household (Eisenman et al. 2016), building (Lin et al. 2016), social community (Houston 2015), district (Graham et al. 2016) and infrastructural scales (NIST 2015).

To this end, Community Resilience in North America is much more oriented towards planning and implementation and is much less concerned with matters of transition theory (Wilson 2012) or development (McNamara and Buggy 2016) that have framed predominately European interpretations of Community Resilience—often referenced as Community-based Adaptation (CBA). This could be partial explained by the fact that North American countries have struggled for years to reduce poverty and socioeconomic vulnerability and have moved beyond top-down ideological program development in favor of practical, feasible and scalable solutions that aren't entirely reliant on state support. This very point has been a source of critical commentary by those who view this brand of resilience as leading to a neoliberal state domination of vulnerable elements of society (Evans and Reid 2014) through its veneration of "decentralization, contextualization, autonomy and independence" (Haldrup and Rosen 2013, p. 143). Others have suggested that by adding 'community' to resilience, it can "address the widespread concern that resilience thinking appears to be value-free… or tainted with a negative anti-community individualism" (Mulligan et al. 2016, p. 348).

Citing Anguelovski et al. (2016) and Sovacool et al. (2015), Graham et al. (2016, p. 114) note that research has increasingly identified cases where, "leaders may be using climate resilience discourse to exacerbate or perpetuate unsustainable, speculative, and exclusionary decisions which might reproduce and exacerbate historic inequities associated with infrastructure and land use development". In this sense, existing hegemonic actors have been observed to utilize the rhetoric of community resilience as a mechanism to justify actions who benefits do not, or are not likely to, inure to a 'community' that is otherwise abstractly, if not obtusely, defined. It could be argued, in explanation of these observations, that the indeterminacy of the scope and boundaries of a community are sufficiently vague so as to cloud the identity of the true parties in interest and hence shield their actions from greater public scrutiny.

Viewed in a more positive light, community resilience has often followed a parallel track to that of adaptation mainstreaming to that extent that both concepts seeks to better integrate various aspects of social service delivery and public administration. Community resilience has sought integrate assessment and planning methodologies that incorporate models from economic development, public health, mental health, natural resource conservation, urban planning and disaster planning (Davidson 2015). For instance, Community Resilience in and of itself could be viewed as a hybrid domain by and between disaster risk reduction and adaptation science. While environmental attributes were historically referenced as sources of risk under disaster risk reduction, Community Resilience has incorporated adaptation science references to expand the application of environmental services (Scholes 2016). Part of this shift in perspective could be explained by the fact that hard-edge environmental engineering interventions have not only been costly to build and maintain, but they have also been increasingly recognized for the limitation of their design performance and useful life. For example, increased downpours in more powerful convection storm events have exceed the storm water management capacity of a variety of systems throughout the continent.

Another area of substantive overlap is between post-disaster public health planning and social adaptive capacity. Public health engagement and assessment methodologies and strategies have increasingly given recognition to concepts such as social capital (Ebi and Semenza 2008; Huang et al. 2011; Portinga 2012), adaptive management (Hess et al. 2012), and integrated long-term program management (Petkova et al. 2015). An additional overlapping theme been the role of civil service organizations in promoting public health (Poutianinen 2013). Disaster risk reduction scholarship has expanded the role of these civic organizations from first responders to critical actors of institutional adaptive capacity (O'Sullivan et al. 2013). Some have even go so far to say that measure of community resilience are representative indicators of a community's social sustainability (Magis 2010). Finally, community resilience has also been extended to examine the resilience and adaptive capacity of public health facilities and systems (Wulff et al. 2015).

In methodological terms, community resilience has likely had multiple impacts on North American scholarship. First, it is helped narrow down boundaries of jurisdictions, organizations and institutions that are manageable for observing and identifying relevant behaviors and capacities that speak to localized adaptation processes (Fox-Lent et al. 2015; Sharifi 2016). The outputs of these assessment inquiries are tools, scorecards, models and indices (Cutter 2016). As Cutter notes, "[the] common elements in all [of] the assessment approaches can be divided into attributes and assets (economic, social, environmental, and infrastructure) and capacities (social capital, community functions, connectivity, and planning)" (2016, p. 741).

Second, these research outputs have, however, provided little resolution for the appropriate metrics of resilience and adaptation by focusing on proxy indicators. While proxies are necessary intermediate steps of evaluation based on the fact that scholars are working within the limitations of existing and accessible data sets, there has been little research on new modes and models of measurement. While these indicators may lead to certain inferences as to the resilience of communities before and after shock events, their non-specificity to the dynamic nature of the processes of resilience and/or adaptation likely undermines their probative value for establishing causal or correlative connections between the underlying observed vulnerabilities, capacities and/or interventions or non-interventions. In cases where correlations have been shown, the indicators don't offer much insight in that communities with more robust economies and more stable social attributes are more likely to *recover* (Burton 2015)—not necessarily in the promotion of adaptation or resilience (Leichenko et al. 2015). For other indicators, such as the percent (%) of the population employed in emergency services, it is highly debatable as to the nature of their validity and/or applicability (Cutter et al. 2008b). For example, in this case, a higher percentage of emergency workers in a small community may bias the results that otherwise suggest a greater level of community resilience because wages are generally higher for emergency workers relative to other occupations in small communities. To this end, a percent of emergency workers may just be a confounding variable for other more conventional economic stability and/or durability indicators.

Likewise, the validation of these indicators has focused on case studies where the exogeneity associated with a broad range of social, economic and environmental attributes undermines any cohesive pattern recognition that defines any one catchment area as more or less 'resilient' than another. To this end, it can be argued that validation is limited by and between indicators and alternative data sets for a single jurisdiction (Sherrieb et al. 2010). In order to compare indicators, one has to create weighted indices, which is in contradiction to the established notion that communities weight their own priorities for managing their vulnerabilities. While indices are helpful in identifying complex issues and interrelationships, they offer little insight into operationalizing resilience or adaptation planning exercises, investments or interventions.

Finally, this body of assessment research raises some epistemological and ontological problems specific to the defined status of being 'resilient' or 'adaptive.' As both resilience and adaptation are processes, it raises the questions as to the utility and validity of measuring a single snapshot of an object in isolation to processes whose duration, intensity and robustness are likely to be either unknown, unobservable or only identified or contextualized in historical terms. In addition, one's resilience is theoretically limited to the parameters of known risks and it is deemed impossible to be show resilience to all risks. Therefore, the notion of specific state is not only philosophically questionable, it lacks qualification and/or quantification as to the parameters of risk (e.g., extreme event, shock, stimulus) that are purportedly in direct relationship to the status. This methodological inconsistency is then compounded by the inconsistent usage of resilience across various disciplines which in the aggregate have worked to thwart measurements, metrics and evaluation criteria for community resilience (Ensor et al. 2016). As adaptation begins to operationalize adaptive capacity research, these unresolved aspects of measurement science will need to be addressed in order to evaluate and/or design strategic and technical interventions and capacities. Adaptation may very well present a unique challenge to that extent that pre- and post-transformational domains may be especially difficult to identify, much less evaluate. To this end, thresholds themselves have been difficult to observe (Carpenter et al. 2005).

Conclusions

The North American experience is very much defined by the great diversity of people, geography, and institutions that shape the continental parameters of adaptation research. With well-developed public institutions and professional capacities in disaster risk reduction and response, the adaptation discourse has been largely shaped by aspects of post-disaster resilience with little to no acknowledgement of the conceptual or practical necessity to draw distinctions between resilience, adaptation and sustainability. This conceptual conflation has undermined legislative intent of emerging laws and policies and has steered the scholarship in favor of inquires that have not always benefited from the emerging global academy in adaptation science. While disaster risk reduction has provided meaningful risk mitigation and impact assessment techniques, there has been little incorporation of adaptive management and governance frameworks for scaling up and incorporating these techniques in the advance of decision making and public policy development. Likewise, while community resilience has begun to conceive of planning exercises and processes that operate at finite bottom-up scales across a wide variety of systems, they are often biased by the assumption that resilience and adaptation lead to future states that are qualitatively defined as absolute goods where such good (i.e., benefits or co-benefits) is equally distributed to members of the community. Utilizing the fiction of 'resilient' is convenient for public policy and observational research designs, but it distracts from the methodological and theoretically understanding of resilience and adaptation as processes.

The challenge moving forward is to give greater recognition to the intersection of resilience and adaptation to the extent that the applications and evaluations of these concepts acknowledge their associated synergies and conflicts. Likewise, adaptation research has the potential to inform parallel lines of scholarship in resilience with disciplinary-specific bodies of literature in everything from anthropology to zoology. With its origins in ecology, adaptation research, particularly socioeconomic adaptation, has sought to explore much broader horizons than those currently binding resilience thinking. With climate change, it will likely be the unknown and unanticipated incremental changes that offer the greatest challenge to human and natural systems. At the same time, adaptation research is challenged to develop knowledge which is operationalizable for advancing current and known conditions. While resolution of various academies will be many years, if not generations, in the making, such resolution does not necessary imply uniform or static frameworks or models. However, in order for adaptation to fulfil its capacity as a paradigmatic science, it must develop coherent theories, methods and terminology that offer potential pathways for the objective consensus as to the identity, validity and utility of new knowledge. This chapter and book represent a small step forward for advancing the science, while at the same time identifying those barriers that represent critical obstacles and challenges for moving forward. The diversity of North American highlights the notion that we may agree as to the challenges that we face, but we do not yet agree as to what the problems are or will be.

References

Adger WN, Arnell NW, Tompkins EL (2005) Successful adaptation to climate change across scales. Glob Environ Change 15(2):77–86

Agrawala S, Carraro M, Kingsmill N, Lanzi E, Mullan M, Prudent-Richard G (2011) Private sector engagement in adaptation to climate change: approaches to managing climate risks. OECD working papers #39. doi:10.1787/19970900

Alexander DE (2013) Resilience and disaster risk reduction: an etymological journey. Natural Hazards Earth Syst Sci 13(11):2707–2716

Anguelovski I, Shi L, Chu E, Gallagher D, Goh K, Lamb Z, Teicher H (2016) Equity impacts of urban land use planning for climate adaptation critical perspectives from the global North and South. J Plann Edu Res 36(3):333–348. doi:10.1177/0739456X16645166

Antle JM, Capalbo SM (2010) Adaptation of agricultural and food systems to climate change: an economic and policy perspective. Appl Economic Perspectives Policy. doi:10.1093/aepp/ppq015

Bahadur A, Tanner T (2014) Transformational resilience thinking: putting people, power and politics at the heart of urban climate resilience. Environ Urbanization 26(1):200–214

Becker A, Inoue S, Fischer M, Schwegler B (2012) Climate change impacts on international seaports: knowledge, perceptions, and planning efforts among port administrators. Clim Change 110(1–2):5–29

Beckford CL, Rhiney K (eds) (2016) Globalization, agriculture and food in the Caribbean: climate change, gender and geography. Springer Nature, London

Béné C, Wood RG, Newsham A, Davies M (2012) Resilience: new utopia or new tyranny? Reflection about the potentials and limits of the concept of resilience in relation to vulnerability reduction programmes. IDS Working Papers, 2012(405):1–61

Berkes F, Ross H (2013) Community resilience: toward an integrated approach. Society Natural Res 26(1):5–20

Bonen A, Loungani MP, Semmler W, Koch S (2016) Investing to mitigate and adapt to climate change: a framework model. Working Paper 16/165. International Monetary Fund, Washington, DC

Brand FS, Jax K (2007) Focusing the meaning (s) of resilience: resilience as a descriptive concept and a boundary object. Ecology Soc 12(1): 23. Available at http://www.ecologyandsociety.org/vol12/iss1/art23/

Burton CG (2015) A validation of metrics for community resilience to natural hazards and disasters using the recovery from Hurricane Katrina as a case study. Ann Assoc Am Geogr 105 (1):67–86

Carpenter SR, Westley F, Turner MG (2005) Surrogates for resilience of social–ecological systems. Ecosystems 8(8):941–944

Carr ER, Abrahams D, De la Poterie AT, Suarez P, Koelle B (2015) Vulnerability assessments, identity and spatial scale challenges in disaster-risk reduction. Jàmbá: J Disaster Risk Studies, 7(1): 1–17. doi:10.4102/jamba.v7i1.201

Childers DL, Cadenasso ML, Grove JM, Marshall V, McGrath B, Pickett ST (2015) An ecology for cities: a transformational nexus of design and ecology to advance climate change resilience and urban sustainability. Sustainability 7(4):3774–3791

Coaffee J (2013) Rescaling and responsibilising the politics of urban resilience: from national security to local place-making. Politics 33(4):240–252

Coaffee J, Rogers P (2008) Rebordering the city for new security challenges: from counter-terrorism to community resilience. Space Polity 12(1):101–118

Cutter SL (2016) The landscape of disaster resilience indicators in the USA. Nat Hazards 80(2): 741–758

Cutter SL, Barnes L, Berry M, Burton C, Evans E, Tate E, Webb J (2008a) A place-based model for understanding community resilience to natural disasters. Glob Environ Change 18(4): 598–606

Cutter SL, Barnes L, Berry M, Burton C, Evans E, Tate E, Webb J (2008) Community and regional resilience: Perspectives from hazards, disasters, and emergency management. CARRI Research Report, Oakridge
Davidson RA (2015) Integrating disciplinary contributions to achieve community resilience to natural disasters. Civil Eng Environ Syst 32(1–2):55–67
Davidson J, Jacobson C, Lyth A, Dedekorkut-Howes A, Baldwin C, Ellison J, Holbrook N, Howes M, Serrao-Neumann S, Singh-Peterson L, Smith T (2016) Interrogating resilience: toward a typology to improve its operationalization. Ecology Society 21(2):27. doi:10.5751/ES-08450-210227
Dawson RJ (2015) Handling interdependencies in climate change risk assessment. Climate 3(4): 1079–1096
Dias P (2015) Is toughness a better metaphor than resilience? Civil Eng Environ Syst 32(1–2): 68–76
Dupuis J, Biesbroek R (2013) Comparing apples and oranges: the dependent variable problem in comparing and evaluating climate change adaptation policies. Glob Environ Change 23 (6):1476–1487
Durkalec A, Furgal C, Skinner MW, Sheldon T (2015) Climate change influences on environment as a determinant of Indigenous health: Relationships to place, sea ice, and health in an Inuit community. Soc Sci Med 136:17–26
Ebi KL, Semenza J (2008) Community-based adaptation to the health impacts of climate change. Am J Prev Med 35(5):501–507
Eisenman DP, Adams RM, Rivard H (2016) Measuring outcomes in a community resilience program: a new metric for evaluating results at the household level. PLOS Currents Disasters. doi:10.1371/currents.dis.15b2d3cbce4e248309082ba1e67b95e1
Ensor JE, Park SE, Attwood SJ, Kaminiski AM, Johson SE (2016) Can Community-based Adaptation Increase Resilience? Climate Develop 001:0001. doi:10.1080/17565529.2016.1223595
Evans B, Reid J (2014) Resilient life: the art of living dangerously. Wiley, Malden
Feld B, Galiani S (2015) Climate change in Latin America and the Caribbean: policy options and research priorities. Latin Am Economic Rev 24(1):1–39
Felgenhauer T (2015) Addressing the limits to adaptation across four damage–response systems. Environ Sci Policy 50:214–224
Feng S, Krueger AB, Oppenheimer M (2010) Linkages among climate change, crop yields and Mexico–US cross-border migration. Proc Natl Acad Sci 107(32):14257–14262
Folke C, Carpenter SR, Walker B, Scheffer M, Chapin T, Rockström J (2010) Resilience thinking: integrating resilience, adaptability and transformability. Ecology Society 15(4): 20. Available at http://www.ecologyandsociety.org/vol15/iss4/art20/
Ford JD, Keskitalo ECH, Smith T, Pearce T, Berrang-Ford L, Duerden F, Smit B (2010) Case study and analogue methodologies in climate change vulnerability research. Wiley Interdisciplinary Rev: Climate Change 1(3):374–392
Ford JD, Smit B (2004) A framework for assessing the vulnerability of communities in the Canadian Arctic to risks associated with climate change. Arctic 57(4):389–400
Fox-Lent C, Bates ME, Linkov I (2015) A matrix approach to community resilience assessment: an illustrative case at Rockaway Peninsula. Environ Syst Decisions 35(2):209–218
Geels FW (2010) Ontologies, Socio-technical Transitions (to sustainability), and the Multi-level Perspective. Res Policy 39(4):495–510
Geels FW (2011) The Multi-level Perspective on Sustainability transitions: Responses to Seven Criticisms. Environ Innovation Societal Trans 1(1):24–40
Graham L, Debucquoy W, Anguelovski I (2016) The influence of urban development dynamics on community resilience practice in New York City after Superstorm Sandy: Experiences from the Lower East Side and the Rockaways. Glob Environ Change 40:112–124
Haraguchi M, Kim S (2016) Critical infrastructure interdependence in New York City during Hurricane Sandy. Int J Disaster Resilience Built Environ 7(2):133–143. doi:10.1108/IJDRBE-03-2015-0015

Hess J, McDowell J, Luber G (2012) Integrating climate adaptation into public health practice: using adaptive management to increase adaptive capacity and build resilience. Environ Health Perspect 120(20):171–179
Haldrup SV, Rosen F (2013) Developing resilience: a retreat from grand planning. Resilience 1(2):130–145
Hamin EM, Gurran N (2009) Urban form and climate change: Balancing adaptation and mitigation in the US and Australia. Habitat Int 33(3):238–245
Houston JB (2015) Bouncing forward assessing advances in community resilience assessment, intervention, and theory to guide future work. Am Behavioral Sci 59(2):175–180
Huang C, Vaneckova P, Wang X, FitzGerald G, Guo Y, Tong S (2011) Constraints and barriers to public health adaptation to climate change: a review of the literature. Am J Prev Med 40(2): 183–190
Ireland P (2010) Climate change adaptation and disaster risk reduction: contested spaces and emerging opportunities in development theory and practice. Climate Development 2(4): 332–345
Johnson C, Toly N, Schroeder H (eds) (2015) The Urban Climate Challenge: Rethinking the Role of Cities in the Global Climate Regime (vol. 4). Routledge, New York
Keenan JM (2014) Material and social construction: a framework for the adaptation of buildings. Enquiry: A J Architectural Res 11(1): 18–32
Keenan JM, King DA, Willis D (2016) Understanding conceptual climate change meanings and preferences of multi-actor professional leadership in New York. J Environ Planning Policy Manage 18(3):261–285
Keim ME (2008) Building human resilience: the role of public health preparedness and response as an adaptation to climate change. Am J Prev Med 35(5):508–516
Kuhn T (1970) The structure of scientific revolutions. University of Chicago Press, Chicago
Kythreotis AP, Bristow GI (2016) The 'resilience trap': exploring the practical utility of resilience for climate change adaptation in UK city-regions. Regional Studies 1–12. doi:10.1080/00343404.2016.1200719
Leichenko R, McDermott M, Bezborodko E (2015) Barriers, limits and limitations to resilience. J Extreme Events 2(01):1550002. doi:10.1142/S2345737615500025
Lin P, Wang N, Ellingwood BR (2016) A risk de-aggregation framework that relates community resilience goals to building performance objectives. Sustainable and Resilient Infrastruct 1–13. doi:10.1080/23789689.2016.1178559
Lockwood M, Raymond CM, Oczkowski E, Morrison M (2015) Measuring the dimensions of adaptive capacity: a psychometric approach. Ecology Society 20(1):37
Magis Kristen (2010) Community resilience: an indicator of social sustainability. Society Natural Res 23(5):401–416
Mann ME, Gleick PH (2015) Climate change and California drought in the 21st century. Proc Natl Acad Sci 112(13):3858–3859
Manuel-Navarrete D, Pelling M (2015) Subjectivity and the politics of transformation in response to development and environmental change. Glob Environ Change 35:558–569
McKee JJ, Rose AN, Bright EA, Huynh T, Bhaduri BL (2015) Locally adaptive, spatially explicit projection of US population for 2030 and 2050. Proc Natl Acad Sci 112(5):1344–1349
McLeman RA (2013) Climate and human migration: Past experiences, future challenges. Cambridge University Press, New York
McNamara KE, Buggy L (2016) Community-based climate change adaptation: a review of academic literature. Local Environ doi:10.1080/13549839.2016.1216954
Meerow S, Newell JP (2016) Urban resilience for whom, what, when, where, and why?. Urban Geography 1–21. doi:10.1080/02723638.2016.1206395
Mendelsohn R (2015) Climate change demands we change. why aren't we?: an economic perspective. Social Res Int Q 82(3):727–747
Mileti D (1999) Disasters by design: a reassessment of natural hazards in the United States. Joseph Henry Press, Washington

Mulligan M, Steele W, Rickards L, Fünfgeld H (2016) Keywords in planning: what do we mean by 'community resilience'? Int Planning Studies 21(4):348–361. doi:10.1080/13563475.2016.1155974

National Institute of Standards and Technology (NIST) (2015) Community resilience planning guide for buildings and infrastructure systems, vol 1. U.S. Department of Commerce, Washington. doi:10.6028/NIST.SP.1190v1

Nawrotzki RJ, Riosmena F, Hunter LM (2013) Do rainfall deficits predict US-bound migration from rural Mexico? Evidence from the Mexican census. Popul Res Policy Rev 32(1):129–158

Neumann JE, Emanuel K, Ravela S, Ludwig L, Kirshen P, Bosma K, Martinich J (2015) Joint effects of storm surge and sea-level rise on US Coasts: new economic estimates of impacts, adaptation, and benefits of mitigation policy. Clim Change 129(1–2):337–349

Olsson P, Galaz V, Boonstra WJ (2014) Sustainability transformations: a resilience perspective. Ecology Society 19(4):1. doi:10.5751/ES-06799-190401

Olsson L, Jerneck A, Thoren H, Persson J, O'Byrne D (2015) Why resilience is unappealing to social science: theoretical and empirical investigations of the scientific use of resilience. Sci Adv 1(4):e1400217. doi:10.1126/sciadv.1400217

O'Sullivan TL, Kuziemsky CE, Toal-Sullivan D, Corneil W (2013) Unraveling the complexities of disaster management: a framework for critical social infrastructure to promote population health and resilience. Soc Sci Med 93:238–246

Park SE, Marshall NA, Jakku E, Dowd AM, Howden SM, Mendham E, Fleming A (2012) Informing adaptation responses to climate change through theories of transformation. Glob Environ Change 22(1):115–126

Pelling M, O'Brien K, Matyas D (2015) Adaptation and transformation. Clim Change 133(1): 113–127

Petkova EP, Ebi KL, Culp D, Redlener I (2015) Climate change and health on the US gulf coast: Public health adaptation is needed to address future risks. Int J Environ Res Public Health 12 (8):9342–9356. doi:10.3390/ijerph120809342

Poortings W (2012) Community resilience and health: the role of bonding, bridging, and linking aspects of social capital. Health Place 18(2):286–295

Poutianinen C, Berrang-Ford L, Ford J, Heymann J (2013) Civil society organizations and adaptation to the health effects of climate change in Canada. Public Health 127(5):403–409

Presidential Directive—Critical Infrastructure Security and Resilience PPD-21 (2013) Available at https://www.whitehouse.gov/the-press-office/2013/02/12/presidential-policy-directive-critical-infrastructure-security-and-resil

Preston BL, Dow K, Berkhout F (2013) The climate adaptation frontier. Sustainability 5(3): 1011–1035

Revi A, Satterthwaite D, Aragón-Durand F, Corfee-Morlot J, Kiunsi RB, Pelling M, Sverdlik A (2014) Towards transformative adaptation in cities: the IPCC's fifth assessment. Environ Urbanization 26(1):11–28

Resurreccion J, Santos JR (2012) Multiobjective prioritization methodology and decision support system for evaluating inventory enhancement strategies for disrupted interdependent sectors. Risk Anal 32(10):1673–1692

Rigg J, Oven K (2015) Building liberal resilience? A critical review from developing rural Asia. Glob Environ Change 32:175–186

Rodin J (2014) The resilience dividend: being strong in a world where things go wrong. PublicAffairs, New York

Rose A (2016) Capturing the co-benefits of disaster risk management on the private sector side. World Bank Policy Research Working Paper (7634). World Bank Group, Washington, DC

Scavia D, Field JC, Boesch DF, Buddemeier RW, Burkett V, Cayan DR, Fogarty M, Harwell MA, Howarth RW, Mason C, Reed DJ (2002) Climate change impacts on US coastal and marine ecosystems. Estuaries 25(2):149–164

Schipper ELF (2009) Meeting at the crossroads?: Exploring the linkages between climate change adaptation and disaster risk reduction. Clim Develop 1(1):16–30

Scholes RJ (2016) Climate change and ecosystem services Climate change. Wiley Interdisciplinary Reviews, UK. doi:10.1002/wcc.404
Sharifi A (2016) A critical review of selected tools for assessing community resilience. Ecol Ind 69:629–647
Shepard CC, Agostini VN, Gilmer B, Allen T, Stone J, Brooks W, Beck MW (2012) Assessing future risk: quantifying the effects of sea level rise on storm surge risk for the southern shores of long island New York. Natural Hazards 60(2):727–745
Sherrieb K, Norris FH, Galea S (2010) Measuring capacities for community resilience. Soc Indic Res 99(2):227–247
Sims REH (2004) Renewable energy: a response to climate change. Sol Energy 76(1):9–17
Sovacool BK, Linnér BO, Goodsite ME (2015) The political economy of climate adaptation. Nature Clim Change 5(7):616–618
Spencer B, Lawler J, Lowe C, Thompson L, Hinckley T, Kim SH, Voss J (2016) Case studies in co-benefits approaches to climate change mitigation and adaptation. J Environ Planning Manage 1–21:0001. doi:10.1080/09640568.2016.1168287
Smith A, Stirling A (2010) The politics of social-ecological resilience and sustainable socio-technical transitions. Ecology Society 15(1): 11. Available at http://www.ecologyandsociety.org/vol15/iss1/art11/
Steffen W, Richardson K, Rockström J, Cornell SE, Fetzer I, Bennett EM, Biggs R, Carpenter SR, de Vries W, de Wit CA, Folke C (2015) Planetary boundaries: guiding human development on a changing planet. Science 347(6223):1259855
Tebaldi C, Strauss BH, Zervas CE (2012) Modelling sea level rise impacts on storm surges along US coasts. Environ Res Lett 7(1):014032
Weis SWM, Agostini VN, Roth LM, Gilmer B, Schill SR, Knowles JE, Blyther R (2016) Assessing vulnerability: an integrated approach for mapping adaptive capacity, sensitivity, and exposure. Clim Change 136:615. doi:10.1007/s10584-016-1642-0
Whelchel AW, Beck MW (2016) Decision tools and approaches to advance ecosystem-based disaster risk reduction and climate change adaptation in the twenty-first century. In: Ecosystem-based disaster risk reduction and adaptation in practice, Springer International Publishing, pp 133–160
Williams AN (2016) Towards a deeper understanding of the Caribbean water supply crisis. Caribbean Water Transshipment Company Ltd, Dominica
Williams C, Fenton A, Huq S (2015) Knowledge and adaptive capacity. Nature Climate Change 5(2):82–83
Wilson GA (2012) Community resilience and environmental transitions. Routledge, NewYork
Wulff K, Donato D, Lurie N (2015) What is health resilience and how can we build it?. Annu Rev Public Health 36:361–374

Erratum to: A Region Under Threat? Climate Change Impacts, Institutional Change and Response of Local Communities in Coastal Yucatán

Lysann Schneider and Tobias Haller

Erratum to: Chapter 10 in: W. Leal Filho and J.M. Keenan (eds.), *Climate Change Adaptation in North America*, Climate Change Management, https://doi.org/10.1007/978-3-319-53742-9_10

The original version of the book was inadvertently published with incorrect author name "M.A. Lysann Schneider" which has been corrected to read as "Lysann Schneider" in Chapter 10.

The updated online version of this chapter can be found at https://doi.org/10.1007/978-3-319-53742-9_10

W. Leal Filho and J.M. Keenan (eds.), *Climate Change Adaptation in North America*, Climate Change Management,
https://doi.org/10.1007/978-3-319-53742-9_22

MIX
Papier aus verantwortungsvollen Quellen
Paper from responsible sources
FSC® C105338

If you have any concerns about our products,
you can contact us on
ProductSafety@springernature.com

In case Publisher is established outside the EU,
the EU authorized representative is:
Springer Nature Customer Service Center GmbH
Europaplatz 3, 69115 Heidelberg, Germany

Printed by Libri Plureos GmbH
in Hamburg, Germany